REACTIVE
INTERMEDIATES

REACTIVE INTERMEDIATES

A Serial Publication

VOLUME 2

Edited by

MAITLAND JONES, JR.
Princeton University

ROBERT A. MOSS
Rutgers University

A WILEY-INTERSCIENCE PUBLICATION

JOHN WILEY & SONS

New York • Chichester • Brisbane • Toronto • Singapore

Library of Congress Cataloging in Publication Data:

Reactive intermediates. v.2
  New York, John Wiley & Sons, c1978

    (2) v. ill. 24 cm.

  "A Wiley-interscience publication."
  Key title: Reactive intermediates, ISSN 0190–8375.

    1. Chemistry, Physical organic–Collected works.   2. Chemical reaction, Conditions and laws of–Collected works.

  QD476.R414   547'.1'39   79-640411
  ISBN 0-471-01875-9   MARC-S

Printed in the United States of America

10 9 8 7 6 5 4 3 2 1

CONTRIBUTORS

D. Bethell
Robert Robinson Laboratories
University of Liverpool
Liverpool, England

Weston Thatcher Borden
Department of Chemistry
University of Washington
Seattle, Washington 95195

Charles P. Casey
Department of Chemistry
University of Wisconsin
Madison, Wisconsin 53706

Charles K. Dustman
Department of Chemistry
University of Nebraska-Lincoln
Lincoln, Nebraska 68588

Peter P. Gaspar
Department of Chemistry
Washington University
Saint Louis, Missouri 63130

Maitland Jones, Jr.
Department of Chemistry
Princeton University
Princeton, New Jersey 08540

Leonard Kaplan
Union Carbide Corporation
South Charleston, West Virginia 25303

Ronald H. Levin
IBM
Office Products Division
Boulder Colorado 80302

Walter Lwowski
Department of Chemistry
New Mexico State University
Las Cruces, New Mexico 88003

Robert A. Moss
Wright and Rieman Laboratories, Rutgers,
The State University of New Jersey
New Brunswick, New Jersey 08903

Stuart W. Staley
Department of Chemistry
University of Nebraska-Lincoln
Lincoln, Nebraska 68588

D. Whittaker
Robert Robinson Laboratories
University of Liverpool
Liverpool, England

PREFACE TO VOLUME 2

Our intent for this second volume of *Reactive Intermediates: A Serial Publication* is unchanged from that for Volume 1. We hope to provide authoritative, critical, and selective analyses of the recent literature for each major type of reactive organic intermediate. Once again, the volume is intended for scientists who are actively involved or contemplating involvement in the chemistry of reactive intermediates.

Authors have been urged to select critically and evaluate those recent contributions that should be brought to the close attention of researchers and students. Literature coverage in the present volume focuses on 1977–1979, but important contributions from earlier years, as well as some references to 1980 work, have been included at the authors' discretion. As a convenience to the reader, the authors have marked with an asterisk (*) those references that they consider most significant for detailed discussion, analysis, and rereading.

Inevitably there have been changes in the coverage and cast of authors. New chapters on "Metal–Carbene Complexes" and "Diradicals" have been added. These enlarge upon matters included in the earlier chapter on carbenes. More important, the new chapters bring into proper perspective new material essential to these rapidly developing subfields. Unfortunately, we have been unable to include the chapter on "Theory of Reactive Intermediates and Reaction Mechanism." We hope to reestablish such a chapter in future volumes.

Changes in authorship include the debuts of W. T. Borden ("Diradicals") and C. P. Casey ("Metal–Carbene Complexes"). The province of "Carbanions" has passed from W. J. le Noble to S. W. Staley and C. K. Dustman. K. N. Houk has left, and D. Whittaker has joined D. Bethell in writing the "Carbocations" chapter.

MAITLAND JONES, JR.
ROBERT A. MOSS

Princeton, New Jersey
New Brunswick, New Jersey
June 1981

CONTENTS

REACTIVE INTERMEDIATES

1

ARYNES

RONALD H. LEVIN

IBM, Office Products Division, Boulder, Colorado 80302

I. INTRODUCTION

During the period from 1977 to 1979, the pace of meaningful contributions in the field of aryne chemistry slackened somewhat. However, a number of full papers that further amplify earlier discoveries, as well as several articles offering new and perspicacious observations, did appear. The highlights from these publications are discussed in the sections below.

II. *ortho*-BENZYNE

A. Structure

A number of theoretical papers dealing with the energies and geometries of arynes have appeared during the current reporting interval. Most notable among these was a contribution by Noell and Newton.[1] These authors applied the GVB formalism to *o*-, *m*-, and *p*-benzyne and obtained results that were at variance with the earlier findings from restricted Hartree-Fock (RHF) calculations, especially with respect to *m*- and *p*-benzyne. These two arynes were found to possess diyl structures at their energy minima and *o*-benzyne was found to be the only true aryne displaying a significant bonding interaction between the radical centers.

As we discussed in our chapter in Volume 1 of this series, Laing and Berry[2] have performed a normal coordinate analysis using the IR spectrum of matrix-isolated *o*-benzyne obtained by Chapman et al.[3] This analysis, along with several explicit assumptions led Laing and Berry to propose a geometry for *o*-benzyne. As a test of their method, they used their structure to predict the IR spectrum of tetradeuterio-*o*-benzyne. The validity of their model may now be examined, as tetradeuterio-*o*-benzyne has been matrix isolated and its IR spectrum recorded.[4] The vibrational frequencies for protio-*o*-benzyne based on Berry's structure were within 1.1% of the frequencies experimentally observed by Chapman. However, in the case of the deuterio analogue, significant differences between Berry's predictions and the experimental findings exist. Not only were some of the predicted bands absent (this may simply be due to low band intensities) but differences in excess of 2.0% between the predicted and observed frequencies were common. With experimental IR data on *o*-benzyne-d_0 and *o*-benzyne-d_4 now in hand, it would be of interest to define a structure that is more compatible with the available information.

B. Reactivity Patterns

During the past few years, several papers have appeared that deal with the relative rates and stereoselectivity of reactions involving *o*-benzene. Among these

TABLE 1.1.   Relative Rates of o-Benzyne Cycloaddition

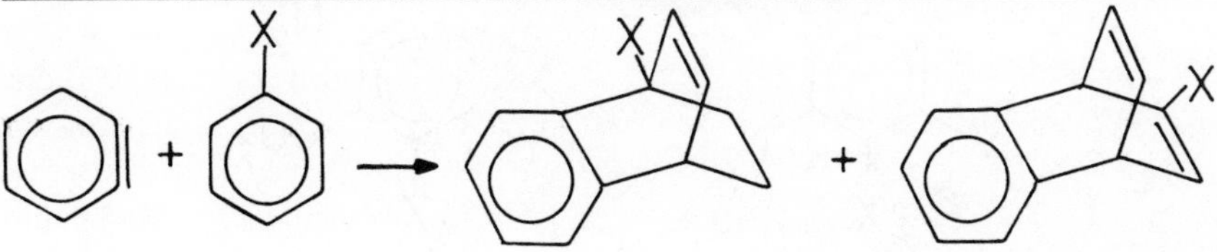

	OMe	Me	H	Cl	CO$_2$Me	CF$_3$
K_{rel} 1	19.1	2.3	1.0	0.44	0	0
K_{rel} 2	2.4	2.9	1.0	0.97	1.5	0.87

was an article by Oda and co-workers in which the reaction of o-benzyne with monosubstituted benzenes was discussed.[5] Two basic modes of Diels-Alder cycloaddition are possible and after correction for statistical factors and secondary reactions, the results presented in Table 1.1 were obtained. As expected, the type

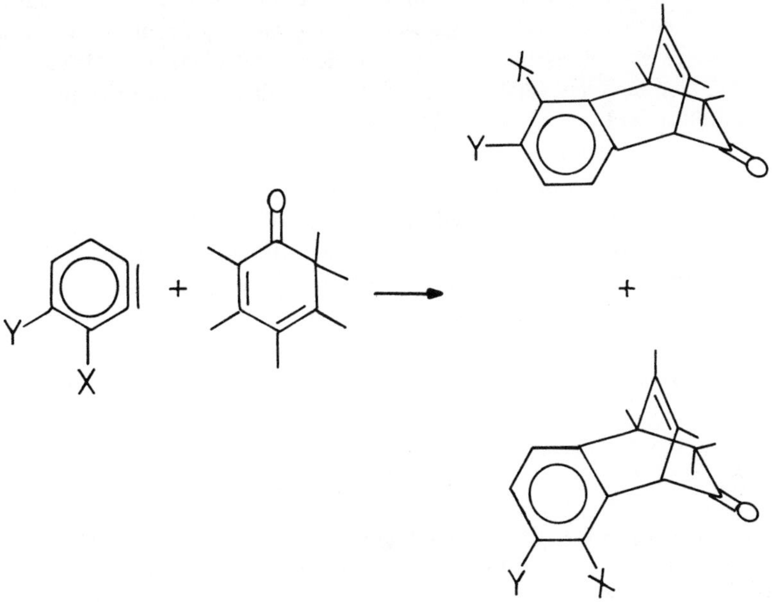

FIGURE 1.   Selectivities of o-Benzyne cycloadditions.

$$X = Me; NO_2, Cl; Y = H$$

$$X = H; Y = Me, NO_2$$

$$X = Me; Y = Me$$

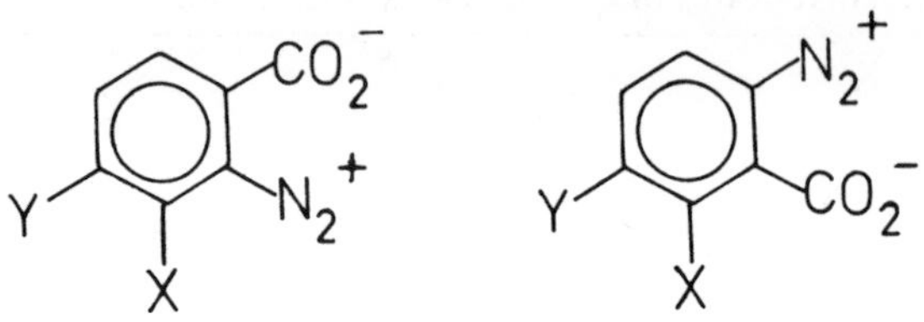

FIGURE 2.   A test for free o-benzyne.

1 products displayed the largest change in relative rate as a function of the substituent X. In accord with the electrophilic nature of o-benzyne (the type 1 reaction correlated with σ_p^+ and yielded $\rho = -1.8$), type 1 reactivity decreases as the substituent becomes more electron withdrawing. This, in turn, allows the type 2 process, the rate of which is only mildly affected by substituents, to compete more effectively.

In a sequel, these same authors probed the effect of silver ion upon this reaction.[6] The ratio of type 1 to type 2 products was only slightly altered, but biphenyls and benzocyclooctatetraenes were among the major projects. These latter compounds as well as the effect of the substituent upon their yields, were anticipated from and lent further credence to the earlier work of Friedman[7] and Vedejs and Shepherd[8] on silver-mediated o-benzyne reactions.

Oku has explored the reverse situation, where substituted o-benzynes react with dienes to produce isomers 3 and 4 (Figure 1).[9] Over the range of substituents explored, the 3/4 ratio varied by little more than a factor of 2. While it is always of interest to argue over such changes, and the authors do so at length, it seems likely that these small differences result from factors currently beyond our ken. One noteworthy observation served to further establish the intermediacy of a free o-benzyne species in anthranilic acid diazotizations. It was found that identical 3/4 ratios were obtained when o-benzyne precursors 5 and 6 were employed (Figure 2).

C. Polymer Bound

A full paper has appeared detailing Mazur and Jayalekshmy's investigation of polymer-bound o-benzyne.[10] Their method entailed the covalent attachment of an o-benzyne precursor, namely, a 1-aminobenzotriazole, to a variety of polystyrene resins. Subsequent lead tetraacetate oxidation produced a polymer-anchored aryne as judged by trapping experiments with tetraphenylcyclopentadienone (TPC). Delayed trapping experiments in which the TPC was added at varying intervals after oxidation, permitted estimates to be made of the aryne lifetime. Reduction of the data led to a "persistence time" for the bound aryne of 53 ± 15 sec!

It then became of interest to determine which kinetic process is limiting this lifetime. Dimerization to yield biphenylene analogues was a likely possibility, especially in view of the fact that the nonpolymer-attached aryne produced such dimers, under otherwise identical conditions, in up to 65% yield. However, while as little as 0.3% biphenylene would have been detected, none was observed in the case of the polymer-supported aryne. Further investigation ultimately revealed

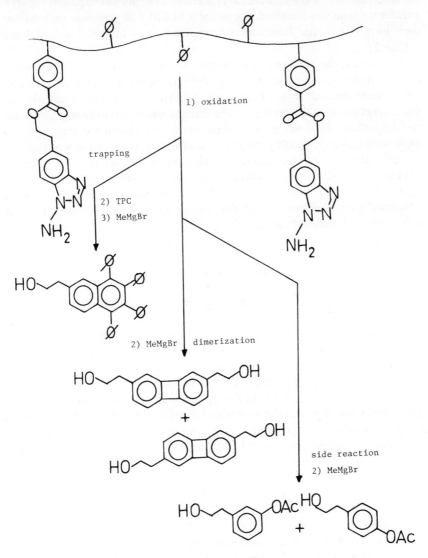

FIGURE 3.   Possible reactions of polymer-bound o-benzyne.

that phenyl acetates were being produced, such a result being understandable in the presence of lead tetraacetate. Iodobenzene diacetate is another oxidant that can transform the aminotriazole into *o*-benzyne, but without the production of acetate as a side product. Hence it was reasoned that use of this material in the generation step should lengthen the aryne lifetime by removing the otherwise limiting decay process of aryl acetate formation. Indeed, when iodobenzene diacetate was employed, aryl acetates were not detected and the aryne persistence time was increased dramatically to 810 ± 80 sec. Now biphenylenes were formed and this dimerization process served to limit the aryne lifetime (Figure 3).

In assessing the verity of their analysis, the authors bring up the possibility that *o*-benzyne may not be involved in the various reactions they observed, but rather some masked arynic form might be participating. They conclude that direct spectroscopic observation of the aryne would be the only way to rule out such hypotheses. However, it would seem that an experiment involving diastereotopic transition states could serve as a valid substitute for direct observation. For example, the ratio of cycloadducts obtained upon the reaction of 1,4-dimethoxyanthracene with free *o*-benzyne[11] and the ratio of ene products obtained upon reaction of 2-methyl-1-butene with free *o*-benzyne could serve as useful benchmarks for comparison with the corresponding results obtained from the polymer-bound, presumed aryne.

III.   *meta*-BENZYNE

A.   Structure

Noell and Newton's[1] GVB examination of *m*-benzyne led them to conclude that this molecule exists as a ground-state singlet diyl, the geometry of which is only moderately distorted from that of benzene. The authors further note that this open structure lies 32 kcal/mol below its bicyclic counterpart (whether the bicyclic structure exists as a local energy minimum was not ascertained). It was also pointed out that the RHF theoretical approach artificially enhances the stability of closed-shell bicyclic species with respect to their open-shell isomers, in this particular case placing bicyclo[3.1.0]hexatriene 17 kcal/mol below the open 1,3-dehydrobenzene structure.

B.   Bicyclo[3.1.0]hexatriene

Washburn has continued his efforts to define the mechanism by which the dibromobicyclohexene **7** is transformed into the fulvene **9** (Figure 4). As discussed in our Chapter in Volume 1 of this series a panoply of deuterium, carbon-13, and halogen labeling studies pointed toward the involvement of the

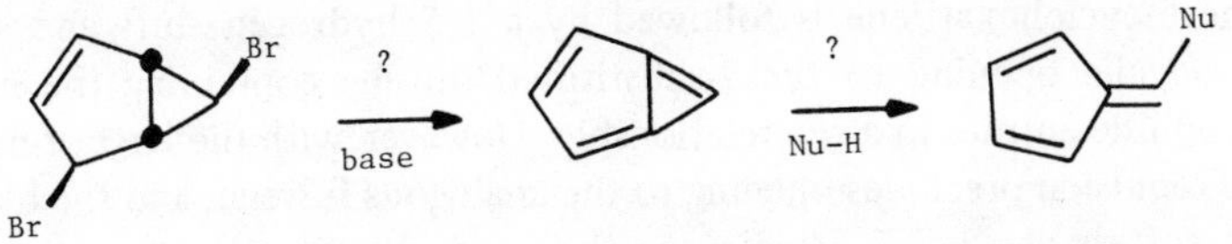

FIGURE 4.   The involvement of *m*-benzyne in fulvene formation.

m-benzyne isomer **8**. However, as also noted, isotopic loss and randomization hampered a definitive interpretation. In an effort to circumvent these difficulties, Washburn has employed alkyl markers and has reported results that are at variance with all proposed mechanisms other than the aryne pathway.[12]

When the methyl derivative **10** was subjected to the standard reaction conditions, only fulvene **12** was isolated. The *t*-butyl analogue, on the other hand, produced and positional fulvene isomer **13** as the kinetic product (Figure 5). This latter result was particularly instructive when combined with the fact that the *syn*- and *anti*-isomers **13** and **14** had equal thermodynamic stability. Clearly then, when a small alkyl group such as methyl is present, nucleophilic addition

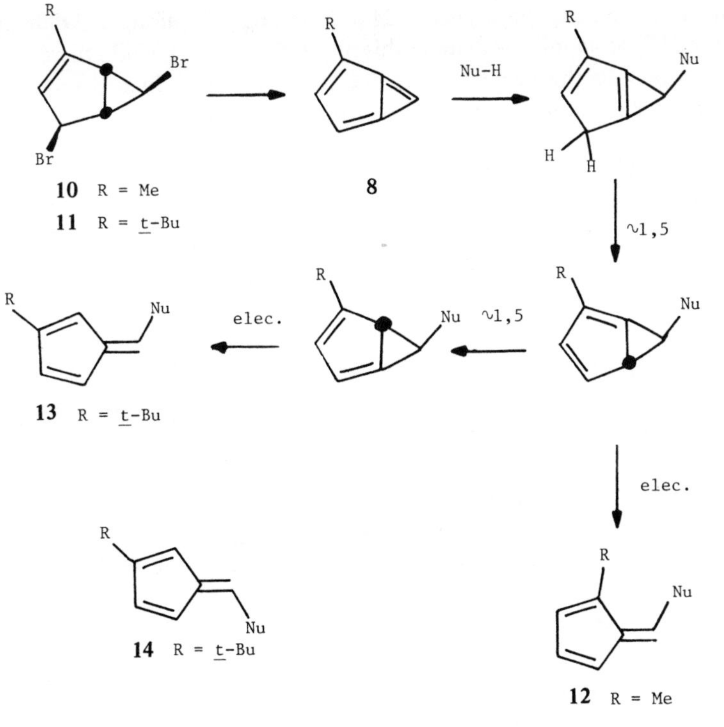

FIGURE 5.   Reactions of alkylated bicyclohexatrienes.

to the bicyclohexatriene is followed by a 1,5 hydrogen shift and subsequent electrocyclic opening to the 2-substituted fulvene containing the methyl and nucleophilic entities in a *syn* relationship. However, with the larger *t*-butyl group, steric repulsion precludes opening to the analogous fulvene, and the bicyclohexadiene suffers another 1,5 hydrogen shift and ultimate electrocyclic opening to **13**. Again the stereospecificity of this transformation must be noted. Alternate mechanisms would have generated both **13** and **14**; only the pathway involving bicyclohexatriene **8** can account for these observations.

A detailed account of these and the earlier experiments is presented in Washburn's account of the subject.[13] Therein he also mentions his lack of success in capturing **8** with diphenylisobenzofuran, furan, and other trapping agents.

C. The 1,3-Diyl

Compound **15**, a benzoannelated relative of **7**, has recently been prepared by Billups and his co-workers.[14] When this substance was treated with potassium *t*-butoxide in THF-d_8 at 0°C, no less than 10 products were formed. Most notable was the presence of compounds **18–20**. Although they were formed in low yield (combined yield ~10%) and the stereochemistries of **18** and **19** were not completely secured, the available data do strongly implicate 1,3-dehydronaphthalene (**17**) as an intermediate in this process (Figure 6). While there is even less data relevant to the occurrence of the other products, their formation can be explained by direct interception of benzobicyclohexatriene **16**.

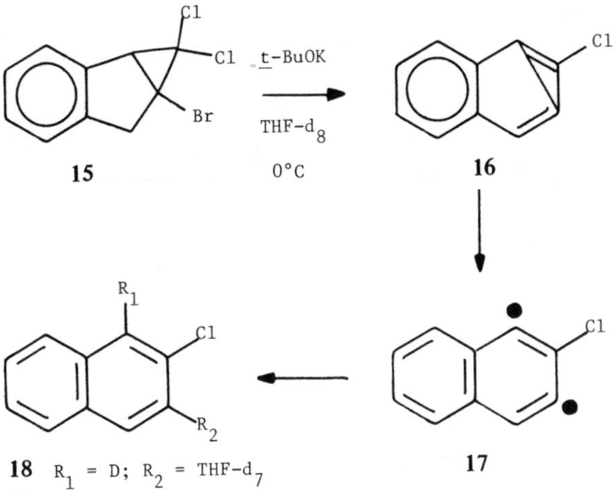

18 R_1 = D; R_2 = THF-d_7
19 R_1 = THF-d_7; R_2 = D
20 R_1 = R_2 = D

FIGURE 6.   Generation of 1,3-dehydronaphthalene (**17**).

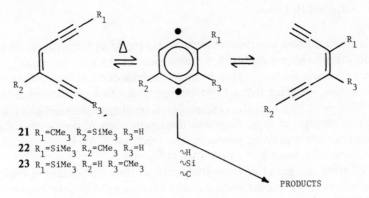

21 R_1=CMe$_3$ R_2=SiMe$_3$ R_3=H
22 R_1=SiMe$_3$ R_2=CMe$_3$ R_3=H
23 R_1=SiMe$_3$ R_2=H R_3=CMe$_3$

FIGURE 7. Intervention and self-trapping of a *p*-benzyne diyl.

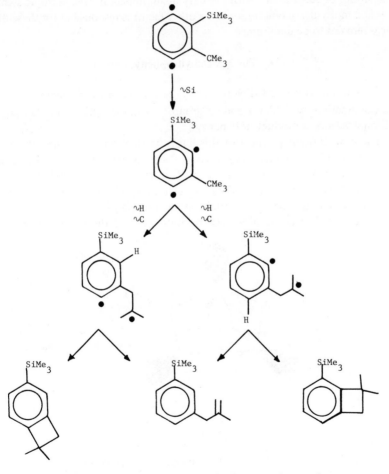

FIGURE 8. The rearrangement of *p*-benzyne into *m*-benzyne.

9

Germane to these observations is the recent report by Bergman on the pyrolysis of substituted ene-diynes.[15] Stereoisomers **21–23** were pyrolyzed and the myriad products identified. Most of these compounds could be accounted for by a mechanism involving initial Cope rearrangement to a substituted *p*-benzyne and subsequent radical shifts (Figure 7). A labeling experiment along with an assiduous display of logic suggested the intervention of a *m*-benzyne diyl to account for the remaining products. The mechanistic hypothesis advanced by these workers in the case of **23** is shown in Figure 8.

The intriguing aspect of these studies lies in their relationship to Washburn's observations on the parent system. In the latter case the bicyclic form is responsible for the observed chemistry, whereas in Billups' and Bergman's experiments the 1,3-diyl plays a role. Whether the different reaction temperatures, the benzoannelation, or substituents with their attendant impact upon electronic stabilization and molecular geometry, or some other factor is responsible for these differences remains to be determined.

D. The 3,5-Dehydrophenyl Cation

The curious cation **24** (Figure 9) has recently been the subject of a theoretical examination by Schleyer and his group.[16] With molecular formula C_6H_3 it is a formal solvolysis product of *m*-benzyne.

The most intriguing aspect of this ion is its electronic structure. It possesses the benzene-like aromatic array of six π-electrons delocalized above and below the molecular plane. In addition to this feature, however, there is another aromatic system contained within the molecular plane. This unit is comprised of the three σ-orbitals on the three carbon atoms lacking hydrogens. Since there are two electrons delocalized over these orbitals, we have the familiar three-center,

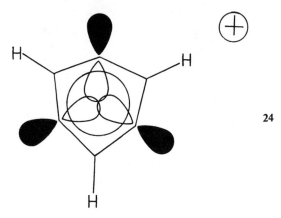

24

FIGURE 9.   The 3,5-dehydrophenyl cation.

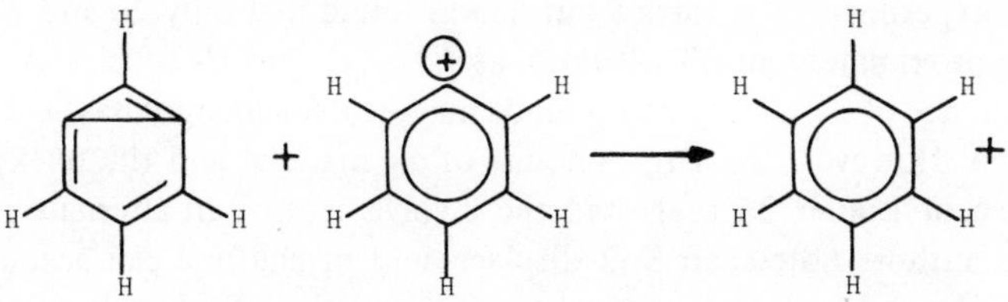

FIGURE 10.   Isodesmic reaction involving **24**.

two-electron aromatic interaction. The net result is two orthogonal electronic frameworks, both of which are aromatic. Hence **24** may be aptly described as a doubly aromatic ion.

The molecular properties attendant such double aromaticity were probed by calculations at various levels of sophistication. In all cases, the bond indices, charge distribution, and overlap populations were supportive of such stabilization. Furthermore, the isodesmic reaction in Figure 10 was found to be exothermic. With respect to the structure of **24**, the D_{3h} form was found to be a local minimum on the $C_6H_3^+$ potential energy surface. The authors point out that $C_6H_3^+$ ions have been known to mass spectroscopists for some time; they go on to note that appropriate mass spectral experiments are currently underway in their laboratory.

IV.   para-BENZYNE

A.   Structure

Noell and Newton's consideration of p-benzyne at the GVB level led these authors to conclude that the open-shell 1,4-diyl, with a geometry similar to benzene, lies considerably below the butalene structure in terms of energy.[1] Their results further suggested that the butalene geometry may correspond to a local energy minimum.

B.   Butalene

Breslow and his co-workers have amassed a considerable amount of experimental evidence supporting the intermediacy of butalene (**25**) in the dehydrochlorination of Dewar chlorobenzene. As a further test of their hypothesis, it was decided to repeat the reaction with an alkyl labeled Dewar chlorobenzene.[17] Because of the nonsymmetric nature of the presumed intermediate, they expected specific isomers (**26, 27**) to be produced (Figure 11). A priori, one would expect **27** to predominate for steric reasons.

When the experiment was carried out, it was found that only **26** and **27** were formed, in approximately an $80:20$ ratio, as $d_0, d_1, d_2,$ and d_3 species. As in the past, the existence of d_2 and d_3 compounds may be explained by base-catalyzed H–D exchange. However, the large amount of d_0 material and the unexpected preponderance of isomer **26** suggested the involvement of an alternate mechanism. As the authors noted, an S_N2 displacement of chlorine can account for this variance. Hence the data can be taken as supportive of a butalene intermediate after the contribution from a significant competing pathway is recognized.

In an effort to generate p-benzyne and/or butalene by an alternate method, Bergman and Johnson prepared the novel propellane **28**.[18] As this molecule is the formal Diels-Alder adduct of butalene and anthracene, there was the expectation that it would serve as a convenient source of p-benzyne or butalene. Unfortunately, **28** is thermally quite stable, requiring temperatures near $700°C$ to

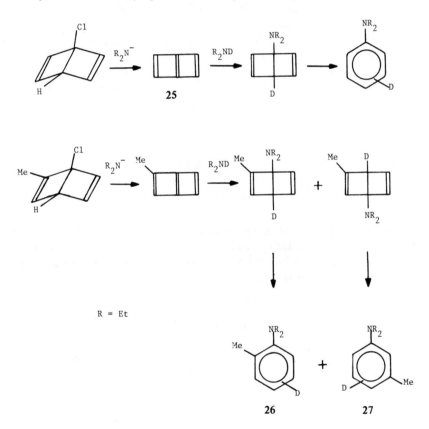

FIGURE 11.   Reaction of butalenes with amines.

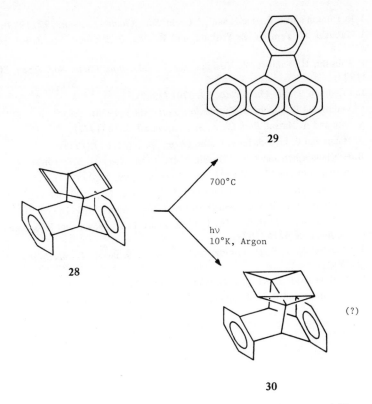

29

28

700°C

hν
10°K, Argon

(?)

30

FIGURE 12.   Thermal and photochemical properties of **28**.

effect substantial conversion. Even at these elevated temperatures, butalene and
p-benzyne do not appear to be released, but rather a deep-seated rearrangement
to benzofluoranthene **29** occurs. Interestingly, it was noted that low-temperature
matrix photoysis of **28** produces a molecule tentatively identified as prismane **30**
(Figure 12). Pyrolysis of **30** may indeed prove intriguing.

V. REFERENCES

1.*   J. O. Noell and M. D. Newton, *J. Am. Chem. Soc.,* **101**, 51 (1979).

2.   J. W. Laing and R. S. Berry, *J. Am. Chem. Soc.,* **98**, 660 (1975).

3.   (a) O. L. Chapman, K. Mattes, C. L. McIntosh, J. Pacansky, G. V. Calder, and G. Orr,
         J. Am. Chem. Soc., **95**, 6134 (1973);
      (b) O. L. Chapman, C.-C Chang, J. Kolc, N. R. Rosenquist, and H. Tomioka, *J. Am.
         Chem. Soc.,* **97**, 6586 (1975).

4.* I. R. Dunkin and J. E. MacDonald, *J. Chem. Soc., Chem. Commun,* 772 (1979).

5.* I. Tabushi, H. Yamada, Z. Yoshida, and R. Oda, *Bull. Chem. Soc. Japan,* **50**, 285 (1977).

6.* I. Tabushi, H. Yamada, Z. Yoshida, and R. Oda, *Bull. Chem. Soc. Japan,* **50**, 291 (1977).

7.* L. Friedman, *J. Am. Chem. Soc.,* **89**, 3701 (1967).

8.   E. Vedejs and R. Shepherd, *Tetrahedron Lett.,* 1863 (1970).

9.* A. Oku and A. Matsui, *Bull. Chem. Soc. Japan,* **50**, 3338 (1977).

10.* S. Mazur and P. Jayalekshmy, *J. Am. Chem. Soc.,* **101**, 677 (1979).

11. B. H. Klanderman and T. R. Criswell, *J. Am. Chem. Soc.,* **91**, 570 (1969).

12.* W. N. Washburn and R. Zahler, *J. Am. Chem. Soc.,* **99**, 2012 (1977).

13.* W. N. Washburn, R. Zahler, and I. Chen, *J. Am. Chem. Soc.,* **100**, 5863 (1978).

14.* W. E. Billups, J. D. Buynak, and D. Butler, *J. Org. Chem.,* **44**, 4218 (1979).

15.* G. C. Johnson, J. J. Stofko, Jr., T. P. Lockhart, D. W. Brown, and R. G. Bergman, *J. Org. Chem.,* **44**, 4215 (1979).

16.* J. Chandrasekhar, E. D. Jemmis, and P. von R. Schleyer, *Tetrahedron Lett.,* 3707 (1979).

17.* R. Breslow and P. L. Khanna, *Tetrahedron Lett.,* 3429 (1977).

18.* G. C. Johnson and R. G. Bergman, *Tetrahedron Lett.,* 2093 (1979).

2
CARBANIONS

STUART W. STALEY and CHARLES K. DUSTMAN

Department of Chemistry, University of Nebraska-Lincoln, Lincoln,
Nebraska 68588

I. INTRODUCTION

The importance of carbanions as synthetic intermediates continues to grow at a significant rate. This class of compounds also continues to provide a fertile testing ground for theories of bonding and delocalization. Studies concerning the interaction of these species with counterion–solvent complexes have become an integral part of research on carbanions in solution. The application of gas-phase techniques promises to provide an increasing amount of new data that, when compared with solution data, will provide new insights into the roles of counterion and solvent in determining reactivity. There is a continuing need for *direct* observation of these unstable species. NMR spectroscopy is the most generally useful technique, but, in our view, the acquisition of direct structural information on alkali metal salts by X-ray diffraction will provide important scientific returns in the near future.

II. SOLID-STATE STRUCTURES

X-Ray diffraction studies of organometallic compounds constitute an important but underutilized source of information concerning cation–anion interactions. Previous work has shown that structural information gained from the solid state can be applied to structures in solution.

1

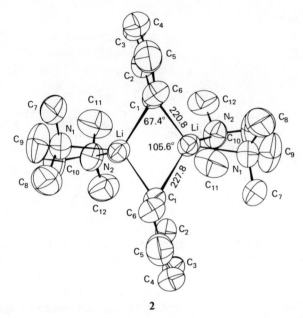

2

Methyllithium (MeLi) has been found to crystallize with tetramethylethylene-diamine (TMEDA) as a complex with stoichiometry $(MeLi)_4(TMEDA)_2$.[1] As shown in 1, the lithium atoms form a tetrahedron, with the carbon atoms of the methyl groups arranged over the faces of the tetrahedron. The lithium–lithium distances are 2.561–2.571 Å, which is considerably larger than the sum of their van der Waals radii (1.36 Å). The TMEDA molecules form bridges between adjacent MeLi tetramers. Methyllithium is known to exist in part as tetrameric species in ether solvents.[2]

The structure of phenyllithium also has been determined as its TMEDA complex.[3] The structure is dimeric, the two phenyl rings being bridged by the lithium atoms, each of which is tetracoordinate as a result of complexation with a TMEDA molecule (2). The lithium–lithium distance is again large (2.49 Å), whereas the lithium–carbon distances are slightly greater (2.208 and 2.278 Å) than those found in ethyllithium (2.188 Å). The incorporation of TMEDA into this structure was essential for structure determination since in its absence crystals of phenyl-lithium suitable for X-ray diffraction analysis have not been obtained.

The sodium atoms in cyclopentadienylsodium TMEDA lie between the cyclo-pentadienyl rings, the whole forming a zigzag chain as shown in 3.[4] The sodium atoms are slightly displaced from a position directly over the center of the five-membered rings, and hence the sodium–carbon distances vary from 2.856 to 2.963 Å. Each sodium atom is coordinated by a TMEDA ligand. Infrared and ^{1}H NMR studies indicate that the system is monomeric in solution.

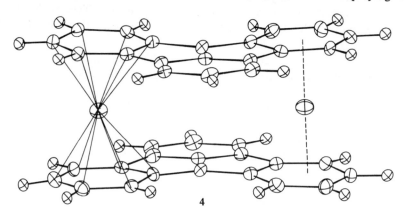

3

Compound **4** crystallizes as a dimer, and some solvent of crystallization (benzene) is also included in the lattice in the ratio of one part **4** to 0.5 part C_6H_6.[5] The two monomers are planar and oriented in an *anti*-arrangement with two lithium atoms coordinated to two of the six-membered rings in a sandwich structure. This is particularly interesting because self-consistent HMO calculations indicate that it is the uncomplexed six-membered ring and the pentalenyl moiety that bear the bulk of the charge (0.5 electron versus 0.25 electron in each of the other two rings). The structure actually adopted reflects the importance of cation–cation repulsions since the cation–cation distance is maximized while stabilizing complexation interactions are retained, even if the latter are not the largest ones possible. The lithium–ring center distance of 1.93 Å is within the sum of the van der Waals radii of lithium and benzene (0.68 + 1.7 = 2.38 Å). The latter distance is comparable to the lithium six-membered ring carbon atom distances, suggesting some covalent interaction. Calculations employing the

4

CNDO/2 method assuming either an electrostatic or a covalent cation–anion interaction gave optimized lithium positions in equally good agreement with experiment.

III. ALKYLLITHIUM COMPOUNDS

Geometries of alkyllithium species and the effects of aggregation on structure and dynamic behavior have been investigated by *ab initio* 4-31G calculations. Planar methyllithium is calculated to be 35.3 kcal/mol less stable than the tetrahedral form, yet the activation barrier for inversion of tetrahedral H_3CLi is known to be 15 kcal/mol.[6] The most stable H_3CLi dimer was calculated to be **5**, and a possible transition state for inversion to **6** was calculated to be 23.3 kcal/mol less stable than **5** at the 4-31G level. These calculations indicate that

aggregation reduces inversion barriers in alkyllithium compounds. Presumably investigation of inversion barriers in trimers or tetramers would improve the agreement between theory and experiment.

Calculations of the structure of dilithiomethane have indicated that the tetrahedral singlet and triplet, as well as the planar triplet, lie within 3.9 kcal/mol of each other.[7] Investigation of the H_2CLi_2 dimer[8] indicated that the head-to-head structure **7**, possessing square planar carbon atoms, is the most stable. The value of $\angle$LiCLi was calculated to be 93.3°. However, the bridged structure **8** was calculated to be only 1.9 kcal/mol less stable than **7**. The occurrence of such small energy differences between calculated structures points out the continued importance of experimental studies.

Aggregation of alkyllithiums in solution has been studied by infrared spectroscopy[9] and by kinetic methods.[10] Some of the most interesting results in this field have arisen from alkali metal NMR data. The natural-abundance ^{6}Li spin lattice relaxation times (T_1) and lithium-proton nuclear Overhauser effect enhancement factors $(\eta_{^6Li-\{^1H\}})$ for methyllithium, n-butyllithium, and phenyllithium were obtained in solvents in which these species are known to be tetrameric, hexameric, and dimeric, respectively.[11] The small solvent-dependent change in $\eta_{^6Li-\{^1H\}}$ for n-BuLi was interpreted as evidence that ^{6}Li relaxation is dominated by intramolecular effects. Quadrupolar relaxation by the ^{6}Li nucleus contributed up to 16.5% to its T_1. It was concluded that the small linewidth of ^{6}Li relative to ^{7}Li could result in a superior spectral resolution despite its lower resonance frequency.

An example illustrating this point is afforded by propyllithium, which has been the subject of an extensive investigation using ^{6}Li and ^{7}Li NMR spectroscopy.[12] The ^{7}Li spectrum of n-propyllithium (90% ^{13}C at C_1) in cyclopentane consists of a single line at higher temperatures. On cooling to below $-40°C$ a new set of signals (heralding the appearance of a new aggregate structure) appears, but no ^{13}C-^{7}Li coupling is observable. Apparently fast intraaggregate cation exchange leads to decoupling at high temperatures, while ^{7}Li quadrupolar relaxation precludes observation of ^{13}C-^{7}Li coupling at lower temperatures. Observation of ^{6}Li and ^{13}C showed a sharp (<1 Hz) ^{6}Li resonance and a nine-line $^{13}C_1$ signal ($J = 3.5$ Hz). From an analysis of the multiplicities and intensities of the resonances, it was concluded that C_1 is bonded to six lithium atoms, implying a hexameric aggregate. The potential value of ^{6}Li NMR spectroscopy in determining alkyllithium structures is readily apparent.

Calculations (MINDO/3 and 4-31G) of cyclopropyl anions bearing chloro or fluoro substituents on the anionic center predict that in the absence of intimate counterion interactions these anions prefer sp^3 hybridization and possess a substantial barrier to inversion[13] (**9**, X = H, 17.4 kcal/mol; X = Cl, 39.7 kcal/mol; X = F, 14.8 kcal/mol). Electron-withdrawing groups can also accelerate the

9

10   **11**

a, X = Cl   a, X = Cl
b, X = Br   b, X = Br
c, X = CH$_3$

rate of formation of cyclopropyl anions since **10a** and **10b** react to afford **11a** and **11b**, whereas **10c** fails to react at all under identical conditions.[14]

Cyclopropyllithium derivative **12** retains its configuration at $-72°C$ in THF,

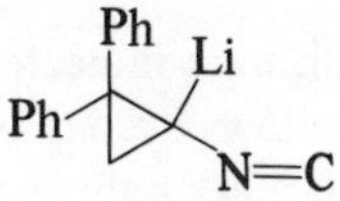

12

whereas racemization occurs on warming to $-5°C$.[15] Racemization was found to be independent of concentration, counterion (Na and K salts were investigated), and solvent. The addition of crown ethers also failed to affect the configurational integrity of **12**. It thus seems that this anion possesses a substantial barrier to inversion even under conditions where it exists as a free ion pair.

A report concerning the possible synthesis of tetralithiotetrahedrane (**13**), a derivative of the elusive tetrahedrane, has been published.[16] The product was obtained as a white powder from photolysis of dilithioacetylene. ^{13}C NMR spectroscopy demonstrated the magnetic equivalence of the carbon atoms, and field desorption mass spectrometry gave two peaks of m/e 76 ($C_4{}^7Li_4$) and m/e 75 ($C_4{}^6Li^7Li_3$). Hydrolysis of **13** afforded acetylene. Tetralithiotetrahedrane is predicted to prefer the face-centered structure **14** by *ab initio* STO-3G calculations and illustrates the tendency for lithium substitution to reduce strain energy. Results of a detailed structure determination of **13** should prove most interesting.

13    **14**

IV. ALLYL ANIONS

The allyl anion has been investigated theoretically at the STO-3G level and the optimized geometry was found to be that in **15**.[17] The $C_1C_2C_3$ bond angle is larger than that calculated for the allyl cation ($123°$) and radical ($127°$). The opening of this angle is attributed to an unfavorable 4π-electron interaction across the termini of the allyl system. Compression of the internal angle to $115°$

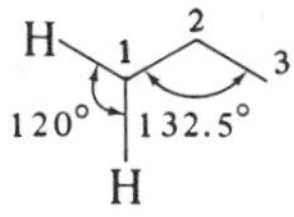

15

results in a 1,3-overlap equivalent to 15% of that of a single bond. The calculated inversion barrier (conversion of a pyramidal to a planar anion) was found to be small (ca. 0.1 kcal/mol), with the energy minimum occurring for a methylene out-of-plane angle of about 10°.

Similar calculations for allyllithium indicated that **16** is the minimum energy conformation.[18] This geometry allows overlap of the allyl HOMO with the empty p_x orbital on lithium, leading to a net stabilization. The bridged species was calculated to be 16 and 18 kcal/mol more stable than **17a** and **17b**, respectively. The barrier to rotation in **16** was found to be 16 kcal/mol at the 6-31G* level.

were found to exist in the Z-form; only the lithium and sodium salts provided measurable amounts of the E-isomer.[19] Substitution of Na and K for Li leads to upfield shifts of the γ-carbon ^{13}C resonance concurrent with a downfield shift of the α-carbon resonance, indicating enhanced delocalization as the cation radius increases. Potassium represents the limit of cation influence on delocalization (insofar as it is detectable by ^{13}C NMR spectroscopy) as the K, Rb, and Cs salts possess essentially identical spectra.

Metalation of 1,3-di-*n*-butoxypropene (**19a**) was found to occur exclusively at a vinylic position rather than at an allylic one.[20] *Ab initio* calculations revealed that the four-electron interactions of an alkoxy substituent with the allyl anion raises the energy of the allyl species to within 2-3 kcal/mol of that of the vinyl anion.[21] This may be compared with an approximately 25 kcal/mol difference between the unsubstituted species. Theoretical kinetic acidities were evaluated

Geometries of organolithium compounds obtained by *ab initio* calculations are often exotic and must be accepted with reservations. These are gas-phase structures, and coordination with solvent or aggregation in solution could drastically alter the geometries of these systems. Indeed, the unusual geometries calculated for these species may in part reflect an absence of opportunities for external coordination of the lithium atoms.

In an extensive investigation of substituted allyl anions (**18**) most of the salts

by a Mulliken population analysis and it was found that the presence of a hydroxy group lowered the population density of the α-vinylic carbon–proton bond below that of the allylic carbon–proton bond, a reversal of the situation in the propenyl system (**19b**).

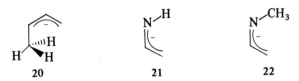

19a   19b

Ab initio 4-31G calculations of the crotyl anion[22] predict the *cis*-configuration **20** to be stabilized by 1.5 kcal/mol relative to the *trans*. This is supported by experimental data for the potassium salt indicating that the *cis* form is preferred by 2.2 kcal/mol over the *trans* form. PMO considerations suggest that the methyl group prefers the staggered conformation, which permits a stabilizing cyclic six-electron through-space interaction.

A recently published paper argues instead that conformational preferences can be explained by electrostatic interactions.[23] This argument is based on 4-31G calculations of acetaldehyde imine **21**, which is found by both experiment and calculation to prefer the *syn*-conformation because it minimizes the lone pair–C_3 electrostatic repulsion and maximizes the attractive interaction between the negatively charged C_3 and the positively charged N—H proton. The N-methyl derivative **22** also prefers the *syn*-conformation for similar reasons. Extension of this reasoning to the crotyl anion is not as straightforward, however, as the lone pair–C_3 interaction is lacking and the positive charge on the methyl protons of **20** is undoubtedly less than that in **22**. Although these considerations would lead to a reduced preference for the *cis*-crotyl anion as observed (i.e., *cis*-crotyl anion is 2.2 kcal/mol more stable than the *trans*-isomer, whereas *syn*-**22** is 6.2 kcal/mol more stable than *anti*-**22**), the relative importance of electrostatic effects and through-space attractive interactions is not clear.

20   21   22

V. OTHER ACYCLIC CARBANIONS

Studies of the rate of deuterium exchange and epimerization have led to the conclusion that the inversion barrier separating **23** and **24** must be at least 22 kcal/mol.[24] It was suggested that the barrier in the parent allenes is probably lower than this value.

NMR (^{1}H and ^{13}C) and optical spectroscopic studies of **25** indicated that this anion exists predominantly in a "planar" structure.[25] Interestingly, protonation occurs primarily at the less highly charged benzylic position to afford a 3:1 mixture of **26** and **27**, respectively. The foregoing result once again demonstrates the complexity of the factors that influence the position of protonation and alkylation in delocalized carbanions.

Metalation of isobutene with potassium *t*-butoxide and *n*-butyllithium in hexane at −20°C afforded the potassium 2-methallyl anion **28**, which was characterized by IR spectroscopy.[26] The low carbon–potassium stretching frequency (207 cm^{-1}) was taken as evidence for the ionic nature of the C−K bond. A force constant of 0.55 mdyn/Å was derived for the C−K stretch (compared to 0.31 mdyn/Å for allylsodium and about 0.44 mdyn/Å for allyllithium). Further treatment of **28** with KOt-Bu/*n*-BuLi led to a decrease in the intensity of absorptions due to **28** and the appearance of a new set of bands that were attributed to the trimethylenemethane dianion **29**.

The conformations of acyclic delocalized anions have long been an area of interest. Oxidation [FB(OR)$_2$ then H$_2$O$_2$] of **30**, **33**, and **35** afforded the product mixtures shown.[27] Oxidation occurs only at the terminal carbons and does not interconvert isomers. The anions probably exist as contact ion pairs. It was concluded that several factors affect the anion conformations. The W form (exemplified by **33**) would be preferred for unsubstituted pentadienyl anions because of the minimization of interactions among C−H groups. Introduction of a methyl group at the 2- or 4-position renders the U form (see **30** and

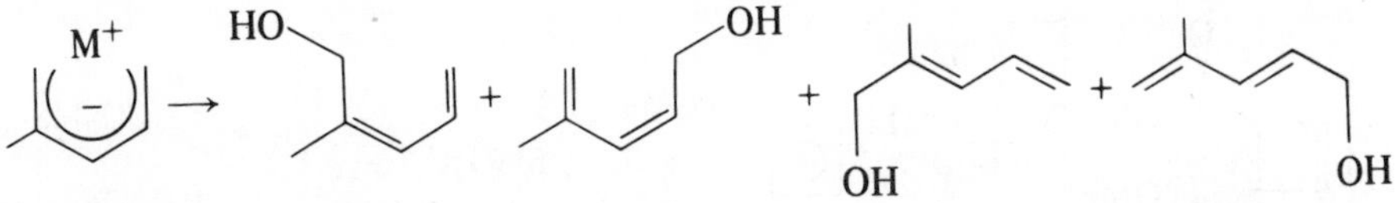

30   31a   31b   32a   32b

M = Li   12%   10%

M = K   22%   41%

55%

2%

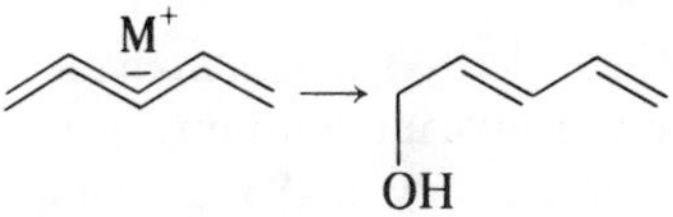

33, M = Li, K   34

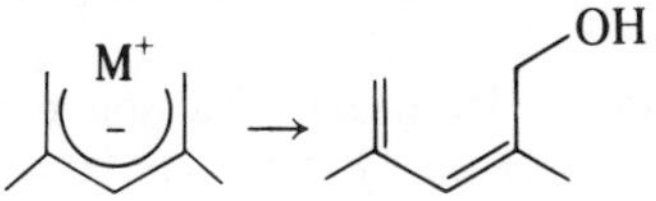

35, M = Li, K   36    37

35) almost as stable as the W form. This would be the expected order of stability for free ions. Consideration of the cation–anion interaction shows that lithium favors η^3 coordination, and hence W and S (or sickle, see 37) geometries, whereas potassium favors η^5 coordination and a U geometry.

In the unsubstituted pentadienyl case, the W form is so sterically favored that η^5 coordination via the U form with potassium is relatively energetically unfavorable, and hence both the Li and K salts of 33 exist in the W form. In methylated pentadienyl anion 30, the lithium salt produces a mixture of U and W conformations, whereas the larger potassium cation so favors η^5 coordination that the U form predominates. Finally, the interactions between methyl groups in 35 destabilize the W form so much that only the U form is observed for both counterions.

VI. CYCLIC DELOCALIZED CARBANIONS

The cyclobutadienyl dianion is predicted to be an aromatic system by HMO theory, yet this ion has not been observed. In a recent investigation of the 1,2-dicarbomethoxy derivative 38, it was concluded on the basis of ^{13}C NMR spectroscopy that only 40% of the excess charge resides in the four-membered ring.[28] As can be seen by the fractional charges given in the diagram, the bulk of this charge is localized on the substituted carbon. The pK_a of 38 was found to be similar to that of 39, suggesting that 38 is about as stable as its open chain

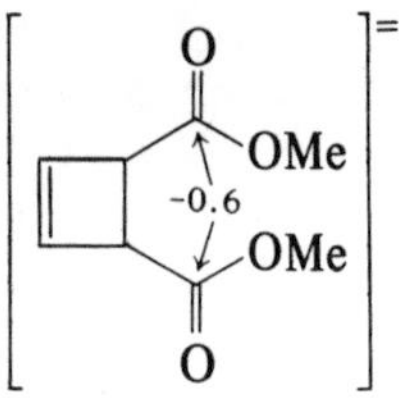

38 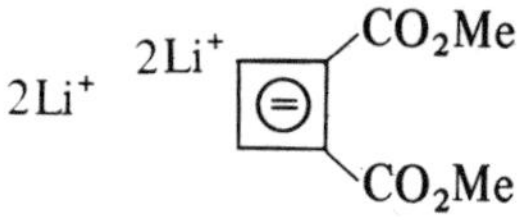   38a 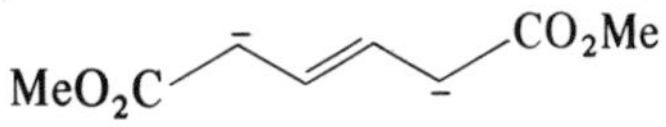   39

analogue. These results led the authors to conclude that **38** is nonaromatic, that is, resonance forms such as **38a** do not make a significant contribution to the ground state. However, the lack of evidence for aromaticity in **38** should not be taken as evidence for the nonaromaticity of the parent cyclobutadienyl dianion, since the charge localization induced by the carbomethoxy substituents is lacking in the parent dianion.

Ion-pairing effects have been shown to exhibit a significant effect on the equilibrium (Eq. 1) of the acetylcyclononatetraenyl anion (**40** $\rightleftharpoons$ **41**).[29] At $-52°C$ in THF, only **41** is observed by ^{1}H NMR spectroscopy. Upon warming, both **40** and **41** are observed, and at $25°C$ only signals due to **40** are present. The $\Delta G^{\ddagger}$ for this process was found to be $\leqslant 10.2$ kcal/mol. High temperatures favor contact ion pairs, which leads to **40** being the more stable structure, whereas lowering the temperature favors solvent-separated ion pairs and leads to the delocalized anion **41** as the more stable species. Support for this interpretation is provided by the observation that in dimethoxyethane (a stronger coordinating solvent than THF) only **41** is observed.

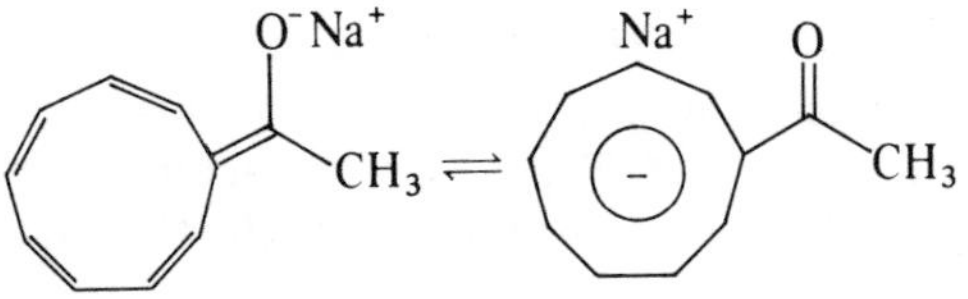

40   41

A systematic study of ion-pairing effects on the ^{13}C NMR spectra of anions has appeared.[30] Variable-temperature ^{13}C NMR spectra of the triphenylmethyl (**42**), fluorenyl (**43**), diphenylmethyl (**44**), trimethylsilylbenzyl (**45**), and benzyl

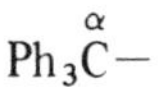

42 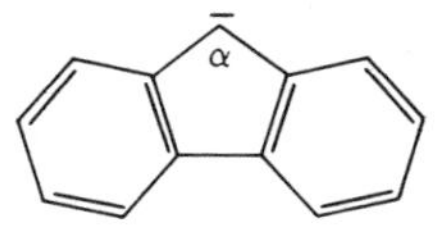   43 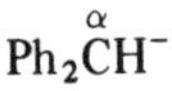   44 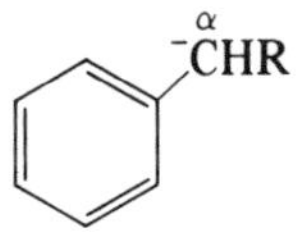   45, R = Me$_3$Si
46, R = H

(46) anions with various cations were obtained in a variety of solvents and indicate that raising the temperature displaces the α-carbon chemical shifts of the Li and Na salts to higher fields. This is attributed to a change from solvent-separated to contact ion pairs as the temperature is increased. The K, Rb, and Cs salts did not display a significant temperature dependence, either because no appreciable change in solvation occurred under the experimental conditions or because ^{13}C NMR is an insensitive probe of solvation effects in these cases. It was also noted that the α-carbon chemical shifts decrease as the cation size increases. This was attributed to the decreased ability of cations of large radius to polarize charge density toward the α-carbon.

It has been suggested that cations influence the ^{1}H NMR chemical shifts of anions in two ways. A direct effect arises from polarization of C—H electrons by the electric field of the cation. An indirect effect is caused by redistribution of π-electron density due to the cation. The altered electron density at a carbon atom produces a polarization of the C—H electrons. These predictions have been investigated in the case of the indenyl anion 47.[31]

Variable temperature ^{13}C NMR spectroscopy revealed that at -60°C in DME the chemical shifts of 47a and 47b were almost equal, indicating that both exist

47

a, M = Li

b, M = Na

primarily as solvent-separated ion pairs. The resonances change with temperature in a nonlinear fashion. In both cases the five-membered ring signals move to higher fields with increasing temperature, while C_4, C_5, C_6, and C_9 shift to lower fields. This is taken to mean that charge becomes concentrated in the five-membered ring upon contact-ion-pair formation, a result in accordance with conclusions based on optical and ^{1}H NMR spectroscopy.[32] The observed shifts allow prediction of the magnitude of the indirect effect on ^{1}H NMR spectra since proton and carbon charge-induced shifts are related by a factor of about 1/16. It was found that the indirect and direct effects are similar in magnitude.[33] Although all the ^{1}H resonances move downfield with increasing temperature, the direct effect is largest for H_1, H_2, and H_3, in consonance with the cation becoming localized over the five-membered ring in the contact ion pair. Although the uncertainty in the value of the factor of 1/16 is well known, it does not seriously affect the conclusions of the studies.

Dianions of pyrene 48 and its isomers 49 and 50 have been studied by ^{1}H

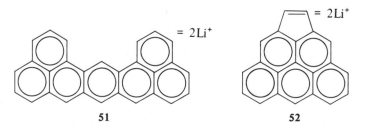

48   49   50

NMR spectroscopy.[34] The centers of gravity of the resonances shift upfield on formation of the dianion from the neutral hydrocarbons by 7.2, 3.2, and 1.3 ppm, for **48**, **49**, and **50**, respectively. The predicted charge-induced upfield shift for dianion formation is 1.3 ppm, in agreement with the shift for **50** only. The ethylene bridge lies on a nodal plane of the HOMO in **48** and **49**, whereas in **50** the corresponding orbital possesses large coefficients there. The excess charge in **48** and **49** is therefore restricted to the perimeter of the dianion, producing a 16π-electron antiaromatic system. The paramagnetic ring current thus induced gives rise to the upfield shifts in excess of those predicted for charge effects alone. PPP and CNDO/2 calculations support the perimeter charge model and also predict a higher degree of bond alternation in **49** than in **48**, which leads to a reduced paramagnetic ring current in **49**. It should be noted that this represents a zeroeth order approach, since the paratropism in **48** actually arises from the magnetic dipole-allowed mixing of low-lying excited states with the ground state.

Several polycyclic dianions have been reported for whose parent hydrocarbons no Kekulé structures can be drawn. Both **51**[35] and **52**[36] are stable species in

$= 2Li^+$

$= 2Li^+$

51   52

THF under the experimental conditions. Plots of ring-current-corrected ^1H chemical shifts versus HMO charge densities were linear in both cases.

Benzannelated anions continue to be extensively investigated. Dianion **53** possesses a paramagnetic ring current as determined by ^1H NMR spectroscopy and is best described as possessing a 16π-electron periphery.[37] The center of gravity of the proton chemical shifts for **53** appears about 2.5 ppm upfield of that for **54**. This has been taken to reflect the inverse dependence of the magnitude of paramagnetic ring currents on the HOMO-LUMO energy difference, since that calculated for **53** is less than that for **54**.

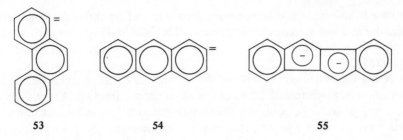

53   54   55

The charge-induced upfield shift of the five-membered ring protons in penta-lenyl dianion **55** is canceled by the development of a diamagnetic ring current, suggesting that **55** is best regarded as an 18π-electron annulene.[38] Experimentally determined charge densities were found to be in good agreement with those calculated from the HMO ω-technique.

Treatment of tetraene **56** with n-BuLi in THF-d_8 produced a deep-red solution that was considered to contain the nonplanar delocalized dianion **57** on the basis of its NMR spectrum.[39] While the product seems reasonable enough, the reported coupling constants (δ 2.92 (t, 2, $J = 7.0$ Hz, H_6, H_{14}), 4.62 (d, 4, $J = 12$ Hz, H_4, H_8, H_{12}, H_{16}), 5.86 (t, 4, $J_1 = J_2 = 7.0$ Hz, H_5, H_7, H_{13}, H_{15}) are not internally consistent. This point must be cleared up before the structure can be considered to be secure.

The spectrum of **57** slowly disappeared at room temperature and was replaced with a new spectrum (δ 6.35; t, 2, $J = 4.0$ Hz, H_6, H_{14}), 6.67 (d, 4, $J = 5.0$ Hz, H_4, H_8, H_{12}, H_{16}), 7.01 (t, 4, $J = 5.0$ Hz, H_5, H_7, H_{13}, H_{15}). This was formulated as a planar delocalized dianion (**58**). Vicinal coupling constants of 4 Hz are

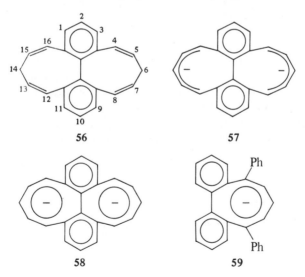

56   57

58   59

unprecedented for a planar nine-membered ring and we therefore do not consider the assigned structure to be correct. This problem clearly requires further investigation. The same reservations apply to the reported generation of a related anion (59).[40]

It was concluded on the basis of the relatively low field chemical shift of the methano bridge protons (δ 2.64) of the anion derived from 60, that this species does not possess a delocalized homophenalanide structure (61a), but instead is best represented as a heptatrienide ion (61b). However, this analysis does not

60	61a	61b

take into consideration the paramagnetic component of the anisotropy of 61a. This effect might very well account for the low field chemical shift.

Compounds 62a and 62b gave 63a and 63b, respectively, on deprotonation.[42]

62	63
a, R = H	a, R = H
b, R = Me	b, R = Me

The bridgehead carbons in 63 exhibit ^{13}C resonances far downfield from those in 62, suggesting a decrease in s-character upon anion formation. The benzo protons are also shifted downfield, indicating the development of an enhanced diamagnetic ring current in the six-membered ring.

An interesting example of charge delocalization in a three-dimensional framework is afforded by dianion 65, obtained by lithium metal reduction of 64

64	65

in THF-d_8.[43] Upon reduction, the phenyl proton resonances shift upfield, as do those of the aliphatic protons. This is interpreted as resulting from charge delocalization over the entire carbon skeleton. Oxidation (O_2, $-78°C$) causes the regeneration of **64** along with some polymer. An ion-pairing effect is indicated, as the potassium and sodium salts of **65** decompose at $-78°C$ in ether to give stilbene and the naphthalene radical anion.

Reports of anions possessing more than two charges have been appearing in the literature and we expect this trend to increase in the near future. The bis-

66   **67**

cyclooctatetraene tetraanion **66** exists as two more or less orthogonal cyclooctatetraene rings and is unexceptional.[44] The tetraanion of octalene (**67**) displays a diamagnetic ring current, as would be expected for a planar 18π-electron periphery.[45]

Evidence has been obtained for the existence of the cycloheptatrienyl trianion **68**, a 10π-electron aromatic system that formally possesses a remarkable 1.43π-electrons on each carbon atom.[46] The "aromatic" stability of this system is illustrated by the fact that trismetalation of **69** affords **68**, probably by elimina-

68   **69**   **70**

tion of LiH from dianion **70**, followed by electrocyclic ring closure and further deprotonation. The existence of the highly resonance-stabilized 1,2,3-triphenyl-allyl trianion (**71**) has been inferred from trapping experiments.[47] This trianion is stable at room temperature as the Li, Na, K, or Cs salt.

71

E = D(Me)

One of the key problems with regard to the structure of carbanions concerns the extent to which the requirement for charge dispersal or electron delocalization competes with the need to minimize bond angle strain and "antiaromatic" destabilization in cyclic $4n\pi$-electron carbanions. The first factor promotes a planar structure, whereas the latter two promote a folded structure. To date, no direct structural evidence, such as X-ray diffraction data, is available. However, in one case the indirect evidence of vicinal H—H coupling constants has indicated that a cyclic 12π-electron anion is nonplanar.[48] The relative importance of the above factors, as well as of others, such as the effect of interaction with the counterion, continues to be an important problem. In this regard, several recent studies are of interest.

Dicyclooctatetraeno[1,2:4,5]benzene (72) has been reduced by electrochemical methods and by treatment with Na/K alloy in THF-d_8 to afford the corresponding dianion (72^{2-}).[49] It was noted that the eight-membered ring protons were shifted upfield by about 1 ppm in the NMR spectrum on reduction, whereas the corresponding protons in 74 were all shifted downfield on reduction to dianion 75. This contrasting behavior was taken as evidence against 72^{2-} existing as 73a ⇌ 73b in rapid equilibrium. It was concluded that the available data are in accord with planar structure 73c.

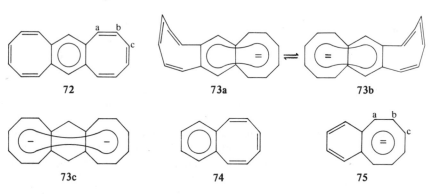

Interestingly, reduction of 76 to the corresponding dianion in THF-d_8 leads to an upfield shift of the eight-membered ring protons.[37] An examination of ESR data, as well as ^1H and ^{13}C NMR spectra led to the conclusion that the dianion approaches the planar or nearly planar "border line" structure 77.

What is interesting about the foregoing results is that the upfield shift of the eight-membered ring protons was deemed to be compatible with structure 77 in one study but incompatible with equilibrating structures 73a ⇌ 73b in another. The difference cannot be due to the different counterions employed because the chemical shifts of the lithium, sodium, and potassium salts of cyclooctatetraene are nearly the same (to within ± 0.06 ppm) in THF.[50]

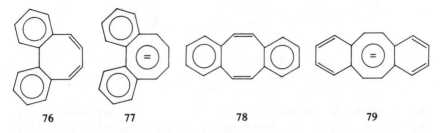

| 76 | 77 | 78 | 79 |

Another parameter beside ^{1}H and ^{13}C chemical shifts that bears on the problem of determining the structure of these dianions is the value of $J_{^{13}C-H}$ in **72** and **76** and their corresponding dianions. Typically there is a decrease of about 10 Hz in $J_{^{13}C-H}$ on going from a cyclooctatetraene to its dianion, as exemplified by the values of 155 and 145 Hz for cyclooctatetraene (COT) and COT^{2-}, respectively.[49] Similar results have been reported for **77** and **79** and for H_c of **75**. Decreases of only 6.5 and 8 Hz occur for H_a and H_b of **75**. Two factors that influence this decrease can be identified. One is the addition of negative charge to the molecule; this has been estimated to cause a decrease of about 2.5 Hz in COT^{2-}.[51] The second is the decrease in s character in the carbon orbital of the C—H bond as $\angle CCC$ is increased from about $120°$ to about $135°$. (It has been shown by X-ray diffraction that COT^{2-} is planar.[52]) It is this influence of hybridization on $J_{^{13}C-H}$ that provides structural information.

In view of the foregoing it is of considerable interest that $J_{^{13}C-H}$ values for the eight-membered rings protons in **72** decrease from 156–158 Hz to 145, 150 (J_a and J_b), and 151 Hz (J_c) on reduction to the dianion. The latter two values are somewhat higher than other values reported for cyclooctatetraene dianions (143–148 Hz). One possible explanation for this observation is that 72^{2-} is not planar. A structure not previously considered is one in which *both* eight-membered rings in 72^{2-} are bent simultaneously, as in structure **73d**. Such an arrangement might provide a reasonable compromise between the requirement for charge delocalization in 72^{2-} and the need to minimize the relative destabilization of 20π-electron perimeter. Such a distortion is consistent with the following facts: (1) The b—c bond in **73c** is calculated (by the Hückel ω-technique) to have the lowest π bond order; (2) the previously mentioned values for J_{C-H_c} and possibly also for J_{C-H_b} are slightly higher than those for other eight-membered ring protons on a time-averaged basis; and (3) the signals for the eight-membered ring carbons in **74**, **76**, and **78** move upfield by about 40–50 ppm on reduction to the corresponding dianion, whereas C_a and C_b in the less densely charged 72^{2-} move upfield by 25 ppm, but C_c shifts by only 14 ppm.

While a "deconjugation" of position c in 72^{2-}, coupled with $\angle CCC$ at C_b and C_c being $< 135°$, is consistent with structure **73d**, caution must be exercised. The influence of charge and structure on ^{13}C chemical shifts requires much fur-

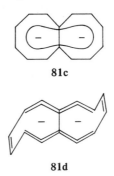

73d

ther elucidation and the influence of charge on J_{13C-H} is not well understood. [For example, J_{13C-H} for *all* of the benzo protons decreases (from 5 to 11 Hz) on reduction of **76** to **77** even though there is presumably no change in planarity.] It is clear that a detailed study of the relationship between structure and J_{13C-H} would be of great value.

In a related study, the dianion of octalene (**80**) has been generated by reduction with lithium in THF-d_8 and observed by ^{1}H and ^{13}C NMR spectroscopy.[45] As in the case of **72**$^{2-}$, the structure of **80**$^{2-}$ is in doubt. It was suggested that the ^{1}H NMR spectrum supports equilibrating structures **81a**⇌**81b**, but it was also observed that the chemical shifts of **80**$^{2-}$ (C$_a$ in **80**$^{2-}$ shifts upfield by only 2 ppm

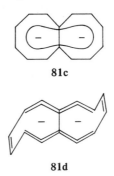

80   81a   81b

compared to the average for C$_a$ and C$_{a'}$ in **80**) can be predicted on the basis of π charge densities. However, the latter prediction would require a structure more akin to the fully delocalized **81c**. It is possible that the ^{1}H and ^{13}C NMR spectra of **80**$^{2-}$ can be reconciled by postulating a folded structure such as **81d**. Such a structure would be less paratropic than **81c** and might also accommodate the

81c

81d

relative downfield position of C$_a$. This leads to the prediction that J_{13C-H} would be greater at position c (and possibly also at b) compared to a. Clearly, more research is required but precedent already exists for a folding of the ring in cyclic $4n\pi$-electron anions.[48]

VII. HOMOCONJUGATION

Homoaromaticity in delocalized anions has been demonstrated in only a small number of cases. Hence much effort has been expended on discovering new anions that possess this property. The most important experimental method used in searching for homoaromatic behavior is NMR spectroscopy. Homoaromaticity is expected to result in ring currents (detectable by ^{1}H NMR spectroscopy) and in a partial equalization of π charge densities relative to those observed for acylic delocalized anions.

Examination of the ^{13}C NMR spectra of a series of cyclohexadienyl anions, including 82–85, failed to reveal compelling evidence for homoaromaticity in this

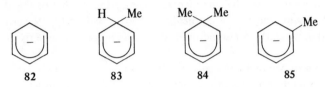

82   83   84   85

series of compounds.[53] Although the chemical shifts of the terminal sp^2 carbons are sensitive to substitution at C_6, there is little deviation from pentadienyl anion-type shifts in the rest of the π-system.

The conclusion has been reached that 86 and 87 are homoaromatic based on

86   87

the large NMR chemical shift differences between H_a and H_b (1.60 and 1.95 ppm, respectively).[54] We view this conclusion as somewhat tenuous because of the lack of evidence for homoaromatic character in the parent cyclohexadienyl anion (82). Inspection of molecular models suggests that the distance between C_1 and C_5 would be too large for significant π-overlap. An alternative explanation for the above chemical shift differences is that the eight-membered ring becomes conformationally rigid on anion formation, resulting in different anisotropic effects of the benzo rings on H_a and H_b.

Other potentially homoaromatic anions recently have been reviewed.[55] By and large, there is relatively little evidence for mono-, bis-, or trishomoaromatic anions in the literature.

A second type of homoconjugation is spiroconjugation. The first spiroanion with π-bonding in both rings (88) recently has been reported.[56] Interestingly, the carbon resonances of C_3 and C_5 are shifted upfield from those of 89, while those of the α- and β-carbons are deshielded and shielded, respectively, relative to those of 90. These complex changes may not be simply interpretable in terms

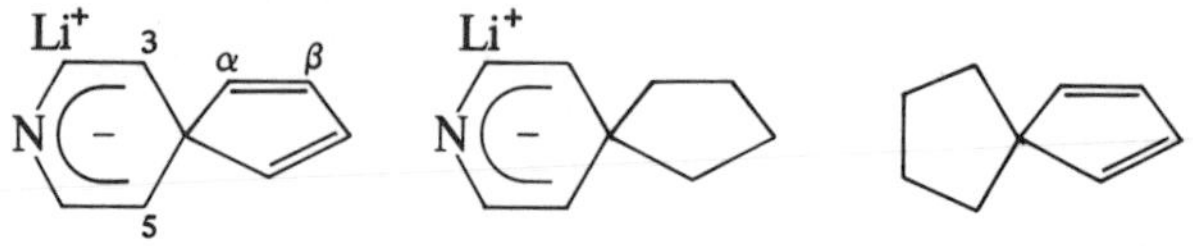

88   89   90

of charge transfer from the cyclopentadiene π-system to that of the charged ring since this would lead to the deshielding of C_β as well as C_α. Nevertheless, it is possible that evidence for some interesting interactions will emerge from studies of 88.

A third area of homoconjugation is that of longicyclic stabilization and destabilization. An interesting series of methylation results, exemplified by the enolate anion of barrelenone (91), is summarized in Table 2.1.[57] It is readily seen that

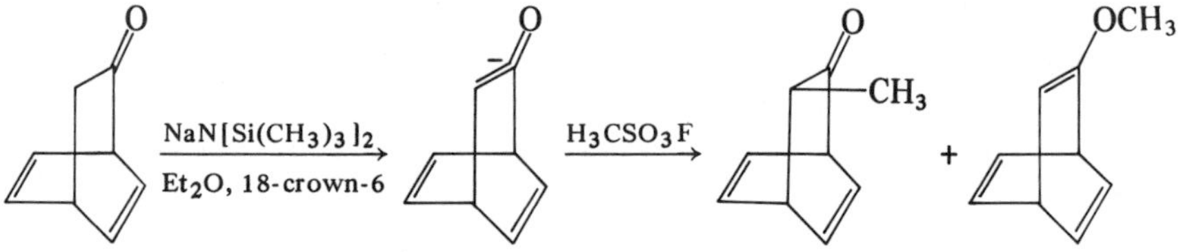

91

TABLE 2.1.   Methylation of the Na Salts of Bicyclic Ketones[a] with Methyl Fluorosulfate in the Presence of 18-Crown-6 at 78°C in Ether

Ketone	$P_{O/C} = l_O/l_C{}^{b}$	$\varrho_O/\varrho_C{}^{c}$
Barrelenone (91)	1.63	0.05
Benzobarrelenone	2.02	0.10
Dibenzobarrelenone (95)	2.33	0.35
Dihydrobarrelenone (96)	2.06	0.53
Dihydrobenzobarrelenone	2.39	0.87
Tetrahydrobarrelenone	2.78^{d}	1.90

[a] Prepared with $NaN[Si(CH_3)_3]_2$ in Et_2O at $-78°C$.
[b] l = HMO charge, calculated with $h_O = 1.0$, $k_{CC} = k_{CO} = 1.0$, homoconjugation $k_{CC} = \pm 0.5$.
[c] Yield ratio percent O-methylation to percent C-methylation.
[d] Calculated for the anion of 2-butanone.

the enolate anions calculated to be least polar are also the ones that are least reactive at oxygen compared to carbon. This trend is also supported by the observation that 92 is alkylated to afford a 9 : 1 mixture of 93 and 94.[58]

92

93   94

The foregoing results can be explained by a decrease in the "bicycloaromatic" stabilization in the transition state leading to C alkylation as the polarity parameter $P_{O/C}$ increases. This rationale appears reasonable, even though the homoconjugative value for β (± 0.5) is somewhat exaggerated, and other factors such as steric effects may also play a role (see 95 and 96, Table 2.2). It would be most interesting to see how the product ratio (Q_O/Q_C) changes with ion-pairing effects.

The cyclopentadienyl anion and tropyl cation analogues of triptycene (97 and 99) have been studied in order to assess the effects of through-space interactions between the aromatic rings.[59] The UV maxima of 97 at longer wavelengths showed small hypsochromic shifts relative to those for triptycene (98), whereas the spectrum of 99 exhibited a strong charge-transfer band. Furthermore, the proton and ^{13}C chemical shifts in these three compounds also provide evidence for charge interaction. Two observations are of particular interest. First, the values of $J_{^{13}C-H}$ for the benzo rings increase in the order 97 $<$ 98 $<$ 99. A parallel observation is that the $J_{^{13}C-H}$ values in the five-membered ring of 97 increase relative to that in the cyclopentadienyl anion (157 Hz), whereas those in the seven-membered ring of 99 decrease relative to that in the tropyl cation (171 Hz). Since, as previously discussed,[51] negative charge may increase $J_{^{13}C-H}$, whereas positive charge decreases it, these changes are all consistent with a transfer of charge from the ionic rings to the benzo rings in 97 and 99.

A second set of observations of interest is that 97 is protonated to afford 100

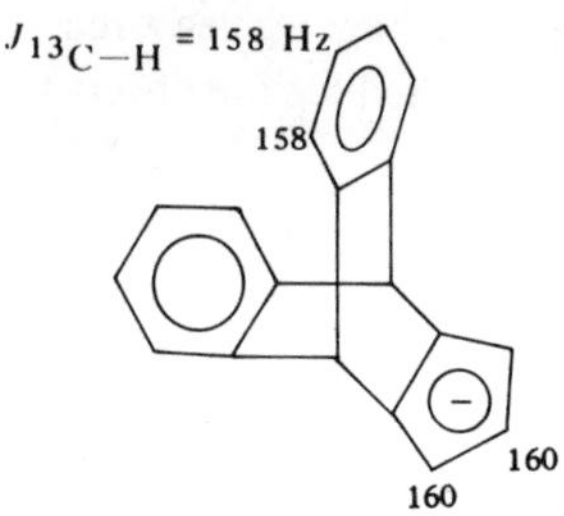

97 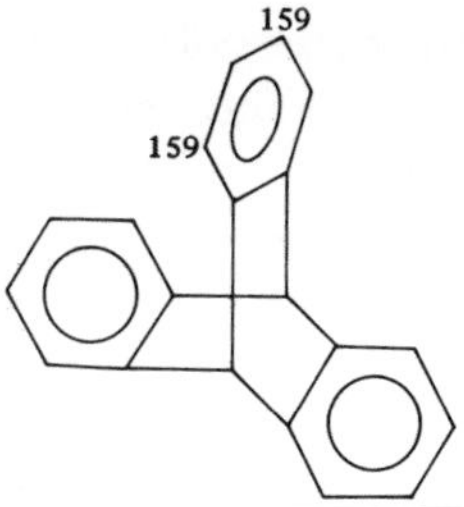   98 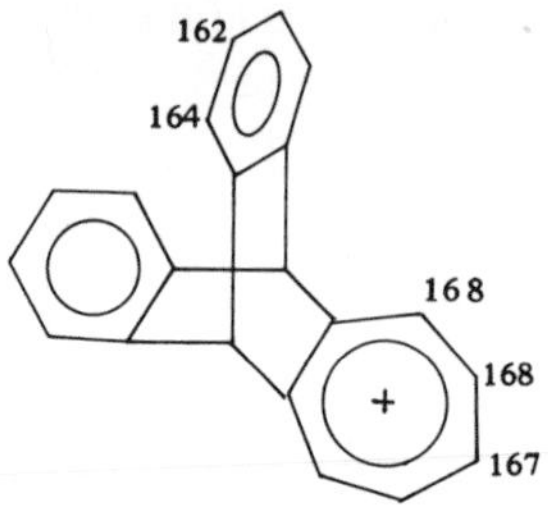   99

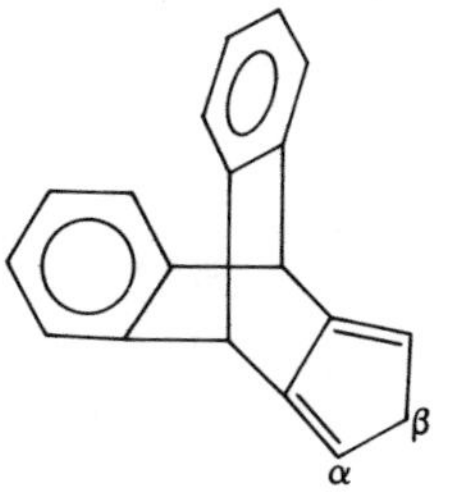

100 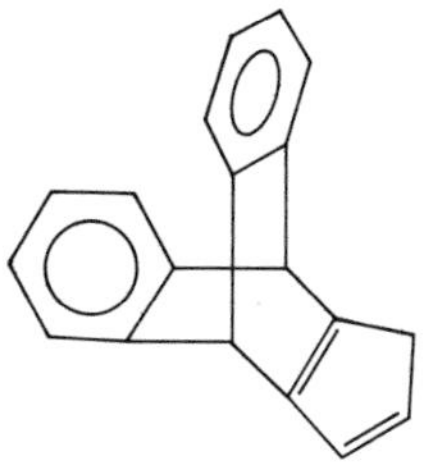   101

and H–D exchange in the five-membered ring of **100** occurs much more readily at the β- than at the α-position.[59] The latter result might be interpreted as reflecting a reluctance of the protonated product to adopt an energetically unfavorable "antibicycloaromatic" structure (**101**), since **100** appears to be a thermodynamic as well as a kinetic product.

The "bicycloaromatic" anion **103** has been proposed as an intermediate in the reduction of **102** by lithium followed by methanol at −78°C to afford **104**.[60]

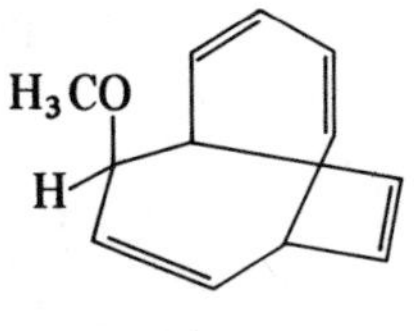

102 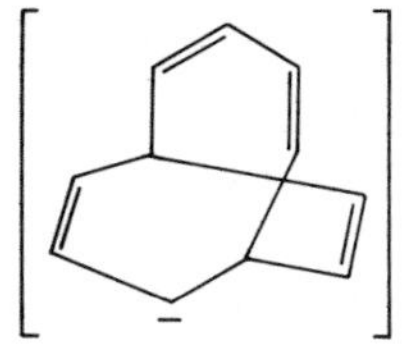   103 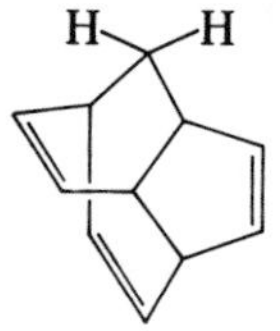   104

It was concluded that the magnitude of longicyclic stabilization in **103** is not sufficient to prevent rearrangements under these conditions.

VIII. DYNAMIC BEHAVIOR

Rotational barriers in delocalized carbanions are quite often low enough to be investigated by NMR spectroscopy. The allyl anion was found to have a maximum rotational barrier in **105a** of 18.0 ± 0.3 kcal/mol when M^+ is cesium.[61] This represents a new lower limit to the rotational barrier in the free allyl anion, which has been calculated to be about 25–29 kcal/mol. The rotational barrier

about the C_2-C_3 bond in *Z*-**105c** (M = K) was found to be 17.0 kcal/mol, while that for the *E*-isomer was $\leqslant 14.0$ kcal/mol. The equilibrium isomer distribution indicated the *E*-isomer to be less stable than the *Z*-isomer by 0.4 kcal/mol, implying that the transition state for rotation of the *Z*-isomer is 2.6 kcal/mol less stable than that of the *E*-isomer. The authors propose that because of steric crowding in the *syn* transition state **106a**, rotation in the *Z*-isomer occurs via an anti arrangement **(106b)** of the cation with the allyl moiety. The *E*-isomer rotates via the *syn*-conformation while maintaining cation overlap with both charged centers of the allyl anion, leading to a net stabilization of 2.6 kcal/mol.

Line shape analysis of the ^{13}C NMR spectra of **107** and **108** at various temperatures allowed the determination of the activation parameters for rotation

about the C_1-C_2 bond.[62] The free energies of activation are identical within experimental error. The ground state of **108** is more sterically crowded than that of **107**, but the transition state for rotation of **108** is less stabilized by resonance than that of **107**. These two effects apparently cancel, resulting in activation barriers of similar magnitude.

An interesting substituent effect on the barrier to rotation in the substituted diphenylmethyl anion **109** has been reported.[63] Variable temperature ^1H and ^{13}C NMR investigations reveal that the methyl-substituted ring rotates about the exocyclic bond at a faster rate than does the deuterium-substituted ring. The authors

D—⟨ ⟩—CH⁻ (109)   D—⟨ ⟩—CH⁻ (110)   D▬▬—CH⁻ (111)

109   **110**   **111**

argue that the inductive effect of the methyl group polarizes electron density away from the 1'-7' bond, leading to the lower π-bond order and a more facile rotation. An equivalent argument can be formulated in transition state terminology, namely, that the methyl group destabilizes transition state **111** more than it does transition state **110**.

Cross-conjugated dianion **112** possesses a variety of rotational barriers.[64] The activation energies, as determined by ^{1}H NMR spectroscopy, are $\Delta G^{\ddagger}_{12} = 11.5$ kcal/mol and $\Delta G^{\ddagger}_{45} = 14.5$ kcal/mol. Rotation about the C_2–C_3 bond was too rapid to be measured, and hence **112** exists as a mixture of all possible conformers. The order of activation barriers follows the HMO π-bond orders ($\rho_{12} =$

112   **113**

0.56; $\rho_{23} = 0.54$; $\rho_{45} = 0.78$). Monoanion **113** was also synthesized and is believed to exist in the indicated conformation.

IX. PHOTOCHEMISTRY AND ION PAIRING IN EXCITED STATES

The nature of cation–anion interactions in excited states can be investigated by optical absorption and emission spectroscopy. Absorption spectra of contact ion pairs generally exhibit blue shifts with decreasing cation radius because of the inability of the cation to change its position so as to stabilize the more delocalized excited state during the very short excitation process. This behavior has been reported for the di- and triphenylmethyl anions[65] and for **114** and **115**.[66] Counterion-dependent red shifts were noted in the emission spectra of ions **114** and **115**. These are attributed to stabilization of the excited state relative to the ground state of the anion. The magnitudes of these shifts were smaller than those of the blue shifts of the absorption spectra, presumably because cation stabiliza-

114   115

tion of the delocalized excited state is not as effective as stabilization of the more localized ground state. Plots of the reciprocal of the cation radius versus emission wavenumber were linear, providing evidence for excited state ion pairs. Some systems, such as Li^+ 114 in tetrahydropyran, did not fall on the line. These deviations were found to be due to appreciable concentrations of solvent-separated ion pairs.

The structures of the ground and excited states of the indolyl anion (116) have been the subject of an extensive investigation.[67] Of the first three UV maxima (representing the first three singlet excited states) S_1 and S_3 were sensitive to the cation, solvent, and temperature, whereas S_2 was relatively unaffected.

116   117

Generally, such shifts are attributed to a change in the ion-pairing equilibrium, with decreased contact ion pairing producing a bathochromic shift. The authors point out, however, that the cation–anion interaction in 116 has two causes: first, an electrostatic π-cloud–cation interaction, and second, a nitrogen lone-pair dipole–cation interaction. Increasing delocalization will allow the dipole interaction to become relatively more important, and at the limit will favor nitrogen-cation coordination.

The ^7Li chemical shift is known to be a function of the nature of ion pairing, being 6-9 ppm upfield from $1M$ aqueous LiCl for a contact ion pair and 1-3 ppm upfield for a solvent-separated ion pair. Lithium–nitrogen σ-complexes lead to a further downfield ^7Li shift. In the carbazolyl anion (117) the ^7Li resonance appears at -0.10 ppm in DME. Since the indolyl anion ^7Li resonance is found at -0.16 ppm, the authors conclude that the cation forms a σ-complex with the nitrogen atom. The authors propose that the cation-dependent shifts in fluorescence maxima are due in part to a σ–π complexation equilibrium.

The oscillator strengths for these transitions are dependent on the amount of charge redistribution upon excitation. In S_0 charge is localized in the five-membered ring, whereas in S_1 and S_3 charge is localized in the six-membered ring. In S_2 the charge is rather evenly distributed. Hence the oscillator strength is found to be S_3 (0.13) > S_1 (0.08) > S_2 (0.02). Calculations including counter-ion effects predicted oscillator strength and ion-pair shift orders in agreement with experiment.

Isomer **119** is produced almost exclusively and in high yield on photolysis of **118**, and subsequent photolysis and/or thermolysis leads to formation of **120**.[68] Although this is the orbital symmetry-allowed product, other processes may be operative. When **118** was irradiated in the presence of *cis*-stilbene, the olefin isomerized at a rate ten-fold greater than that in the absence of **118**, suggesting

electron transfer from the latter. Oxidation of **118** with $CdCl_2$ at $-70°C$ led, however, to 10–20% of dimeric and ring-opened products. It seems likely then that photoisomerization of **118** occurs by a concerted mechanism. Interestingly, photolysis of the *cis*-2,3-diphenyl-1-cyanocyclopropyl anion reveals that ring opening is only a minor pathway.

The photoisomerization of the *trans,trans*-diphenylallyl anion (**121**) produces *cis,trans*-isomer **122**, which reverts to **121** in the dark.[69] The Li salt of **122** was

characterized by absorption and ^{1}H NMR spectroscopy. The Na, K, Rb, and Cs salts of **121** exhibit similar behavior, and the kinetics of **122** → **121** are found to be first order. The rate constants decrease with increasing cation radius, becoming too small to measure in the Rb and Cs cases. The lack of a concentration effect or a common ion effect for the Na salt of **122** indicated that isomerization does not occur via the free ion. Increasing the percentage of tight ion pairs by addition of ether led to an increase in the isomerization rate. Presumably the greater ability of the small lithium counterion to form strong bonds with a localized carbanion in the transition state accounts for its relatively more rapid isomerization from **122** to **121**.

Steric effects on photoisomerization were noted for **123** and **125**.[69] The 1,3-phenyl–CH_3 interaction in **124b** renders **124a** the only accessible photoisomer, whereas **125** fails to photoisomerize because of 1,2-phenyl-CH_3 interactions in **126**.

125   126

Photolysis of a variety of arylmethyl carbanions in DMSO leads to homologous methylation of the carbanion.[70] The mechanism is suspected to involve electron transfer from the carbanion to DMSO followed by radical reactions, but a detailed investigation has not been reported.

X. REACTIONS

Several recent reviews[71, 72] of carbanion rearrangements show that these species provide a rich and challenging field for mechanistic investigations. An example is provided by the conversion of 1-lithio-2,2-diphenylpent-4-ene (**127**) to its isomers **128** and **129**, produced by allyl and phenyl migration, respec-

127   128   129

tively.[73] When M = Li, the **128/129** ratio is 97:3, whereas replacement of Li with K and Cs changes the ratio in favor of phenyl migration (32:68 and 16:84, respectively). Treatment of the potassium salt of **127** (from Li^+ **127** and KOt-Bu) with 18-crown-6 at $-75°C$ gave rapid rearrangement to **128**. These reactions are controlled by ion-pairing phenomena, with loose or solvent-separated ion pairs favoring allyl migration, and tight or contact ion pairs favoring phenyl migration. Phenyl migration is postulated to occur via **130**, in which charge is restricted

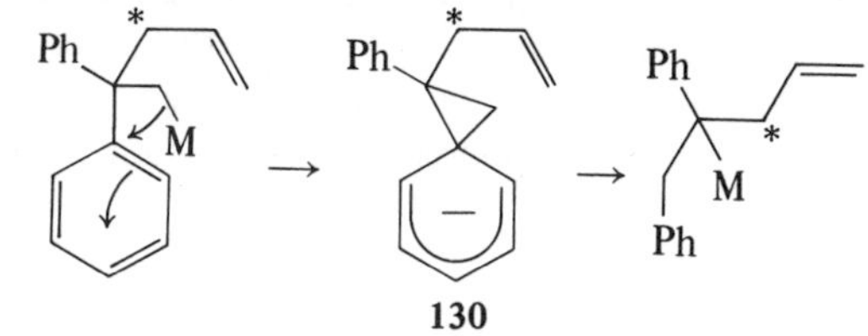

130

to the phenyl ring. The allyl migration presumably proceeds via a more delocalized transition state such as **131**, which is favored by loose ion pairing (Eq. 2).

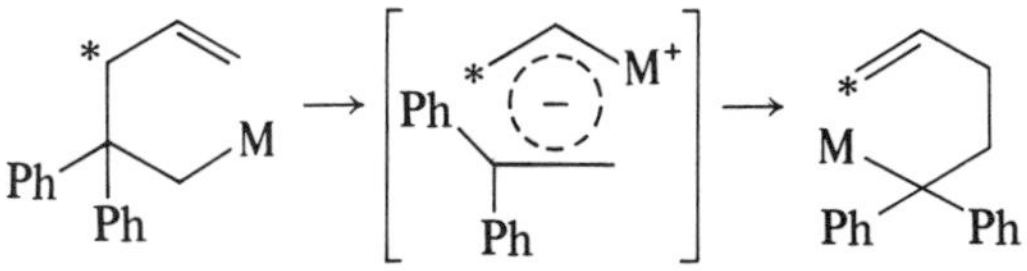

$$\tag{2}$$

131

These results provide an interesting contrast to those of a previous study, in which phenyl migration was promoted by the lithium counterion.[74]

Analysis of the carboxylic acids obtained from carboxylation of the rearrangement products of ^{14}C labeled **127** gave the results summarized in Table 2.2. The amount of label found in the position predicted by Eq. 2 (inversion) decreases with increasing temperature, concomitant with an increase in label at position 3. A disproportionation–recombination mechanism was proposed to account for the origin of scrambled label (Eq. 3). This mechanism is given experimental

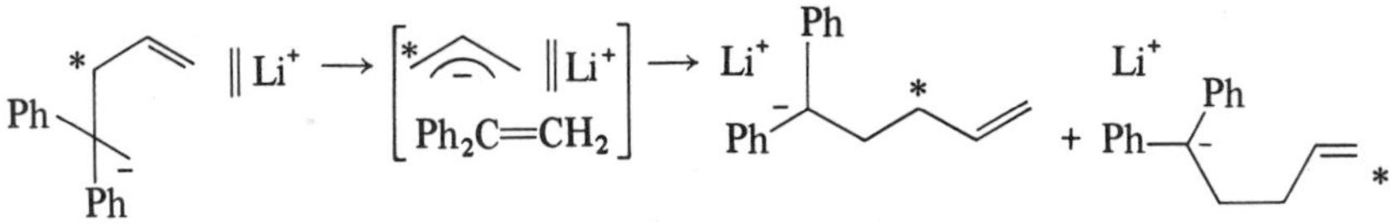

$$\tag{3}$$

TABLE 2.2.   Carbon-14 Distribution in the Rearrangement Products of **127**

Temperature (°C)	% Inversion (label found at *)	% Scrambled (label found at 3)
-20 to 0	33 ± 6	67
-32	54 ± 12	46
-50	63 ± 6	37
-70	100 ± 8	0

support by the observation that rearrangement of unlabeled **127** in the presence of labeled allyllithium at -20 to $0°C$ produces 14% incorporation of the label in the rearranged product.

Hexenyl anion **132** gave unexpected products upon quenching.[75] The product

| 132 | 133a | 133b | 133c |

| 133d | 133e |

distribution was sensitive to the counterion and solvent, as illustrated in Table 2.3. The authors proposed the intramolecular hydrogen transfer mechanism of Eq. 4 to account for the products, all of which (except **133b**) arise from the 1-propylallyl anion **134**. Intimate participation of the cation is, evidently important in the formation of **134** as conditions favoring contact ion pairs (Cs^+, THF) yield a higher percentage of products derived from **134** than do conditions

(4)

favoring solvent-separated ion pairs.

Several anionic cycloreversions have been reported.[76] Reaction with base (2 equiv of lithium tetramethylpiperidide) of **135** and **136** under identical conditions afforded cyclopentadienyl anion and the cyanocyclopentadienyl anion via a [4 + 2] cycloreversion mechanism. Compounds **137** and **138** react similarly under the same conditions.

A possible example of an anionically accelerated [2 + 2] cycloreversion is

TABLE 2.3. Product Distribution (%) from Quench of **132**

M	Solvent	133a	133b	133c	133d,e
Na	THF	85	3	3	9
Na	DME	94	2	4	1
Cs	THF	7	18	18	57
Cs	DME	10	44	12	34

represented by the base-promoted rearrangement and cleavage of **139**.[77] Although intriguing, no support for this mechanism other than the characterization of the products was advanced.

The regioselectivity of the base-promoted cyclopropyl ring cleavage in **140** and in four other methyl-substituted isomers coupled with the absence of deuterium exchange on treatment with KOt-Bu/DMSO-d_6, has led to the conclusion that the rate-limiting transition state for the overall reaction involves the deprotonation of **140** and its derivatives with little distortion from its tublike ground-state conformation.[78] As the foregoing results indicate, studies of regioselectivity represent a potentially powerful probe in mechanistic investigations.

The role of single electron transfer (SET) pathways versus nucleophilic dis-

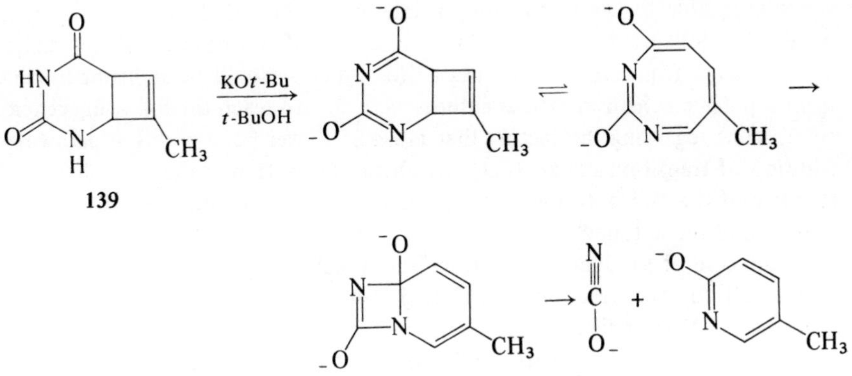

139

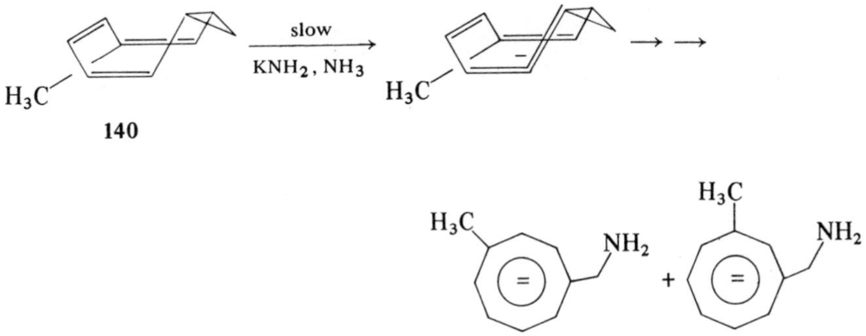

140

100%   0%

placement as the mode of Grignard attack is an ongoing subject of investigation. The isomerization of enone **141** has been used as a probe for electron transfer

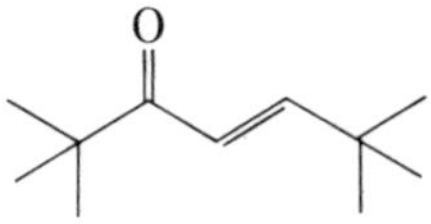

141

since the ketyl of **141** isomerized rapidly.[79] Based on the absence of significant isomerization of configurationally pure **141**, both H_3CMgBr and allyl magnesium bromide were determined to add to the enone by a polar mechanism, while t-BuMgCl addition produced isomerization of **141** and was therefore considered to proceed by a SET route. Addition of p-dinitrobenzene (DNB) as a radical trap did not affect the rate of H_3CMgBr attack on *cis*-**141** but did quench the isomerization of **141** caused by SET reactions catalyzed by transition metals. The authors concluded that isomerization occurs via the free ketyl, which is intercepted by DNB, while addition arises from a solvent-caged anion pair that is not accessible to DNB. It is not possible to rule out a SET mechanism accompanied by collapse to products too rapidly to allow diffusion of the radical intermediates from the solvent cage. This process would be indistinguishable from a polar mechanism. These authors were able to reach the following general conclusions regarding the factors that favor SET over polar addition pathways: addition of transition metals ($FeCl_3$), enhanced solvation of the ketyl, increased stability of the radical formed from the Grignard reagent, and lowered reduction potential of the ketone.[80]

A mixture of products arising from both polar addition and SET mechanisms was noted for the reaction of 142 with benzophenone in ether.[81] Although addition of DNB failed to affect product ratios, the SET mechanism seems securely based on the isolation of radical products such as those shown and

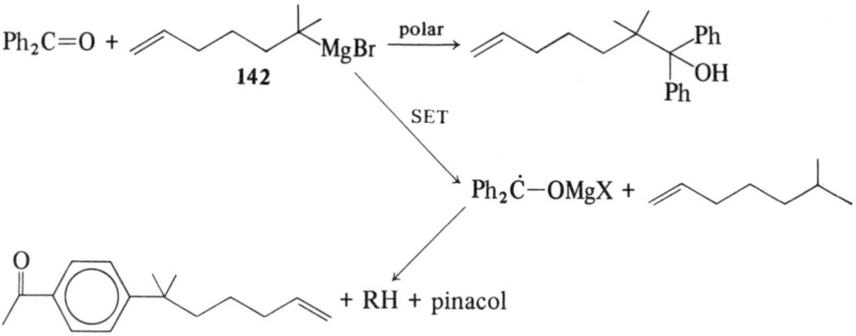

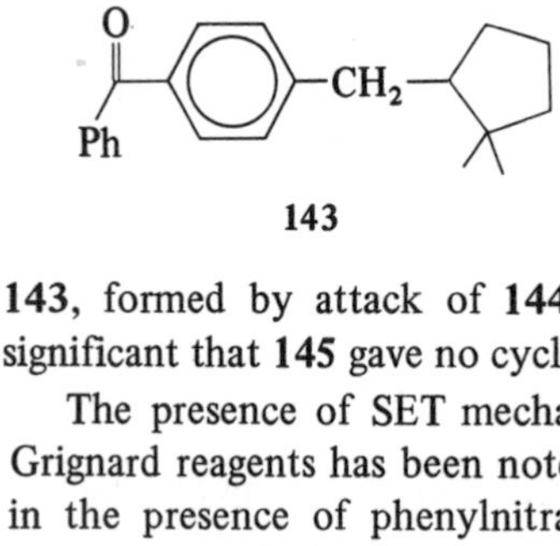

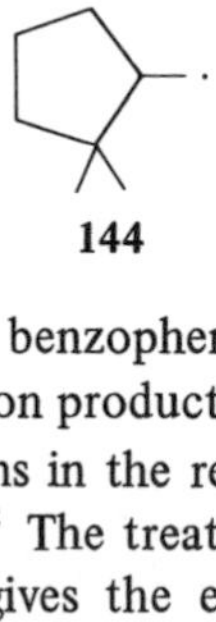

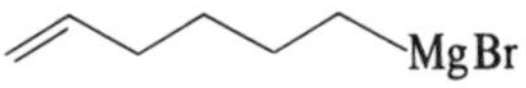

143, formed by attack of 144 on benzophenone. In the latter instance it is significant that 145 gave no cyclization products.

The presence of SET mechanisms in the reactions of carbanions other than Grignard reagents has been noted.[82] The treatment of 146 with methoxide ion in the presence of phenylnitrate gives the equilibrium shown in Eq. 5. The

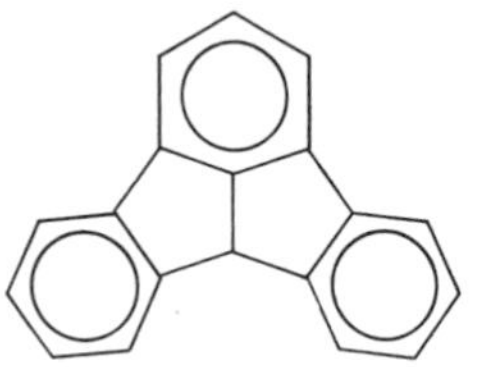

$$\underset{k_h}{\overset{k_h\ MeO^-}{\rightleftharpoons}} 146^- \underset{k_e}{\overset{k_a PhNO_2}{\rightleftharpoons}} 146^{\cdot} + PHNO_2 \cdot \quad \underset{dimer}{\downarrow k_1} \quad \underset{azoxybenzene}{\downarrow k_2}$$

(5)

146

ratio k_e/k_{-e} was found to be dependent on the counterion. Large counterions (Cs$^+$) stabilize **146**$^-$ more effectively than do smaller counterions (K$^+$), and hence shift the equilibrium toward contact ion pairs. The larger counterion also stabilizes **146**$^-$ relative to PhNO$_2^{\div}$ because there is appreciable hydrogen bonding of solvent to the latter. A large counterion seriously disrupts the solvation sphere upon complexation with PhNO$_2^{\div}$. A cation of large radius/charge ratio also stabilizes the delocalized anion (**146**$^-$) relative to the more localized PhNO$_2^{\div}$.

147   148

The 1,1-diphenylethylene radical anion (**147**) dimerizes to give the dianion **148**.[83] Interestingly, in the presence of electron acceptors (A) such as anthracene pyrene, perylene, and tetracene, **148** disproportionates by an SET mechanism as illustrated by Eq. 6.

$$\text{Na}_2^+\,\textbf{148} + \text{A} \underset{}{\overset{k_1}{\rightleftharpoons}} \text{Na}^+\textbf{148}^{\div} + \text{Na}^+\text{A}^{\div}$$

$$\text{Na}^+\textbf{148}^{\div} \xrightarrow{k_2} \text{Na}^+\textbf{147} + \text{Ph}_2\text{C}{=}\text{CH}_2$$

$$\underset{k_3}{\overset{\text{A}}{\longrightarrow}} 2\,\text{Ph}_2\text{C}{=}\text{CH}_2 + \text{Na}^+\text{A}^{\div} \tag{6}$$

$$\text{Na}_2^+\textbf{148} \xrightarrow{h\nu} \text{Na}^+\textbf{148}^{\div} + \text{Na}^+,\text{e}^- \tag{7}$$

The spontaneous decomposition of Na$^+$ **148**$^{\div}$ has a lower activation energy pathway than that assisted by A $(k_2 > k_3)$.[83] The driving force for this reaction is the delocalization energy gained by the H$_2$C—CH$_2$ bond electrons and the formation of a double bond upon dissociation of Na$^+$ **148**$^{\div}$.

Flash photolysis of **148** leads to bleaching of the solution followed by reformation of **148** in the dark.[84] The authors determined that photoejection (Eq. 7) occurred upon photolysis, followed by spontaneous disproportionation and recombination.

150 R$_1$ = Br, R$_2$ = H   152   153
151 R$_1$ = H, R$_2$ = Br

Alkylation of $2Na^+(Ph_2CNPh)^{2-}$ (**149**) with bromides **150–153** which are resistant to nucleophilic displacement, also proceeds by a SET mechanism.[85] The rate of alkylation of **149** by tertiary halides is faster by at least a factor of 6 than the rate of alkylation by primary halides, a strong argument in favor of a difference in mechanism.

It has been shown that **154b** does not react with di-t-butylketone at $-20°C$ but that formation of **155** does occur on treatment of **154a** and the ketone with

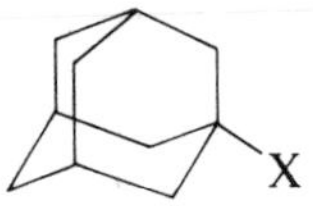

154
a, X = Cl, Br
b, X = Li

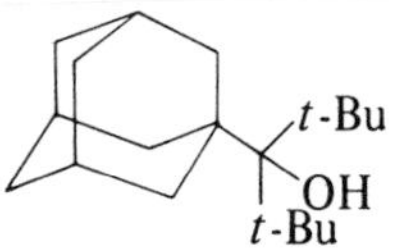

155

lithium powder.[86] This provides evidence for an SET mechanism in the Barbier reaction.

Quite often the isolation of dialkylated products is taken as evidence for the existence of a dianonic nucleophilic intermediate. Thus in Eq. 8 formation of **157** would normally be supposed to occur via **156**. However, such is not the case. The observation that rapid addition of Me_3SiCl to a mixture containing the supposed **156** led to a decreased yield of **157** suggested that the two-step

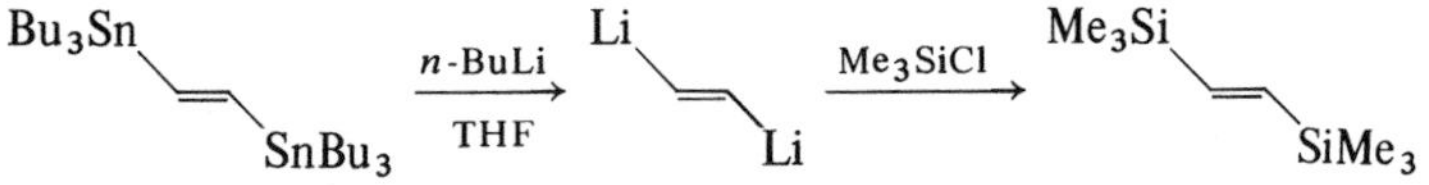

$$Bu_3Sn\diagdown\diagup SnBu_3 \xrightarrow[THF]{n\text{-BuLi}} Li\diagdown\diagup Li \xrightarrow{Me_3SiCl} Me_3Si\diagdown\diagup SiMe_3 \qquad (8)$$

156   **157**

mechanism of Eq. 9 was occurring.[87] Reaction of Me_3SiCl with n-BuLi reduces

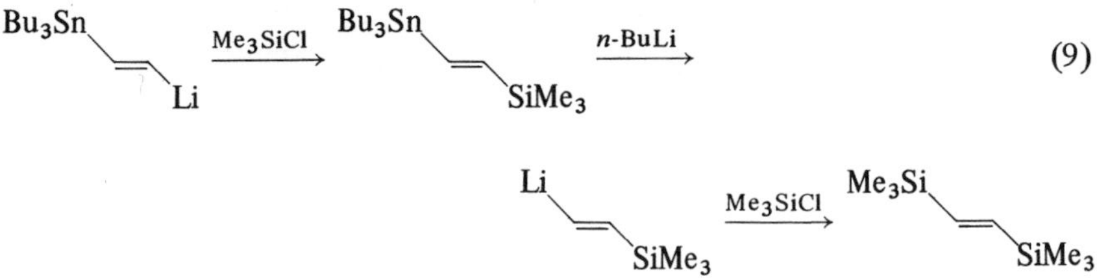

$$Bu_3Sn\diagdown\diagup Li \xrightarrow{Me_3SiCl} Bu_3Sn\diagdown\diagup SiMe_3 \xrightarrow{n\text{-BuLi}} \qquad (9)$$

$$Li\diagdown\diagup SiMe_3 \xrightarrow{Me_3SiCl} Me_3Si\diagdown\diagup SiMe_3$$

158

the amount of **158** (and hence **157**) that will eventually be formed. It is thus important when using this method to be sure that the trapping agent reacts with the purported polyanion more rapidly than with the base used to generate the polyanion. It should be emphasized that in cases where the polyanion is stable enough there is no substitute for its direct spectral characterization.

XI. REGIOSELECTIVITY AND COMPLEXATION

Control of regioselectivity by variations of the equilibrium between tight and loose or free ion pairs is an attractive strategem, but is useful as a synthetic tool only if the effect is sufficiently general. The stereospecific deuteration of 10-lithio-9-*t*-butylanthracene (**159**) illustrates the usefulness of this approach.[88] In

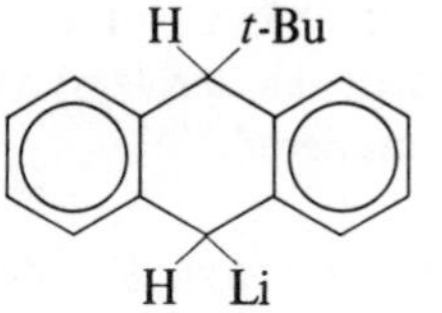
H
t-Bu
H
Li

159

highly solvating solvents, free ion pairs predominate (as determined by optical spectroscopy) and deuteration occurs to give the *cis*-product. Deuteration of contact ion pairs results in the exclusive formation of the *trans*-product. When THF is the solvent, both types of ion pairs exist and thus a mixture of products is formed.

A second example of complexation control of product distribution is afforded by the cyclization of **160a**.[89] In benzene, THF, and CH_2Cl_2, the formation of *cis*-cyclopropane **161a** is favored, whereas in DMF and acetonitrile *trans*-isomer **161b** is favored. Addition of the chelating agents 18-crown-6 and [2.2.2] cryptand enhances the amount of *trans*-isomer. These results suggest that intermediate **160a** is favored by tight ion pairs, and **160b** by loose ion pairs or free ions. One possible explanation is that the cation complexes with the *β*-carbomethoxy

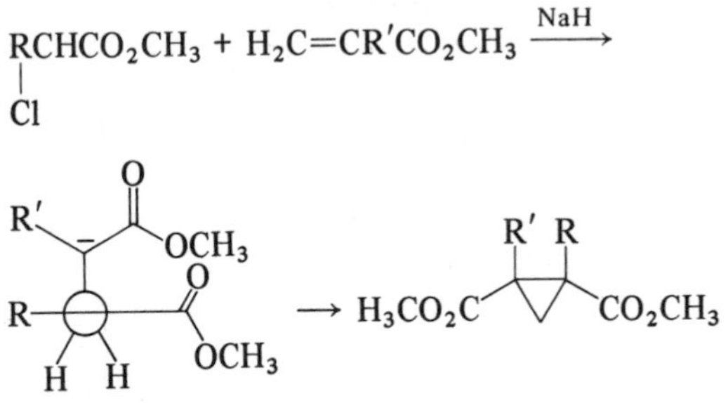
RCHCO2CH3 + H2C=CR'CO2CH3
Cl
NaH
R'
O
OCH3
O
R
H H
OCH3
R' R
H3CO2C
CO2CH3

160a **161a**

⇅

group as well as with the anionic sites in tight ion pairs, thereby favoring **160a**. In the loose or free ion, **160a** is destabilized relative to **160b** because of the unfavorable dipole–dipole interaction of the carbomethoxy groups.

Examples such as the above are intriguing, but the generality of the approach can be questioned. A detailed analysis of the regioselectivity of addition of nucleophiles to enones (in particular acrolein) based on ion pairing suggests that three limiting cases may be considered:[90]

1. Free nucleophile adds to a carbonyl. This occurs when cation–substrate and

$$Nu + \text{\Large$>$}\!\!=\!\!O \xrightarrow{k_0} Nu\text{\large$\dashv$}OH$$

nucleophile interactions are weak (i.e., large cations, presence of crown ether, strongly solvating media).

2. Free nucleophile adds to a complexed carbonyl. Complexation enhances the

$$Nu + \text{\Large$>$}\!\!=\!\!O\cdots M^+ \xrightarrow{k_M^+} Nu\text{\large$\dashv$}OH \qquad k_M^+ > k_0$$

electrophilicity of the carbonyl by lowering the energy of the LUMO and increasing the positive charge on carbon.

3. Ion pair adds to a free carbonyl. Nucleophilicity of the anion is decreased by lowering its HOMO and decreasing the charge separation.

$$M^+Nu + \text{\Large$>$}\!\!=\!\!O \xrightarrow{k_p} Nu\text{\large$\dashv$}OH \qquad k_p < k_0$$

The majority of reactions are a blend of the limiting cases, and support the concepts of complexation control (strong cation–carbonyl and relatively weak cation–nucleophile interactions) versus association control (strong cation–nucleophile and relatively weak cation–carbonyl interactions). Application of these criteria to the regioselectivity of additions of nucleophiles to acrolein showed that hard–soft interactions, as well as orbital overlap requirements, are at least as important as differences in complexation in controlling product distributions. A simple and broad prediction of regioselectivity based on ion-pairing phenomena alone appears unlikely to be developed, although the reactions of some classes of compounds may be rationalized on this basis.

XII. GAS-PHASE STRUCTURE AND CHEMISTRY

It is clear that one of the frontiers of the 1980s will be the study of the structure and chemistry of organic ions in the gas phase. The initial stimulus for this work was the desire to learn more about the intrinsic properties and structures of ions and their reactivity in the absence of the effects of solvation. More recently, the study of reactions that occur in solvent clusters in the gas phase has permitted investigations in a "fourth phase," which serves as a bridge between the gas and liquid phases. It can be anticipated that there will be an explosion of activity in this area during the next decade.

The investigation of carbanions in the gas phase has resulted from the recent development of the techniques of ion cyclotron resonance (ICR) spectrometry, high-pressure mass spectrometry, and flowing plasma mass spectrometry (flowing afterglow). These are only a few of the powerful new techniques emerging from chemical physics to have an increasing impact on problems in organic chemistry and biochemistry. Indeed, we expect that this trend will accelerate and that it will prove to be one of the dominant forces in the study of reactive intermediates during the next decade.

A variety of reasonably complex carbanions have recently been studied by the flowing afterglow technique. This technique has (among others) the advantages that adsorption–desorption problems can be eliminated and that both the ionic and the neutral species in a reaction possess thermal energy distributions.

One of the more significant observations to be made by using flowing afterglow is that a wide variety of simple carbanions participate in sequential deuterium exchange reactions with D_2O and with deuterated alcohols such as CH_3OD and CF_3CH_2OD. The number of hydrogen atoms that undergo exchange gives important information concerning the structure of the ion in the gas phase. For example, abstraction of hydrogen was found to occur from *both* the acetylene and methyl portions of propyne.[91] This is in contrast to results in solution,

$$HO^- + CH_3C{\equiv}CD \begin{array}{c} \xrightarrow{\ <40\%\ } CH_3C{\equiv}C^- + HOD \\[2ex] \xrightarrow{\ >60\%\ } \overline{CH_2}{-}\overline{C}{-}CD + H_2O \end{array}$$

where no methyl exchange occurred. Only two of the five hydrogens are exchanged in the anion formed by proton abstraction from 2-butyne. This is

$$CH_3\overline{C=C}-CH_2 + D_2O \rightarrow CH_3C\overline{D-C}=CH + HOD \xrightarrow{D_2O} CH_3CD_2C\equiv C^- + HOD$$
$$162$$

explained by water-catalyzed isomerization to form ion **162**.

The anion derived from *cis*-1,3-pentadiene exchanges less readily than that from *trans*-1,3-pentadiene under identical conditions of flow and reaction time,[92] indicating that these isomers do not interconvert. Furthermore, evaluation of the extent of exchange in 1,4-pentadiene indicates that approximately equal quantities of *cis*- and *trans*-anions are formed on deprotonation of this isomer.

Caution must be exercised in interpreting these data, since ambiguous results are sometimes obtained. Double bond migration and *cis-trans* isomerization may occur, producing anions of uncertain structure. However, it has been shown recently that the efficient and unambiguous generation of many organic ions occurs by reaction of their trimethylsilyl derivatives with fluoride ion.[93] Since trimethylsilyl derivatives are easily prepared and handled, this procedure promises to be very useful for the preparation of a variety of anions.

$$F^- + CH_3C\equiv CSi(CH_3)_3 \rightarrow CH_3C\equiv C^- + FSi(CH_3)_3$$

A number of other reactions have been studied, including reactions of anions with nitrous oxide,[94] carbon dioxide,[94] triplet and singlet molecular oxygen,[95] and oxidation and reduction reactions with $C_6H_7^-$, HNO^-, and HO_2^-.[96] It is clear that the studies in the field of "organic physical chemistry," as all too briefly exemplified by the foregoing examples, will grow in importance in the near future. Indeed, the great expansion of new reactions and compounds that the past decade has witnessed, coupled with powerful new physical techniques for studying structure and reactivity, should lead to a growing number of significant advances.

XIII. ACKNOWLEDGMENT

We are pleased to acknowledge the support of the National Science Foundation.

XIV. REFERENCES

1. H. Köster, D. Thoennes, and E. Weiss, *J. Organomet. Chem.*, **160**, 1 (1978).

2. P. West and R. Waack, *J. Am. Chem. Soc.*, **89**, 4395 (1967).

3. D. Thoennes and E. Weiss, *Chem. Ber.*, **111**, 3157 (1978).

4. T. Aoyagi, H. M. M. Shearer, K. Wade, and G. Whitehead, *J. Organomet. Chem.*, 175, 21 (1979).

5. D. Bladauski, W. Broser, H.-J. Hecht, D. Rewicki, and H. Dietrich, *Chem. Ber.*, 112, 1380 (1979).

6. T. Clark, P. v. R. Schleyer, and J. A. Pople, *J. Chem. Soc., Chem. Comm.*, 137 (1978).

7. W. D. Laidig and H. F. Schaefer III, *J. Am. Chem. Soc.*, 100, 5972 (1978).

8. E. D. Jemmis, P. v. R. Schleyer, and J. A. Pople, *J. Organomet. Chem.*, 154, 327 (1978).

9. (a) V. M. Sergutin, V. N. Zgonnik, and K. K. Kalninsh, *J. Organomet. Chem.*, 170, 151 (1979)
    (b) V. M. Sergutin, N. G. Antonov, V. N. Zgonnik, and K. K. Kalninsh, *J. Organomet. Chem.*, 145, 265 (1978).

10. T. Holm, *Acta Chem. Scand. B*, 32, 162 (1978).

11. F. M. Wehrli, *Org. Magn. Reson.*, 11, 106 (1978).

12.* G. Fraenkel, A. M. Fraenkel, M. J. Geckle, and F. Schloss, *J. Am. Chem. Soc.*, 101, 4745 (1979).

13. J. Tyrrell, V. M. Kolb, and C. Y. Meyers, *J. Am. Chem. Soc.*, 101, 3497 (1979).

14. C. Y. Meyers, and V. M. Kolb, *J. Org. Chem.*, 43, 1985 (1978).

15. M. P. Periaswamy and H. M. Walborsky, *J. Am. Chem. Soc.*, 99, 2631 (1977).

16. G. Rauscher, T. Clark, D. Poppinger, and P. v. R. Schleyer, *Angew. Chem., Int. Ed. Engl.*, 17, 276 (1978).

17. D. W. Boerth and A. Streitwieser, Jr., *J. Am. Chem. Soc.*, 100, 750 (1978).

18. T. Clark, E. D. Jemmis, P. v. R. Schleyer, J. S. Binkley and J. A. Pople, *J. Organomet. Chem.*, 150, 1 (1978).

19. S. Bywater and D. J. Worsfold, *J. Organomet. Chem.*, 159, 229 (1978).

20. S. J. Gould and B. D. Remillard, *Tetrahedron Lett.*, 4353 (1978).

21. A. R. Rossi, B. D. Remillard, and S. J. Gould, *Tetrahedron Lett.*, 4357 (1978).

22. P. v. R. Schleyer, J. D. Dill, J. A. Pople, and W. J. Hehre, *Tetrahedron*, 33, 2497 (1977).

23. K. N. Houk, R. W. Strozier, N. G. Rondan, R. R. Fraser, and N. Chuaqui-Offermans, *J. Am. Chem. Soc.*, 102, 1426 (1980).

24. W. J. le Noble, D.-M. Chiou, and Y. Okaya, *J. Am. Chem. Soc.*, 100, 7743 (1978).

25. R. J. Bushby, A. S. Patterson, G. J. Ferber, A. J. Duke, and G. H. Whitham, *J. Chem. Soc., Perkin Trans. II*, 807 (1978).

26. C. Sourisseau and J. Hervieu, *J. Mol. Struct.*, 40, 167 (1977).

27.* M. Schlosser and G. Rauchschwalbe, *J. Am. Chem. Soc.*, 100, 3258 (1978).

28. P. J. Garratt and R. Zahler, *J. Am. Chem. Soc.*, 100, 7753 (1978).

29. G. Boche and F. Heidenhain, *J. Am. Chem. Soc.*, 101, 738 (1979).

30.* D. H. O'Brien, C. R. Russell, and A. J. Hart, *J. Am. Chem. Soc.*, 101, 633 (1979).

31.* H. W. Vos, C. MacLean, and N. H. Velthorst, *J. Chem. Soc., Faraday Trans. II*, 72, 63 (1976).

32.* J. van der Giessen, C. Gooijer, C. MacLean, and N. H. Velthorst, *Chem. Phys. Lett.*, 55, 33 (1978).

33.* C. Gooijer and N. H. Velthorst, *Org. Magn. Reson.*, 12, 684 (1979).

34. K. Müllen, *Helv. Chim. Acta*, 61, 2307 (1978).

35. O. Hara, K. Yamamoto, and I. Murata, *Tetrahedron Lett.*, 2431 (1977).

36. O. Hara, K. Tanaka, K. Yamamoto, T. Nakazawa, and I. Murata, *Tetrahedron Lett.*, 2435 (1977).

37. K. Müllen, *Helv. Chim. Acta*, 61, 1296 (1978).

38. I. Willner, J. Y. Becker, and M. Rabinovitz, *J. Am. Chem. Soc.*, 101, 395 (1979).

39. I. Willner and M. Rabinovitz, *J. Am. Chem. Soc.*, 99, 4507 (1977).

40. M. Rabinovitz, I. Willner, A. Gamliel, and A. Gazit, *Tetrahedron*, 35, 667 (1979).

41. K. Nakasuji, M. Katada, and I. Murata, *Tetrahedron Lett.*, 2515 (1978).

42. R. J. Hunadi and G. K. Helmkamp, *J. Org. Chem.*, 43, 1586 (1978).

43. K. Müllen, *Helv. Chim. Acta*, 61, 1305 (1978).

44. L. A. Paquette, H. C. Berk, C. R. Dagenhardt, and G. D. Ewing, *J. Am. Chem. Soc.*, 99, 4764 (1977).

45.* K. Müllen, J. F. M. Oth, H.-W. Engels, and E. Vogel, *Angew. Chem.*, *Int. Ed. Engl.*, 18, 229 (1979).

46. J. J. Bahl, R. B. Bates, W. A. Beavers, and C. R. Launer, *J. Am. Chem. Soc.*, 99, 6126 (1977).

47. G. Boche and K. Buckl, *Angew. Chem., Int. Ed. Engl.*, 17, 284 (1978).

48. S. W. Staley and A. W. Orvedal, *J. Am. Chem. Soc.*, 95, 3384 (1973).

49.* L. A. Paquette, G. D. Ewing, S. Traynor, and J. M. Gardlik, *J. Am. Chem. Soc.*, 99, 6115 (1977).

50. R. H. Cox, L. W. Harrison, and W. K. Austin, Jr., *J. Phys. Chem.*, 77, 200 (1973).

51. P. Laszlo, *Bull. Soc. Chim. Fr.*, 558 (1966).

52. J. H. Noordik, T. E. M. van den Hark, J. J. Mooij, and A. A. K. Klaassen, *Acta Crystallogr. B*, 30, 833 (1974).

53. G. A. Olah, G. Asensio, H. Mayr, and P. v. R. Schleyer, *J. Am. Chem. Soc.*, 100, 4347 (1978).

54. A. Dagan, D. Bruck, and M. Rabinovitz, *Tetrahedron Lett.*, 2995 (1977).

55.* L. A. Paquette, *Angew. Chem., Int. Ed. Engle.*, 17, 106 (1978).

56. S. Q. A. Rizvi, J. Foos, F. Steel, and G. Fraenkel, *J. Am. Chem. Soc.*, 101, 4488 (1979).

57.* R. Gompper and K.-H. Etzbach, *Angew. Chem., Int. Ed. Engl.*, 17, 603 (1978).

58. R. Gompper and K.-H. Etzbach, *Angew. Chem., Int. Ed. Engl.*, 18, 470 (1979).

59.* D. N. Butler and I. Gupta, *Can. J. Chem.*, 56, 80 (1978).

60. M. J. Goldstein, Y. Nomura, Y. Takeuchi, and S. Tomoda, *J. Am. Chem. Soc.*, 100, 4899 (1978).

61.* T. B. Thompson and W. T. Ford, *J. Am. Chem. Soc.*, 101, 5459 (1979).

62. H. Bauer, M. Angrick, and D. Rewicki, *Org. Magn. Reson.*, 12, 624 (1979).

63. C. H. Bushweller, J. S. Sturges, M. Cipullo, S. Hoogasian, M. W. Gabriel, and S. Bank, *Tetrahedron Lett.*, 1359 (1978).

64. J. Klein and A. Medlik-Balan, *Tetrahedron Lett.*, 279 (1978).

65. E. Buncel and B. Menon, *J. Chem. Soc., Chem. Commun.*, 758 (1978).

66.* T. E. Hogen-Esch and J. Plodinec, *J. Am. Chem. Soc.*, 100, 7633 (1978).

67.* H. W. Vos, C. MacLean, and N. H. Velthorst, *J. Chem. Soc., Faraday Trans. II*, 73, 327 (1977).

68. M. A. Fox, *J. Am. Chem. Soc.*, 101, 4008 (1979).

69. H. M. Parkes and R. N. Young, *J. Chem. Soc., Perkin Trans. II*, 249 (1978).

70. L. M. Tolbert. *J. Am. Chem. Soc.,* **100**, 3952 (1978).

71.* S. W. Staley in *Pericyclic Reactions,* Vol. 2, A. P. Marchand and R. E. Lehr, Eds., Academic, New York, 1977, Chap. 4.

72.* E. Grovenstein, Jr., *Angew. Chem., Int. Ed. Engl.,* 17, 313 (1978).

73.* E. Grovenstein, Jr., and A. B. Cottingham, *J. Am. Chem. Soc.,* 99, 1881 (1977).

74. S. W. Staley and J. P. Erdman, *J. Am. Chem. Soc.,* 92, 3832 (1970).

75. J. F. Garst, J. A. Pacifici, C. C. Felix, and A. Nigam, *J. Am. Chem. Soc.,* 100, 5974 (1978).

76. W. Neukam and W. Grimme, *Tetrahedron Lett.,* 2201 (1978).

77. R. N. Comber, J. S. Swenton, and A. J. Wexler, *J. Am. Chem. Soc.,* 101, 5411 (1979).

78. S. W. Staley, G. E. Linkowski, and M. A. Fox, *J. Am. Chem. Soc.,* 100, 4818 (1978).

79. E. C. Ashby and T. L. Wiesemann, *J. Am. Chem. Soc.,* 100, 3101 (1978).

80. E. C. Ashby and T. L. Wiesemann, *J. Am. Chem. Soc.,* 100, 189 (1978).

81. E. C. Ashby and J. S. Bowers, Jr., *J. Am. Chem. Soc.,* 99, 8504 (1977).

82. R. D. Guthrie and N. S. Cho, *J. Am. Chem. Soc.,* 101, 4698 (1979).

83. S. Lillie, S. Slomkowski, G. Levin, and M. Szwarc, *J. Am. Chem. Soc.,* 99, 4608 (1977).

84. H. C. Wang, E. D. Lillie, S. Slomkowski, G. Levin, and M. Szwarc, *J. Am. Chem. Soc.,* 99, 4612 (1977).

85. J. G. Smith and D. J. Mitchell, *J. Am. Chem. Soc.,* 99, 5045 (1977).

86. P. Bauer and G. Molle, *Tetrahedron Lett.,* 4853 (1978).

87. D. Seyferth and S. C. Vick, *J. Organomet. Chem.,* 144, 1 (1978).

88. M. Daney, R. Lapouyade, and H. Bouas-Laurent, *Tetrahedron Lett.,* 783 (1978).

89. S. Akabori and T. Yoshii, *Tetrahedron Lett.,* 4523 (1978).

90. J.-M. Lefour and A. Loupy, *Tetrahedron,* 34, 2597 (1978).

91. J. H. Stewart, R. H. Shapiro, C. H. DePuy, and V. M. Bierbaum, *J. Am. Chem. Soc.,* 99, 7650 (1977).

92. C. H. DePuy, V. M. Bierbaum, G. K. King, and R. H. Shapiro, *J. Am. Chem. Soc.,* 100, 2921 (1978).

93. C. H. DePuy, V. M. Bierbaum, L. A. Flippin, J. J. Grabowski, G. K. King, and R. J. Schmitt, *J. Am. Chem. Soc.,* 101, 6443 (1979).

94. V. M. Bierbaum, C. H. DePuy, and R. H. Shapiro, *J. Am. Chem. Soc.,* 99, 5800 (1977).

95. R. J. Schmitt, V. M. Bierbaum, and C. H. DePuy, *J. Am. Chem. Soc.,* 101, 6443 (1979).

96. C. H. DePuy, V. M. Bierbaum, R. J. Schmitt, and R. H. Shapiro, *J. Am. Chem. Soc.,* 100, 2920 (1978).

3

CARBENES

ROBERT A. MOSS

Rutgers University, New Brunswick, New Jersey 08903

MAITLAND JONES, JR.

Princeton University, Princeton, New Jersey 08544

I. INTRODUCTION

In the 3 years since Volume 1 of *Reactive Intermediates*, many reviews of carbene chemistry have appeared. Among those that we do not discuss in this volume are descriptions of the ESR spectrum of CH_2,[1] a description of the use of CIDNP in studies of carbene reaction mechanisms,[2] and reviews of fluorocarbenes,[3] unsaturated carbenes,[4,5] and carbenes derived from cyclopropenes.[6] In addition we do not include an examination of the 1,2 hydrogen shift, certainly a common intramolecular reaction of carbenes.[7] There are other reviews of course and we mention these in the discussion as we go along.

II. PHYSICAL MEASUREMENTS

A. Singlet–Triplet Splitting in Methylene

In our chapter in Volume 1 we cited the extraordinary experimental determination of a 19.5 ± 0.7 kcal/mol singlet–triplet gap for methylene by Lineberger and his co-workers.[8] It is no surprise that the ensuing years have seen much comment on this paper and many new theoretical and an occasional experimental determination of this and other energy gaps.[9-22] Most new theoretical values for methylene range between 8.1 and 13.5 kcal/mol, although there are three higher values, two of which[19,20] are quite close to the Lineberger number. However, these two use minimal or admittedly "inferior" basis sets,[19,20] and a third, which yields a very high value of 25.3 kcal/mol, uses an approximate method.[21]

In 1978 two groups photolyzed ketene to generate methylene in the $^1A_1$ state. Laser-induced fluorescence using the $^1B_1$-$^1A_1$ system led to values of 8.1 ± 0.8[22a] and 6.3 ± 0.8[22b] for the singlet-triplet gap of methylene. Both values depend on accurate values for heats of formation of various species, although we find it difficult to argue with the values adopted by the authors. In addition, Frey and Kennedy have elaborated on earlier work and derived a value of 8.7 ± 0.8 kcal/mol.[22c] Their work assumes that the Arrhenius parameters for reactions of triplet methylene and methyl radicals will be similar, but this too seems a reasonable assumption to us. Clearly, these three experimental values are in striking agreement.

Using earlier results of Halevi et al.,[22d] Duncan and Trimble have pointed out that the thermal decomposition of diazomethane must yield both singlet and triplet methylene.[22e] However, no triplet methylene can be detected in the thermal decomposition of diazomethane. Presumably, triplet methylene is activated to the much more reactive singlet in a process that must depend critically on the magnitude of the singlet-triplet separation in methylene. Duncan and Trimble argue that for a gap as large as about 19 kcal/mol, triplets would surely have been detected. They find the lower values much more acceptable.

So we are left with the discrepancy between the 19.5 kcal/mol number of Lineberger and co-workers and nearly every other value. It has been suggested at least four times[9, 16, 17, 22b] that a source of the discrepancy may be the appearance of hot bands in the Lineberger spectra and the consequent misassignment of levels. However, this does not seem to be the case, as generation of CH_2^- from sources at very different temperatures leads to the same differential energy value.[23] So for the moment we are left with no resolution of this problem and we must await further work, as we did at the end of Volume 1.

Two groups have calculated that vinylidene[15a, 15b] and vinylidene carbene[15b] have singlet ground states, the corresponding triplets being about 46-51 kcal/mol higher in energy. Apeloig and Schreiber have calculated the effects of substitution on the singlet states of alkylidene carbenes and found that electropositive groups are stabilizing and electronegative groups are destabilizing.[15c] We deal later with various halocarbenes.

B. Ground State of Carboalkoxycarbenes, Cyanocarbenes, and Vinylcarbenes

Hutton and Roth appear to have resolved the long-standing question of the ground state of carboalkoxycarbenes.[24] As is often the case, analysis of the chemical reactions would lead one to believe that the singlet state is overwhelmingly present, but this says little about the ground state. It is perfectly conceivable that a higher lying singlet state might be thermally populated. The rates of the singlet reactions should be considerably faster than those of the corresponding triplet. Hutton and Roth were able to observe triplet EPR spectra in either frozen solutions or vacuum-deposited matrices. Two species were detected that are presumably the stereoisomers shown below. That the triplet state is indeed the ground state was determined by showing that at least in one case a plot of signal intensity versus $1/T$ was linear over the range 10–40 K; that is, the Curie law was followed.

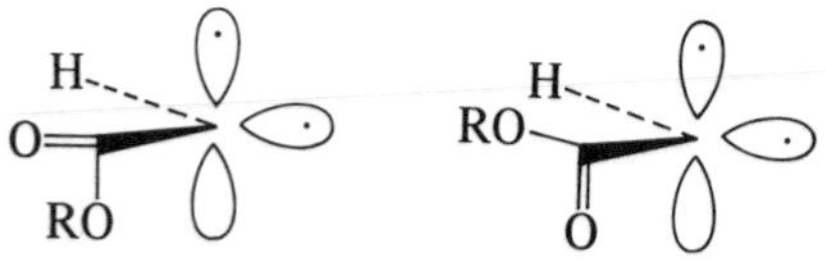

GVB calculations have given descriptions of both singlet and triplet vinyl-carbenes.[24a] The presence of an electron in a σ-orbital in the triplet leads to a carbene-like triplet, lying some 12 kcal/mol below the carbene-like lowest singlet and 14 kcal/mol below the 1,3-diradical-like $^1A''$ singlet state.

In apparent contrast to earlier theoretical and experimental work,[24b] Lucchese and Schaefer have used an *ab initio* theoretical study to reveal a distinctly bent triplet ground state for dicyanocarbene (133°), which lies about 14 kcal/mol below the singlet.[24c]

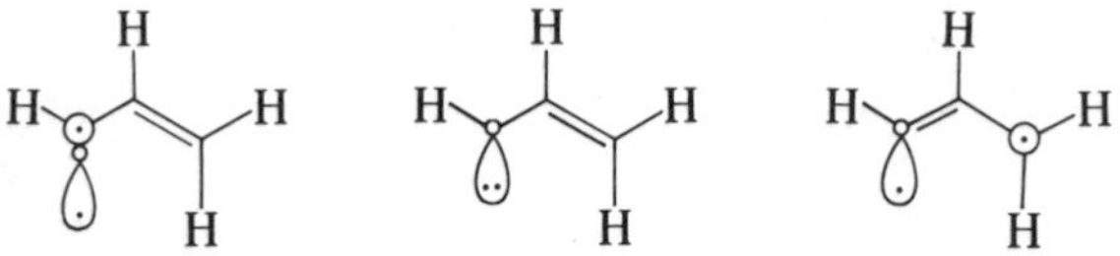

	$^3A''$	$^1A'$	$^1A''$
Relative energy (kcal/mol):	0	12	14

III.   GENERATION OF CARBENES

A.   Can Excited Diazo Compounds Mimic Carbenes?

In a paper expanding on earlier observations of Rando,[25] Tomioka, Kitagawa, and Izawa examined the direct and photosensitized decomposition of diazoamide 1.[26] On direct irradiation substantial amounts of the four- and five-membered ring lactams (2 and 3) were formed. Sensitized decomposition essentially eliminated formation of the four-membered compound 2. At this point a reasonable mechanistic explanation would be that the singlet carbene inserted in the two possible positions to give 2 and 3 but that the triplet carbene, being more selective, gave only the less strained 3 (see Eq. 1). However, addition of methanol, a good singlet trap, succeeded in decreasing *only* the yield of 3. The yields of

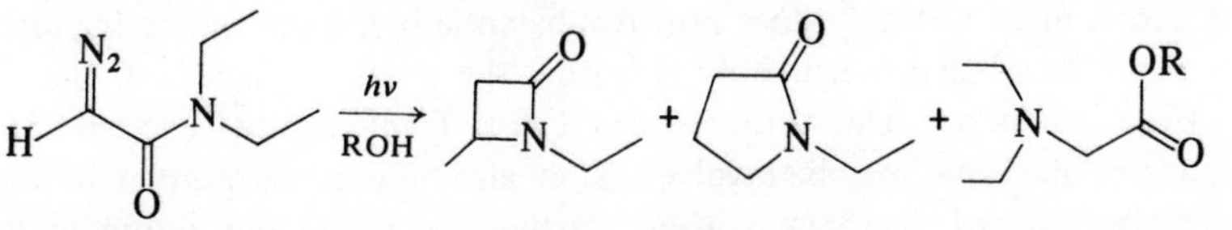

$$(1)$$

compound **2** and of a product **4**, formed by Wolff rearrangement, were insensitive to methanol concentration. Thus it is argued that the four-membered lactam **2** cannot be formed from singlet carbene, and is likely to be formed from the same intermediate, leading to the product derived from Wolff rearrangement. This "mystery source" is surely not the triplet carbene, so what else can it be? The authors argue that it must be the excited singlet diazo compound, thus recalling an argument made by Wulfman, Poling, and McDaniel with respect to diazomalonate decompositions.[27]

The argument of Tomioka does not seem unsound to us and perhaps a substantial effort directed toward proving the intermediacy of excited diazo compounds is warranted. In particular, we suggest another look at the results of Wulfman et al.,[27] which apply to intermolecular insertion reactions of dicarboalkoxycarbenes. Here it should be simple to generate the carbene from a precursor not containing nitrogen and see if the chemistry changes. Others have also suggested that excited carbene precursors may be reactive species. For instance, Hase and Kelley made such a suggestion for ketene photolysis,[28] although this work has been criticized by Richardson and Simons.[29]

The presence of extraordinary reactive intermediates has also been suggested in a most interesting paper by Chang and Shechter.[30] These workers have reinvestigated the intramolecular insertion reactions of *tert*-butylcarbene, first studied by Frey's group 18 years ago.[31] As shown below in Eq. 2, Chang and Shechter find that thermal or photosensitized generation of the carbene leads to exactly

$$(2)$$

$h\nu/\text{sens}/25°\text{C}$	88.4 ± 2.0	11.6 ± 2.0
$180°\text{C}$	88.4 ± 1.0	11.6 ± 1.0
$h\nu/25°\text{C}$	44.7 ± 1.6	55.3 ± 1.6

the same product mixture. This is compatible with the usual mechanistic scheme in which a singlet and triplet carbene equilibrate, the products being determined by the relative magnitudes of the rates of reaction of singlet and triplet carbene. However, in a "normal" state of affairs we would expect that direct photolysis of either the diazo compound or related diazirine would lead to the same product mixture and it most certainly does not. A substantially less discriminating inter- mediate is produced that presumably is neither the usual σ^2-singlet nor the σp- triplet. What can it be? The authors suggest, à la Tomioka, that excited diazo compounds or diazirines may be involved. They also note, however, that another possibility is "excited" carbene without further specifying the nature of the excitation. These observations by Chang and Shechter, and earlier by Frey et al.,[31] are extraordinarily interesting and certainly deserve further study.

Acceptance of intramolecular rearrangements of excited carbene precursors or excited carbenes would add a new level of complication to analysis of carbene chemistry, but the situation may be even more complex. G. R. Chambers has discovered a similar situation in *intermolecular* chemistry.[32] His data, summarized in Table 3.1, recapitulate those of Chang and Shechter, in that the products of thermal or photosensitized generation of the carbene are identical, but direct photolysis yields a different, and less selective intermediate. In Chambers' case, however, the new intermediate, be it excited precursor or excited carbene, must live long enough, without either losing nitrogen or internally converting to the lowest singlet carbene, to undergo *intermolecular* reactions.[32] This is even more remarkable than the observations of Tomioka et al.[26] and Chang and Shecter.[30]

B.   Metal-Mediated Transfers of Carbenes

There are several interesting reports of copper-mediated carbene transfers, including a Simmons-Smith-like cyclopropanation of alkenes that uses CH_2I_2 and copper powder (see Eq. 3).[33] Reaction conditions involved 50 or more hours of heating in an aromatic solvent (benzene or toluene). The conversion is readily

$$\text{(cyclohexene)} + CH_2I_2 \xrightarrow[\text{cat. } I_2]{\text{Cu, 70°C, 50 hr}} \text{(bicyclo product)} \tag{3}$$

85%

performed, is not particularly water or air sensitive, and enjoys most of the advantages of the Simmons-Smith reaction.[34] Although a twofold excess of CH_2I_2 is minimally required, yields are generally good, no insertion products are formed, and additions to stilbene and β-methylstyrene isomer pairs are stereo- specific. Cyclopropanation of 3-methoxycyclohexene occurs *cis* to the methoxyl (72% yield), demonstrating oxygen-based stereochemical control of the type familiar from the Simmons-Smith reaction.

Mechanistic characterization by relative reactivity studies (Table 3.2) impli- cates an organocopper intermediate with strong steric requirements. Mono- alkylated octene-1 is more rapidly cyclopropanated than dialkylated cyclohexene, and styrene is more reactive than either α-methylstyrene or 1,1-diphenylethylene.

TABLE 3.1 Products from 4,4-Dimethylcyclohexadienylidene

Olefin +	Conditions				
trans-2-Butene	hv	0	0	35	65
cis-2-Butene	hv	23	25	0	52
cis-2-Butene	hv, Ph_2CO	29	22	14	35
cis-2-Butene	125°C	34	23	13	30

TABLE 3.2.   Relative Reactivities of
$CH_2I_2 + Cu^a$

Alkene	k/k_0
$C_6H_5CH=CH_2$	3.89
$C_6H_5(CH_3)C=CH_2$	3.12
n-$C_6H_{13}CH=CH_2$	1.80
c-C_6H_{10}	1.00^b
$(C_6H_5)_2C=CH_2$	0.773
c-$C_6H_5CH=CHCH_3$	0.315
t-$C_6H_5CH=CHCH_3$	0.307
t-$C_6H_5CH=CHC_6H_5$	0.058
c-$C_6H_5CH=CHC_6H_5$	0.042

^{a}Selected from Ref. 33; reaction temperature $75°C$ in benzene solution.
^{b}Standard alkene.

These trends oppose those encountered with the "iodomethylzinc" intermediate of the Simmons-Smith reaction,[34] indicating greater steric discrimination by "ICH_2Cu." When, however, the steric environment at the substrate $C=C$ was held constant, the rather strong electrophilic character of the organocopper intermediate emerged: $\rho = -2.4 \pm 0.3$ (versus σ, benzene, $75°C$) for additions to *meta*- and *para*-substituted styrenes.[33] For comparison, $\rho = -1.61$ (versus σ, benzene, $78.6°C$) for the CH_2I_2/Et_2Zn-derived carbenoid,[35] indicating greater charge separation in transition states such as 5, offered for Eq. 3.[33]

5

Transition state 5 is admittedly oversimplified; charge separation is ignored and, more important, the reagent is unlikely to be monomeric. The reaction mixture was heterogeneous and an aromatic solvent (complexation to an organocopper intermediate?) was essential. In saturated hydrocarbon solvent there was no reaction between copper powder and CH_2I_2, whereas in ethereal solvents rapid reaction occurred but CH_2 transfer proceeded in low yield. A dimeric (or more highly associated) "ICH_2Cu" might account for the apparent steric selectivity.

Analogous reactions have been less successfully applied to the generation of other carbenic transfer reagents. Reaction of $Br_2CHCOOCH_3$ and alkenes under conditions similar to those of Eq. 3 gave *syn*- and *anti*-methoxycarbonylcyclopropanes, but yields were generally low.[33] Interestingly, the $BrCHCOOCH_3Cu$ transfer reagent showed a tendency toward *syn*-stereoselectivity, relative to the carbenoid generated by copper-catalyzed decomposition of diazoacetic ester. Additions of the former reagent were *syn*-stereoselective with hexene-1, octene-1, and styrene, whereas additions were less *exo*-stereoselective with cycloalkenes than were the corresponding reactions of the diazo-derived reagent.

Under conditions similar to those of Eq. 3, $FCHI_2$, Cu, and alkenes afforded fluorocyclopropanes in good yields, for example, from cyclohexene, 80%; heptene-1, 61%; styrene, 59%.[36] $ClCHI_2$ and $BrCHI_2$ gave the corresponding chloro- and bromocyclopropanes, respectively, in lesser but acceptable yields. Addition of "CHI" (CHI_3/Cu) to cyclohexene, however, proceeded in only 10% yield. Mechanistically, these CHX transfers appear to be similar to the "ICH_2Cu" reaction. They require an aromatic solvent, proceed without C–H insertion, are stereospecific with *cis*- or *trans*-4-octene, and are generally *syn*-stereoselective. With cyclohexene, for example, *endo*-X/*exo*-X ratios for additions of "CHX" were (X=) F, 2.4; Cl, 2.1; Br, 1.5; and I, 0.5. The transfer reagent was again electrophilic, $\rho = -1.13$ (versus σ^+, benzene, 75°C). A transition state analogous to 5 was offered, but as with the CH_2I_2/Cu system, the $XCHI_2$/Cu reactions are heterogeneous and the organometallic intermediates are probably associated.

Finally, "CH_3CH" transfers were effected with CH_3CHI_2 and Cu under the conditions of Eq. 3. Yields were low (cyclohexene, 31%; heptene-1, 11%), so that this application of the reaction seems of marginal synthetic importance.[37]

Metal–carbene complexes are covered in detail in Chapter 4, but two examples are cited here because of their unusual efficiency in carbene fragment transfer. The shelf-stable iron complex 6 is readily prepared by Meerwein salt methyla-

$$\begin{array}{cc} \overset{\displaystyle CO}{\underset{\displaystyle CO}{\underset{|}{\overset{|}{CpFe-CH_2\overset{+}{S}(CH_3)_2, \, BF_4^-}}}} & \overset{\displaystyle CO}{\underset{\displaystyle CO}{\underset{|}{\overset{|}{CpFe^+{=}CH_2, \, BF_4^-}}}} \\ 6 & 7 \end{array}$$

tion of the corresponding sulfide. Methylene transfer is effected to a variety of alkenes in refluxing dioxane solution,[38] presumably via the intermediacy of iron-methylide 7,[39] formed by extrusion of Me_2S.

Conversions and yields are generally good, although 2 equiv of 6 are required per equivalent of alkene, and it is unclear how efficiently the recovered iron-containing material can be purified and recycled. Mono- and dialkylated substrates were used, including the normally unreactive *trans*-stilbene (64% yield). It is not clear whether any special (steric?) problems would be encountered with tri- or tetrasubstituted alkenes.

A second, important, well-characterized reagent is the phenylcarbene–tungsten pentacarbonyl complex (8), which efficiently transfers the PhCH fragment to alkenes.[40] These reactions are performed in CH_2Cl_2/alkene solutions at $-78°C$,

$$(CO)_5W=CHC_6H_5 \quad Et_4\overset{+}{N}(CO)_5\overset{-}{W}CH(OCH_3)C_6H_5$$
$$\qquad 8 \qquad\qquad\qquad\qquad 9$$

where **8** is generated by *in situ* treatment of **9** with CF_3COOH. Yields range from a low of 36% with tetramethylethylene to a high of 98% with isobutene. Additions to *cis*- and *trans*-2-butene are stereospecific.

There are several important differences between **8** and the "$C_6H_5CHLiBr$" carbenoid **10** generated from benzal bromide and RLi.[41] Complex **8** is very sensitive to steric effects. Relative reactivities for cyclopropanation at $-78°C$ fall in the unusual order: $Me_2C=CMe_2$, 3.5; $MeCH=CMe_2$, 820; $Me_2C=CH_2$, 3500; *cis*-$MeCH=CHMe$, 7.6, *trans*-$MeCH=CHMe$, 3.5; $PhCH=CH_2$, 410; $MeCH=CH_2$, 11; $EtCH=CH_2$, 5.6; $Me_2CHCH=CH_2$, 2.4; $Me_3CCH=CH_2$, 1.0. Note the unusually low reactivity of the tetrasubstituted alkene and the enormous reactivity of isobutene; both observations are atypical of the common electrophilic carbenes and carbenoids, where the tetra > tri > di substrate reactivity ordering is observed.[42a]

Additions of **8** to trimethylethylene and *cis*-2-butene were strongly *syn*-phenyl stereoselective, with *syn/anti* ratios of 94:1 and 41:1, respectively. Additions of **10** to these alkenes were also *syn*-selective, but to much smaller extents (1.3:1 and 2.4:1 at $-10°C$).[41] *Syn/anti* ratios for additions of **8** to $RCH=CH_2$ decreased from 1.8 (R=Me) to 0.01 (R=Me_3C). Relative reactivity studies, combined with these *syn/anti* ratios, allowed partition of the overall reactivities into *syn* and *anti* series. Thus for *syn*-additions of **8** to $RCH=CH_2$ the relative reactivities were: (R=) Me, 7.0; Et, 2.7; *i*-Pr, 0.64; Me_3C, 0. The corresponding *anti*-reactivities sequence was 3.9, 2.9, 1.7, and 1.0. Obviously, very strong steric effects oppose *syn*-"PhCH" transfer. Similar effects are encountered with **10**, but the dependence on R is not nearly as strong.[43]

The selectivity data suggest a very electrophilic addition of **8**, strongly biased toward one end of the substrate's double bond. However, this *stereospecific* reaction is unlikely to proceed through open intermediate **11**. Also rejected[40] is

$$(CO)_5\overset{-}{W}-CH-C_6H_5$$
$$\underset{+}{R_2C-CR_2}$$
$$11$$

reaction via a metallocyclobutane-like transition state,[44] which fails to account for the smoothly decreasing *syn/anti* ratios observed with the 1-alkene substrates, or for the extremely high *syn/anti* ratio observed with trimethylethylene.

Transition states **12** and **13** were offered to account for the data.[40] The former results from attack of the electrophilic carbenic carbon on the less substituted end of $Me_2C=CHMe$ and involves stabilization of the δ^+ induced on the olefinic centers by interaction with the *ipso* phenyl carbon; steric control is

12   **13**

exerted by keeping the substrate's "odd" CH_3 away from the $-W(CO)_5$ group. Rotation of the $-C(CH_3)_2$ fragment toward $W(CO)_5$ converts **12** into a metallo-cyclobutane-like geometry, whence the *syn*-phenyl cyclopropane is formed by "elimination" of $W(CO)_5$. For reactions with more hindered alkenes, such as $Me_3CCH=CH_2$, open transition state **13** is offered.[40] It avoids the destabilizing Me_3C-phenyl interaction that would occur in **12** and leads mainly to *anti*-phenyl product. As the steric demand of the alkene substituent decreases, the reactions are believed to proceed increasingly via **12**.

There is an *ad hoc* character to the postulation of two "competing" transition states for these reactions. Certain features, such as the specific interaction with the *ipso* carbon atom (cf. **12**), appear overinterpreted. Nevertheless, the addition reactions of **8** are of great synthetic and mechanistic interest and they warrant continued study.

C. Unsaturated Carbenes

This area has recently been reviewed,[4, 5] so that only a few papers, touching importantly on our earlier coverage (see Volume 1, pp. 73–76), are noted.

a. Methylene Carbenes

Dimethylmethylene carbene (**14**) appeared to select differently between styrene substrates when generated by different methods [see Eq. 4]. The species derived from triflate **15** by butoxide-induced α-elimination cyclopropanated arylethylenes with $\rho = -0.75$ (versus σ, $-20°C$),[45] whereas the intermediate generated by decomposition of nitrosooxazolidone **16** (presumably via diazo-alkene **17**), cyclopropanated these substrates with $\rho = -3.4$ (σ^+, $40°C$).[46] The origin of this discrepancy was unclear, but we noted in Volume 1 that both ρ values were determined with minimal data sets. Thus we welcome publication of an extensive set of relative reactivities for triflate-generated **14**.[47] Aside from redemonstrating the well-known[46] steric hindrance encountered by **14** in electrophilic addition to highly substituted alkenes, (e.g., $k_{Me_2C=CMe_2}/k_{c\text{-}C_6H_{10}} = 0.027$, cf. **18**), the principal interest of the new results[47] is the close agreement with earlier observations for **14** generated from **16**. Thus, relative to cyclohexene, **14** from **15** ($-20°C$) gave the reactivities: $Me_2C=CMe_2$, 0.027; $c\text{-}C_8H_{14}$, 0.88; and $n\text{-}C_6H_{13}CH=CH_2$, 0.22.[47] For comparison, the corresponding reactivities

$$\underset{\substack{\mathbf{15}}}{\underset{CH_3}{\overset{CH_3}{\diagdown}}C=C\underset{OTf}{\overset{H}{\diagup}}} \xrightarrow{KOt\text{-Bu}}$$

(4)

$$\underset{\mathbf{14}}{\underset{CH_3}{\overset{CH_3}{\diagdown}}C=C:}$$

$$\underset{\substack{\mathbf{16}}}{\underset{CH_3}{\overset{CH_3}{\diagdown}}\underset{O}{\overset{N-NO}{\diagup}}O} \xrightarrow[40°C]{C_2H_5OCH_2CH_2Li} \underset{\substack{\mathbf{17}}}{\underset{CH_3}{\overset{CH_3}{\diagdown}}C=C=N_2} \xrightarrow{-N_2}$$

$$\underset{\mathbf{18}}{} \qquad \underset{\substack{\mathbf{19}}}{\underset{CH_3}{\overset{CH_3}{\diagdown}}C=C\underset{+}{\overset{H}{\diagup}}}$$

for **14** generated from **16** (40°C) were 0.02, 0.7, and 0.2.[46] This is excellent agreement, considering the variation of temperature and precursor, and suggests that both generative procedures do afford **14**.

The methylene carbenes derived from either **16** or **15** therefore behave similarly toward aliphatic alkenes but differently toward arylethylenes. Why the ρ values based on the latter reactions differ so strongly (see above) is still unclear, but the species derived from **16** appears inordinately reactive toward styrene; $k_{styrene}/k_{c\text{-}C_6H_{10}} = 6.2$,[46] whereas the species derived from **15** shows 0.57.[45] The latter species is similar in selectivity to CCl_2 ($\rho = -0.62$, σ^+, 80°C).[48] The close resemblance in reactivities with simple alkenes for alternatively derived **14**, coupled with the wide divergence in reactivities toward arylethenes might mean that the latter olefins react with a strongly electrophilic precursor of **14**, perhaps **19**, derived from **16**. However, the reactivity of **14** (from **16**) toward arylethenes is similar to that of **14** derived from $(CH_3)_2C=CBr_2 + CH_3Li$ (ether, -40°C), or $(CH_3)_2C=CHBr + KOt\text{-Bu}/t\text{-BuOH}$ (-10 to 0°C), for which $\rho = -4.3$ (σ^+).[49] The mystery therefore remains, at least in part.

The situation would be simplified if **14** were readily available from authentic diazoalkene **17**. Importantly, Wittig reaction of dimethyl diazomethylphosphonate with acetone (Eq. 5) appears to form **17**, which decomposes *in situ*, affording **14**, which can be trapped by alkenes.[50] Analogous reactions are observed with

$$(MeO)_2P(O)CHN_2 + Me_2C=O \xrightarrow[THF, -78°C]{KOt\text{-Bu}} (MeO)_2P(O)OK + \underset{\mathbf{17}}{[Me_2C=C=N_2]}$$

$$\xrightarrow[-78°C]{16\ hr} [Me_2C=C\colon] + \quad \bigcirc \quad \rightarrow \quad \text{(bicyclic structure with =CMe}_2\text{)} \tag{5}$$

$$14$$

cyclohexanone in place of acetone and presumably involve carbene **20**. Using 5-nonanone, carbene **21** is produced. It gives cyclopentene **22** (35%) by self-insertion. The latter reaction can occur at $-78°C$, since quenching at this temperature with methanol does not prevent formation of **22** and does not afford **23** (although **23** *is* formed at the expense of **22** when CH_3OH is present at the

$$\bigcirc = C\colon \qquad {n\text{-}C_4H_9 \atop n\text{-}C_4H_9}\!\!\!\diagdown\!\!C=C\colon \qquad CH_3-\!\!\!\bigcirc\!\!\!-n\text{-}C_4H_9 \qquad {n\text{-}C_4H_9 \atop n\text{-}C_4H_9}\!\!\!\diagdown\!\!C=CHOCH_3$$

$$\qquad\qquad 20 \qquad\qquad\qquad 21 \qquad\qquad\qquad\quad 22 \qquad\qquad\qquad\quad 23$$

start of the reation; that is CH_3OH can intercept the diazoalkene precursor of **22**).

Diazoalkene **17** must be very unstable to decompose readily at $-78°C$. Thermochemical calculations suggest that this is not unreasonable.[50] We eagerly await relative reactivity studies for **14** generated according to Eq. 5. A potential complication here is the presence of THF. One hopes that the THF can either be omitted or shown to have no mechanistically significant effect.

In related reactions, tosylazoethylenes **24** and **25** were prepared from $R_2CHCH=O$ and decomposed in cyclohexene or ethylvinyl ether (8-24 hr/25°C) to give 20-40% yields of the anticipated adducts of **14** or **20**, respectively. The mechanism by which **24** affords **14** is unclear. Concerted decomposition via **26**,

$$Me_2C=CHN=NSO_2Ar \qquad\qquad \bigcirc\!\!=CHN=NSO_2Ar$$

$$\qquad 24 \qquad\qquad\qquad\qquad\qquad\qquad\qquad 25$$

$$\text{(structure 26)}$$

$$26$$

and decomposition via ion pair $Me_2C=CHN_2^+ \ ^-O_2SAr$ have been suggested.[51] The intermediacy of **17** is thus problematical. The method of Eq. 5[50] would therefore seem a more promising route to continued exploration of the properties of **14**.

Finally, we note that **14** is also available from silylvinyl triflate **27** by treatment with KF-crown ether complex or $R_4N^+F^-$ (-20 or $0°C/1$-2 hr).[51] Additions to cyclohexene or ethylvinyl ether were essentially quantitative. Precursor 27 is related both to 15[45] and to vinyl chloride 28 (which may be decomposed

$$Me_2C=C\begin{matrix}OTf\\\\SiMe_3\end{matrix} \qquad Me_2C=C\begin{matrix}Cl\\\\SiMe_3\end{matrix}$$

27   28

with $R_4N^+F^-$).[52] Structural considerations suggest that 27 should be the most reactive, mildest, and most useful triflate methylene carbene precursor. Further studies of 14 generated from 27 would be welcome.

b. Vinylidene Carbenes

Two important papers have appeared concerning the freeness of vinylidene carbenes (29)[53] generated from halide precursors.[54,55] The broad conclusion is that either a cation or an anion can be associated with the nascent carbene during various stages of its birth, but that only the cation association is strong enough to influence selectivity toward alkenes. Thus 29 was generated from 30

$$R_2C=C=C: \qquad Me_2C=C=CHBr \qquad Me_2CX-C\equiv CH$$
    29   30   31

under various basic conditions; selectivities of the carbene appear in Table 3.3.

The olefinic selectivities of 29 (R = Me) derived by base-catalyzed elimination from 30 (Cl rather than Br) or 31 (X = Cl or OAc) are very similar,[53] suggesting that the discriminating carbenic intermediate in these cases is anion independent. The data in Table 3.3, however, show that the same conclusion does not follow for the cation. Under the (heterogeneous) "pure" butoxide conditions a species is generated from 30 that shows lower selectivity than the KOt-Bu/18-crown-6

TABLE 3.3. Selectivities of Dimethylvinylidene Carbene (0–5°C)

Alkene	t-BuOK/ 18-Crown-6	NaOH(KOH) Aliquat-336	t-BuOK/ t-BuOH	t-BuOK
$Me_2C=CMe_2$	16.9	19.3	14.7	7.4
$Me_2C=CHMe$	7.6	6.6	5.5	4.3
(cyclic ether)	1.1	2.1	1.5	2.7
(cyclohexene)	1.0	1.0	1.0	1.0
$n\text{-}C_4H_9CH=CH_2$	0.1	0.2	0.2	0.3

generated species. This behavior parallels that encountered with PhCBr generated from $PhCHBr_2$ and KOt-Bu in the absence or presence of 18-crown-6,[56] and a similar analysis has been offered.[54] K^+ is associated with **29** when **30** is decomposed with KOt-Bu alone. Note that **29** (R = Me) is closer in selectivity to the base/crown ("free") carbene when **30** is decomposed with 50% aqueous NaOH/ Aliquat-336 or 1:1 t-BuOK/t-BuOH. The metal cation is presumably solvated in these two cases and unavailable for association with **29**, much as 18-crown-6 prevents such association by sequestering K^+. One also notes that the best yields of adducts from **29** were obtained with the phase-transfer method.

Le Noble examined the t-BuOK/t-BuOH conversion of acetylene **32** to allene **34**, which presumably occurs via vinylidene carbene **33**.[55] Under neutral solvolytic conditions (80% CH_3OH, 50°C), E-**32** and Z-**32** (X = Cl) solvolyze at comparable rates with ∼60% solvolysis to identical 1:3 mixtures of E/Z-**30**-OCH_3 and ∼40% return to E-**34**-Cl (or Z-**34**-Cl, complete retention is observed during Cl^- return). These phenomena are adequately accounted for by cation-anion pairs that undergo competitive return to **34**-Cl or dissociation to solvent-separated species.

t-BuOK/t-BuOH base-catalyzed reactions are much faster than the solvolytic ones, so that the behavior of **33** can be observed. Moreover, t-BuOK-catalyzed D–H exchange (t-BuOD, 30°C) is ∼700 times faster than epimerization with E-**34**-Cl, so that carbanions related to E- (or Z-) **34** are configurationally stable under experimental conditions. Similar results are obtained with Z-**32**-Cl.

E-32   33 E-34

In the key experiments, t-BuO$^-$/t-BuOH decomposition (30°C) of either E- or Z-**32**-Cl afforded ∼35% of *returned* **34**-Cl. From either precursor, **34**-Cl was formed with *net retention*. Thus from E-**32**-Cl, the E/Z ratio of **34**-Cl was 57:43, and from Z-**34**-Cl it was 28:72. Clearly, free **33** was not the sole progenitor of the returned **34** chlorides; a carbene–chloride associated species (**35**) was

35

postulated to rationalize the stereochemical memory effect. Here, Cl⁻ remains principally associated with one face of the carbene on its way from acetylenic precursor to allenic product.

A detailed analysis[55] of the fate of **32** suggests that of the initially formed carbene-Cl⁻ pairs, ~66% dissociate to give **33** (and hence ethers derived from reaction with t-BuOK or t-BuOH), ~5% dissociate and then return externally and stereorandomly, ~10% return internally with retention, and ~19% stereochemically equilibrate Cl⁻ before internal return. The principal conclusion of this elegant work is that the reaction "must traverse a stage in which the carbene is not yet free,"[55] although at the later stage where olefinic cyclopropanation occurs, the perturbing influence of the anion (whether still paired or gone) is not detectable.[53]

D. Phase-Transfer Catalysis (PTC)

This area has been heavily reviewed,[57] so only a few papers of special note require discussion. The prevailing mechanistic interpretation of classical two-phase (50% aq NaOH/CHX₃ + alkene) carbene generation remains, that is, that OH⁻ converts weakly acidic CHX_3 to CX_3^- at the interface; the $R_4N^+Cl^-$ catalyst (which largely resides in the organic phase) then escorts CX_3^-, paired with R_4N^+, into the organic phase where CX_2 is liberated and reacts with the substrate.[57e, 58] Since the various equilibria leading to CX_2 favor starting material, alkene or another acceptor is needed to consume the carbene. The overall success of these reactions, therefore, depends strongly on the substrate's nucleophilicity.

Dehmlow has reported further optimization studies for PTC-CCl_2 cyclopropanations.[58] The most recent prescription calls for 4 molar excesses of $CHCl_3$ and 50% aqueous NaOH and 1 mol-% of the catalyst, all relative to the alkenic substrate. Reagents are mixed at 0-5°C, stirred (>800 rpm) at 25°C for 1-2 hr, and then heated to 50°C to 2-4 hr. Of the 50 catalysts tested, the most effective were $PhCH_2N^+Et_3Cl^-$, $(n\text{-}Bu)_4N^+Cl^-$, $(n\text{-}Bu)_4N^+HSO_4^-$, $(n\text{-}Oct)_3N^+MeCl^-$ (Aliquat-336), and $(n\text{-}Pr)_3N$.

With mixed haloforms such as $CHClBr_2$, carbene–halide exchange under PTC conditions normally leads to products of CCl_2 and CBr_2 capture, as well as to the desired ClCBr adducts.[59] However, dibenzo-18-crown-6 (DBC-6) appears to be a *specific* PTC catalyst for ClCBr generation.[60] Using DBC-6, pure $CHClBr_2$, and typical PTC conditions, Fedorynski obtained good yields of ClCBr adducts with a wide variety of alkenes. The experimental parameters and mechanistic origins of this specific catalysis were studied.[61] For example, with molar ratios of DBC-6/$CHClBr_2$/NaOH/1-hexene of 1:50:285:50 (8 hr, 45°C), 41% of pure ClCBr adduct was obtained.[61] Yields were higher with more reactive alkenes. The use of 18-crown-6, 15-crown-5, dicyclohexyl-18-crown-6, benzo-crown-5, or $R_4N^+X^-$ catalysts gave *mixtures* of CCl_2, CBr_2, and ClCBr adducts. Mixtures were also obtained with DBC-6 and very unreactive substrates (e.g., $Me_3CCH=CH_2$).[60,61]

Dehmlow postulated "ion pair carbene complex" 36 to account for DBC-6 specificity. Here, complexation of the electrophilic "CX_2" (ClCBr) with the

36

benzo substituent and with the X^- $(Br^-) \cdots Na^+ \cdots$ crown unit stabilizes the carbene *and* X^- (Br^-) relative to other PTC situations. In this way, halide exchange is suppressed by 36 and specific transfer of ClCBr to alkene is successfully competitive. As would be anticipated, DBC-6 (via 36, X = Br) is a less reactive catalyst than, e.g., $PhCH_2N^+Et_3Br^-$ and affords a more selective "CBr_2" than does the ammonium ion catalyst.[61] These results are important in their own right and for their suggestive power, pointing toward the rational design of synthetically specific carbene PTC catalysts.

Steinbeck examined PTC/CX_2 insertion reactions (arrows) with acetals 37-39.[62,63] Although 37 afforded little α-insertion product, 1,3-dioxolanes

37   38   39

(38) and 1,3-dioxanes (39) gave α-CHX_2 derivatives in good yields (although with low conversions of starting acetal). Conditions involved the use of CHX_3 and 50% aq NaOH at 0°C for 24 hr, with $PhCH_2\overset{+}{N}Et_3Cl^-$ as catalyst and stirring rates of 800-1000 rpm. Thirty-six examples of CCl_2 insertion were presented for 38, along with nine CBr_2 insertions. R was varied through all principal alkyl substituents, as well as haloalkyl, cyclohexyl, benzyl, and aryl. A limited number of the (less reactive) dioxane substrates were also examined.

Relative reactivities from competition experiments (conditions as above) gave $\rho = -0.63$ $(\sigma^+, 0°C)$ for CCl_2 insertions into 38, R = Ar, and $\rho^* = -0.73$ $(\delta \sim 0)$ for CCl_2 insertions into 38, R = Et, i-Pr, t-Bu, $BrCH_2$, and so on, using the extended Taft equation.[63] In these insertions, CCl_2 apparently functioned as an electrophile with little steric selectivity. It was previously shown that CCl_2

behaved electrophilically in α-benzylic insertion into substituted cumenes, $\rho = -0.89$ (σ^+, 80°C).[64]

Continuing earlier attempts to enhance the selectivity of PTC-generated CCl_2 (see Volume 1, pp. 77-78), "internal C=C competitions" were examined with dienes 40-43.[65] Typical PTC conditions (50% aq NaOH, $CHCl_3$, 34°C, 5 hr) were employed and several t-amine and R_4N^+ catalysts were studied. DABCO (44) afforded the best result, leading to completely regioselective mono-addition (arrow) to 40 in 23% yield. Yields were considerably higher with, e.g., n-Bu_3N and n-$C_{16}H_{33}N^+Me_3Br^-$ (CTABr) catalysts, but substantial quantities of

40   41   42   43   44

diadduct formed. With Bu_3N, 41 and 42 regiospecifically afforded monoadducts in 60 and 36% yields, respectively (99 and 88% corrected for recovered diene), but limonene (43) gave mono- (57%) and diadduct (43%) in 88% overall yield. DABCO gave only monoadduct with 43, but the yield is not stated in the original paper.[65]

Selectivity values are generated by the rate constant ratio (k_1/k_2) for the initial addition of CCl_2 to give monoadduct (k_1) versus the subsequent addition (k_2) to give diadduct. Apparent rate constants were determined for CCl_2 additions to 40 under pseudo-first-order conditions following the disappearance of 40 or its monoadduct over the initial 30-40% of reaction. Experimental conditions were: 10 mmol substrate, 0.5 mmol of catalyst, 5 ml of $CHCl_3$, 10 ml of 33% (w/w) aq NaOH, 34°C. Both k_1 and k_2 depended on the catalyst. Their magnitudes were greatest with CTABr ($k_1 = 7.5 \times 10^{-4}$, $k_2 = 0.83 \times 10^{-4}$ sec^{-1}) and lowest with DABCO (0.83×10^{-4} and 0.0025×10^{-4} sec^{-1}, respectively), leading to k_1/k_2 values of 9.0 (CTABr) and 330 (DABCO). The other catalysts gave intermediate absolute and relative rate constants.

Higher selectivity for the DABCO reactions seems to parallel lower reactivity for this CCl_2 generative system, although the crucial mechanistic factors are unclear. It was suggested that "the reactivity of the amine-catalyzed reactions could be correlated with the stability of a nitrogen ylide which is formed by the interaction between dichlorocarbene and a tertiary amine."[65] However, this is not likely to be a useful suggestion because the CCl_2 precursor in t-amine PTC systems is probably $R_3\overset{+}{N}CHCl_2$, CCl_3^-, where the cation is derived from R_3N and an equivalent of $CHCl_3$.[66] We speculate (perhaps wildly) that the DABCO-derived CCl_2 precursor may resemble **45**, an ion-pair carbene complex, drawn in analogy to **36** (see above). Such a species might also display greater stability and selectivity for the nascent CCl_2. Note that the PTC addition of CCl_2 to **40**, catalyzed by **46**, the *mono*-aza analogue of DABCO, is unselective $(k_1/k_2 = 15)^{65}$

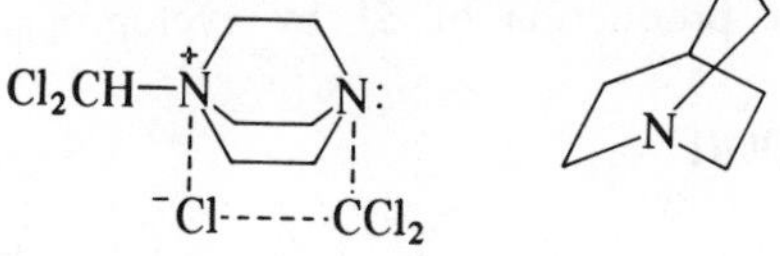

45

46

and resembles typical R_4N^+-catalyzed reactions. The *second* nitrogen atom of DABCO is therefore essential to the apparent CCl_2 selectivity, in accord with the concept expressed in **45**.

E. Other Methods for Generating Dihalocarbenes

A cautionary note emerges from the observation that CCl_2 generation from the Seyferth reagent ($PhHgCCl_3$) may not always be uncomplicated. Iodide-catalyzed CCl_2 generation in 2-methyl-5-*t*-butylpyrrole gave the *mercurial* **47**, in addition to anticipated products **48–50** (see Eq. 6).[67] The yield of **47** is not given, but its structure was secured by X-ray analysis. Compound **47** was stable at 80°C in ethanol or DME, but gave mainly **48** and **50** when treated with KOH, CH_3ONa, or TsOH under these conditions.

The general significance of this report[67] is unclear, but it is plain that one must consider the possible intervention of organomercurials in applications of Seyferth reagents, particularly with uncommon substrates. New reaction pathways for CX_2 should not be postulated without consideration of possible organometallic alternatives (see especially Section V.D).

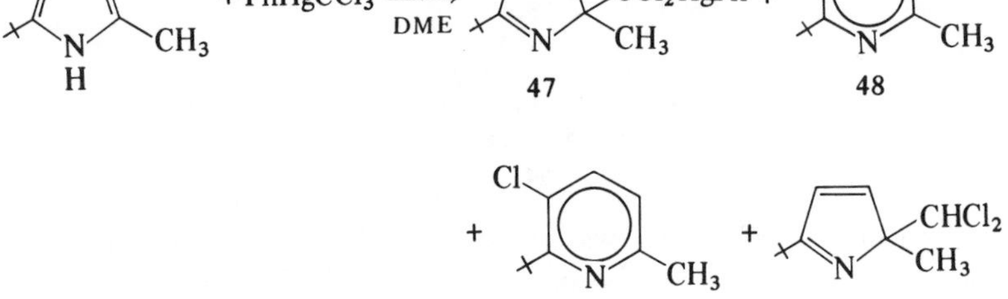

47   **48**   (6)

49   **50**

F. Asymmetric Induction

Syntheses of chrysanthemic acid have been stimulated by the potent insecticidal properties of its naturally occurring esters (pyrethrins). Among the four stereoisomers, esters of the *d-trans* acid (51) showed the greatest biological activity. A careful examination of chiral catalysts and alkyldiazoacetic esters was made to maximize production of 51 by cyclopropanation of 52.[68] An

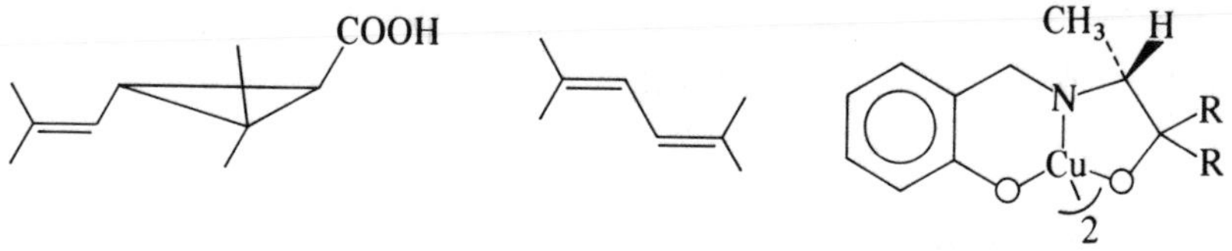

51   52   53

example of the extraordinary results is the addition (75°C, then 20°C, 7 hr) of *l*-menthyl diazoacetate to 52, catalyzed by the copper chelate formulated as 53 (R = 5-*t*-butyl-2-octyloxyphenyl). The resulting mixture of cyclopropanes (72% yield) contained 89.9% of *d-trans*-51 as the *l*-menthyl ester, together with 2.7% of the *l-trans*-51 derivative, and 7.4% of the *cis* epimeric esters. This represents a *trans/cis* ratio of 12.5 and an enantiomeric excess of 94% for the derived *d*-enantiomer of 51.[68] A sterically demanding R group in "HCCOOR" was essential to achieve high *trans/cis* and enantiomeric selectivities.

We also note highly detailed studies of asymmetric carbenoid additions to arylethylenes (see Eq. 7).[69a] With $N_2CHCOOC_2H_5$ and (-)-bis(α-camphorquinonedioximato)cobalt(II) (54) as catalyst, best results were obtained at -15°C.

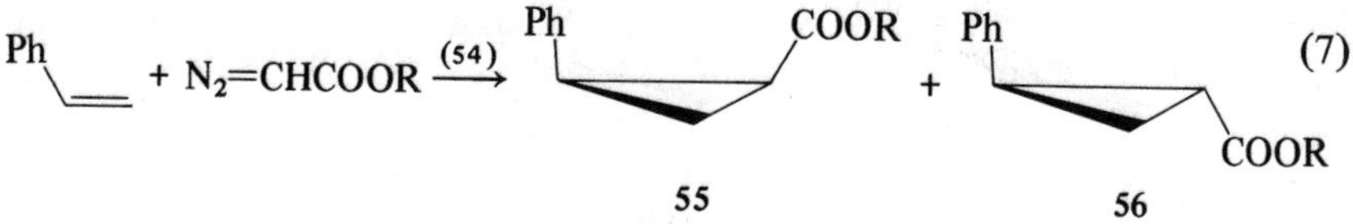

$$Ph\diagdown \!\!=\!\!= \; + \; N_2\!\!=\!\!CHCOOR \xrightarrow{(54)} \quad 55 \quad + \quad 56 \qquad (7)$$

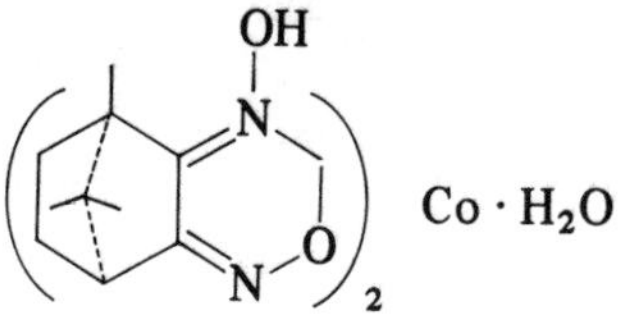

54

An overall chemical yield of 80% was obtained (55/56 = 1:0.95) with enantiomeric excesses of 72 and 76%. Use of a more sterically demanding diazoester enhanced the induction: with R = *neo*-C_5H_{11} in Eq. 7, the overall yield at 0°C was 87% and the optical yields rose to 81 (55) and 88% (56). Catalyst (-)-54 always gave cyclopropanes with the S-configuration at the former carbenic carbon atom.

The reaction appears limited to activated terminal alkenes such as styrene, 1,1-diphenylethylene, butadiene, and isoprene and is regiospecific for the terminal double bond of an extended conjugated system. A prochiral substrate is not required: 1,1-diphenylethylene gave a 70% optical yield (95% chemical yield) of cyclopropane with $N_2CHCOOEt$ and **54**. However, a prochiral carbenoid *is* necessary; the use of diphenyldiazomethane in Eq. 7 provided only a low yield of nearly inactive product.

Further investigation provided mechanistically relevant information.[69b] Above $2M$ styrene, the reaction rate (N_2 elimination *or* product formation) was first order in **54** and $N_2CHCOOEt$, as well as in styrene. Styrenes substituted with p-electron donating substituents showed mildly enhanced reactivity ($k_{p\text{-}CH_3O}/k_{p\text{-}Cl} = 6.6$), and diazoesters with bulky alkyl groups reacted more slowly ($k_{R\,=\,CH_3}/k_{R\,=\,neo\text{-}C_5H_{11}} = 21$), although with greater induction and **55**/**56** selectivity. Coordination of substrate to a complex of diazoalkane and catalyst apparently accelerated nitrogen expulsion, although slower decomposition did occur in the absence of styrene. Bound styrene lost its double bond integrity (cis-1,2-d_2-styrene gave **55** and **56** with cis/$trans$-d_2 distributions of 75:25 and 63:37, respectively), although recovered styrene was not isomerized. Finally, the asymmetric induction was subject to competitive inhibition; addition of basic ligands decreased induction, and added α-picoline actually reversed the sense of induction in the formation of **55**.

The suggested mechanism involved coordination of N_2=CHCOOR to Co(II) catalyst **54**, formation of a carbenoid-**54** complex by loss of N_2, binding of styrene with displacement of water or other ligand, and decomposition to a cobaltacyclobutane intermediate that subsequently furnished the final cyclopropane.[69b] This sequence seems to contradict the observation that the acceleration of N_2 elimination by styrene is first-order. The suggestion that N_2 loss precedes styrene coordination appears to have been inserted to enable construction of a carbenoid-**54** complex that would permit rationalization of the sense of asymmetric induction. As with all *post hoc* constructs, these rationalizations are permitted by the stereochemical facts rather than demanded by them. The importance of this work[69] is less in the detailed mechanism offered than in the detailed studies described and in the very real synthetic potential established for certain asymmetric carbenic cyclopropanations.

IV. CARBENOIDS

An important confluence of theory and experiment has occurred in this old, but still poorly understood, area. Clark and Schleyer have published geometry-optimized *ab initio* calculations for CH_2FLi,[70a] CHF_2Li,[70b] and CCl_3Li.[70c] Simultaneously, Seebach's group has studied $^{13}CBr_3Li$[71a] and several ^{13}C-labeled cyclopropylidene carbenoids[71b,72] by low-temperature ^{13}C NMR spectroscopy.

Three local minima were calculated for CH_2FLi using 4-31G level geometries and 6-31G level energies.[70a] The two lower-energy structures are **57** and **58** (bond angles are shown; see original for bond lengths). Carbenoid **58** is calculated

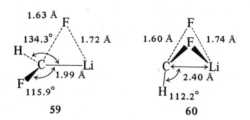

57   58

to be 19-29 kcal/mol less stable than **57**, but separated from **57** by an energy barrier of only 1.4 kcal/mol. Carbenoid **57** has the character of an intimate ion pair of $LiCH_2^+$ and F^-. Its lithium-substituted carbonium ion character is in good accord with the well-documented electrophilic properties of lithium carbenoids toward C=C and C—H bonds.[72] The molecular orbitals of CH_2Li^+ are similar but stabilized relative to those of singlet CH_2. Reactivity similar to that of CH_2 is anticipated for the carbenoid, although insertion reactions should be suppressed in favor of addition reactions. Structure **58** may be regarded as a complex of CH_2 and LiF.

Calculations for CHF_2Li[70b] gave **59** (analogous to **57**) as the most stable structure at the 6-31G level, although by only 4.2 kcal/mol over the unusual

59   60

structure **60**. Carbenoid **59** is again a lithium-substituted carbonium ion–fluoride ion pair, whereas **60** resembles a $CHF_2^-Li^+$ ion pair and should have nucleophilic properties.

Extension of the calculations to CCl_3Li (4-31G level)[70c] gave **61-64**, in order of increasing relative energies (0.0, 1.5, 4.1 and 6.5 kcal/mol, respectively). Of principal interest were the observations that "tetrahedral" **64** was the least stable arrangement, was not a local minimum, and went over to **62** upon removal of enforced C_{3v} symmetry. The most stable arrangement was the unusual **61**, a

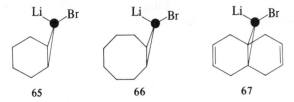

kind of CCl$_3^-$Li$^+$, in which most of the negative charge resides on Cl. Carbenoid **62** is analogous to the presumably electrophilic lithium-substituted carbonium ion forms preferred by CHF$_2$Li (**59**) and CH$_2$FLi (**57**).

How do these calculational results accord with the available experimental data? The calculated structures and energies pertain to gas-phase conditions; no account is taken of solvent or aggregation effects, which could clearly alter unit structures and relative energies. On the other hand, the general occurrence of lithium-substituted carbonium ion structures **57**, **59**, and **62** agrees well with many reactivity studies that demonstrate electrophilic behavior of lithium carbenoids,[42,72] as well as with matrix isolation IR studies of, for example, LiCCl$_3$, which are interpretable in terms of two carbenoids: **63** or **62**, and a C$_{3v}$ species such as **64** or **61**.[73]

The most incisive experimental contributions are afforded by direct NMR studies.[71] Cyclopropylidene carbenoids **65-67** ($\cdot = {}^{13}$C) were prepared in THF by treatment of the corresponding *gem*-dibromides with *n*-BuLi/hexane at

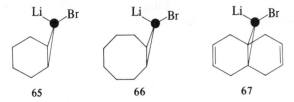

-100°C.[71b] The key ^{13}C observation was substantial *deshielding* (55-56 ppm) of all three carbenic carbon atoms relative to the corresponding H—C—Br compounds. This apparent *decrease* of electron density at the carbenic centers of **65-67** suggests that the structural change H—C—Br → Li—C—Br is accompanied by a weakening of the carbenoids' C—Br bonds; see structure **65a**, which corresponds to an electrophilic lithium cyclopropylidene carbenoid.[71b] This formulation agrees with the calculational results (compare **57**, **59**, and **62**) and with the characteristic chemistry of cyclopropylidene carbenoids.[72]

Even more interesting conclusions attended the generation of ^{13}CBr$_3$Li (THF, -100°C) from ^{13}CBr$_4$ and *n*-BuLi.[71a] Signals corresponding to *three* different species were observed at -100°C in the ^{13}C NMR. Significant properties appear

65a

in Table 3.4. Species A and B displayed $1:1:1:1$ quartets due to strong $^7Li-^{13}C$ coupling and appeared to be monomeric, nonionized organolithium reagents. As with the cyclopropylidene carbenoids, the ^{13}C resonances of A and B were deshielded relative to $HCBr_3$ (91 and 142 ppm, respectively). The higher field ^{13}C signal of A makes its formulation as **68** reasonable, leaving structure **69** or **70** for species B. Note that the latter resemble **62** and **63**, respectively, calculated for CCl_3Li, and represent potentially electrophilic $Br_2CLi^+Br^-$ variants. The observed $Li-C$ coupling establishes the presence of a $Li-C$ bond and probably excludes a bromo analogue of **61** for A or B.

68   **69**   **70**   **71**

Upon warming from -100 to $-80°C$, A was converted to B, which was in turn lost in favor of C at $-70°C$. C persisted in solution up to $-40°C$; its decomposition afforded polymer and $Br_2C=CBr_2$. C was formulated as CBr_2 (**71**),[71a] but could be complexed with THF. Note that the NMR data do not exclude the bromo analogue of **61** as a structure for C. Quenching of A (**68**) or B (**69** or **70**) with methanol gave bromoform; quenching of C did not yield this product.

The ^{13}C NMR spectrum of $CH_3{}^{13}CBr_2Li$ was also studied.[71a] A single carbenoid was detected (-100 to $-40°C$), bearing a deshielded $(Li-)^{13}C$ quartet, and quenchable by CH_3OH to give 1,1-dibromoethane. This species was formulated as a distorted tetrahedron.

By and large, the NMR results are in pleasing agreement with the calculations, especially because both approaches strongly support $X_2CLi^+X^-$ formulations for

TABLE 3.4. $^{13}CBr_3Li$ Carbenoids[71a]

Species	$\delta\ ^{13}C$ (ppm)	Multiplicity (Hz)	Stable Temp. Range (°C)	Assignment
A	101.5	q, $J = 43$	-100 to -80	**68**
B	152.2	q, $J = 40$	-100 to -70	**69** or **70**
C	144.1	s	-100 to -40	**71**

lithium carbenoids, which agree well with typical electrophilic carbenoid chemistry. It is hoped that further spectroscopic results will be forthcoming. Olefin quenching experiments would be particularly welcome. Although species C (**71**?) does not afford bromoform upon methanolic quench, it should afford 7,7-dibromonorcarane when quenched with cyclohexene.

Finally, we note the publication of full experimental procedures for the synthetic use of lithium cyclopropylidene carbenoids in cyclopropane alkylations.[72,74]

V.   REACTIVITY OF CARBENES

A.   1,4-Additions

Since our discussion in Volume 1 of this series, another theoretical treatment by Schoeller and Yurtsever[75] using MINDO/3 has pointed out that even though the symmetry-allowed HOMO–LUMO interaction between π_2 of the diene and the vacant p orbital of the carbene is favorable, the 1,4-addition reaction is thwarted by the obligatory HOMO–HOMO interaction between π_1 of the diene and the filled σ-orbital of the carbene. This unfavorable interaction overwhelms the interaction between σ and the LUMO (π_3) of the diene.

Although we devote substantial space later to other quantitative aspects of carbene reactions, of particular interest are competitive 1,2- and 1,4-additions of CX_2 to norbornadienes.[76] Jefford has argued that 1,4-addition represents a predominantly nucleophilic mode of carbene attack (see **72**) in which the $E_{CX_2}^{HO}$-E_{diene}^{LU} interaction plays a major role, whereas 1,2-addition represents a "normal" electrophilic CX_2 addition.[77] To explore this contention, Jefford

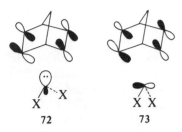

72   73

and Huy added CF_2 (from $Ph\overset{+}{P}CF_2Br,Br^- + KF$, DME, 25°C) to 2-R-substituted norbornadienes and determined overall 1,2/1,4 addition ratios as a function of R (see Eq. 8).[77a] Representative results included: 1,2/1,4 = 8.1 (R = OCH₃), 3.0 (R = Ph), 1.9 (R = H), 1.4 (R = CN), and 1.1 (R = COOCH₃). The trend of decreasing 1,2/1,4 selectivity as R became electron withdrawing was suggested to represent increasing assertion of nucleophilic 1,4-addition as the electrophilic

1,2-addition became increasingly disfavored due to diminished π-electron density.[77a]

This conclusion was offered with reference to Jean's FMO analysis,[78] in which $E_{CXY}^{LU} - E_{diene}^{HO}$ (73) and $E_{diene}^{LU} - E_{CXY}^{HO}$ (72) were evaluated for additions of CCl_2, CFCl, and CF_2 to norbornadiene. Jean pointed out that, in general, interaction 72 was favored over 73 by stronger overlap, but that 73 was preferred over 72 because of a smaller differential orbital energy.[78] STO-3G geometry-optimized calculations of carbenic FMOs showed that E_{CXY}^{LU} rose sharply along the series CCl_2, CFCl, CF_2 (3.24, 4.62, and 6.50 eV, respectively), while E_{CXY}^{HO} rose but modestly (-9.88, -9.55, and -9.30 eV). In passing from CCl_2 to CF_2, orbital interaction 73 thus became 3.26 eV larger in differential energy, while orbital interaction 72 actually diminished by 0.58 eV. The apparent sharp increase in E_{CXY}^{LU} (which also disfavors 1,2-addition), coupled with the small increase in E_{CXY}^{HO}, led to the conclusion that "the dominant electronic factor which augments the relative rate of 1,4 addition (as CXY is changed from CCl_2 to CF_2) is the diminution of the electrophilic character of the carbenic p orbital."[78]

We note two reservations concerning these experimental[77a] and theoretical[78] results. First, the reported enhancement of 1,4-versus 1,2-addition in going from norbornadiene to 2-cyanonorbornadiene (a factor of 1.4) is not very impressive, nor is the selectivity reversed (i.e., 1,4-addition never exceeds 1,2-addition) with any of the substrates.[77a] Second, although E_{CXY}^{LU} does rise in passing from CCl_2 to CF_2, the reported rise in E_{CXY}^{HO} is calculation dependent. Calculations at the 4-31G level on STO-3G geometry-optimized carbenes give $E_{CCl_2}^{HO} = 11.44$ and $E_{CF_2}^{HO} = -13.38$ (!),[79] versus -9.88 and -9.30 eV, respectively, for the STO-3G energies.[78] Thus, at the 4-31G level, $E_{CF_2}^{HO}$ is actually lower[79] than $E_{CCl_2}^{HO}$, rather than higher[78] in energy. The effect on the differential orbital energies for CF_2 addition to norbornadiene (using cited diene HOMO and LUMO energies[78]), given as (electrophilic) interaction 73 versus (nucleophilic) interaction 72, is 13.4 versus 17.2 eV(Ref. 78) and 8.8 versus 21.3 eV (Ref. 79). In both cases interaction 73 has the smaller, dominant differential orbital energy, but it is very strongly dominant with the 4-31G orbital energies.

Indeed, Klumpp and Kwantes[80] have found correlations between the log of the rate of 1,4-addition (log $k_{1,4}$) and the logs of the rates of endo 1,2-addition [log $(k_{endo\text{-}anti})$ and log $(k_{endo\text{-}syn})$] for additions of dichlorocarbene to a series of 7-substituted norbornadienes. Although the correlations are admittedly "crude," much more scatter appears in an attempted correlation of log $(k_{1,4})$

with log $(k_{exo\text{-}anti})$. Thus 1,4- and *endo*-1,2-additions of dichlorocarbene respond rather similarly to the influence of the substituent at C-7. As ρ values of about -2 (σ_I correlation) are observed, *both* reactions appear to be electrophilic.

We conclude that a nucleophilic 1,4-addition of CF_2 to norbornadiene is not yet proven, although it may be correct to consider such an addition "less electrophilic" than that of CCl_2.

The most exciting development in this area, however, is certainly the report by Burger and Gandillon of an intramolecular 1,4-addition (Eq. 9).[81] The authors used the methyl label shown in Eq. 9 to demonstrate that benzvalene is formed from 5-methylcyclopentadienylcarbene by 1,4- rather than 1,2-addition. No mechanistic discussion is included in their paper, but presumably the geometry of the starting material disfavors 1,2-addition and holds the carbene in the proper orientation for conjugate addition. This becomes only the third example

$$(9)$$

of this kind of reaction. We mentioned above one other, the reaction of halocarbenes with norbornadiene,[76-78] and perhaps the remaining example deserves some comment. In 1970 Anastassiou and Cellura[82] reported the addition of triplet dicyanocarbene to cyclooctatetraene and the formation of the product of 1,4-addition (Eq. 10). Both partners must be present for this reaction to succeed.

$$(10)$$

Dicyanocarbene does not generally add to dienes to give 1,4-addition, and cyclooctatetraene gives exclusively 1,2-addition with triplet carbenes such as dicarboalkoxycarbene.[83] The only cogent explanation for this singular behavior comes from M. E. Hendrick, who suggested in 1971 that electron transfer may occur in the originally formed diradical to give **74**, a zwitterion containing both a homotropyllium ion and a well-stabilized carbanion (Eq. 11).[84] If Hendrick is correct, the stabilization afforded the carbanion end of the dipole by the cyano groups is

$$(11)$$

crucial. Why then is the carboalkoxy group ineffective? The pK_a's of the related hydrocarbons may give the answer. Dicyanomethane has a pK_a of 11.2, but malonic ester is more than 10^2 less acidic with a pK_a of 13.5. So the ester group may simply not be good enough at stabilizing the negative charge to permit the electron transfer reaction to occur. On the other hand, for this explanation to be valid, a species such as triplet diacetylcarbene must undergo 1,4-addition, as the related diacetylmethane has a pK_a of 9.0. We would be interested in seeing the results of this reaction.

B. Quantitative Characterization of Carbenic Selectivity

a. The Selectivity Spectrum; HOMO–LUMO Treatments

Progress has been made in the understanding of carbene–alkene cycloaddition reactions.[85] The olefinic selectivity of simple carbenes can be expressed by a "selectivity index," m_{CXY}, obtainable experimentally (for electrophilic carbenes) from a log–log correlation of k_{rel} for CXY additions to a standard set of alkenes versus k_{rel} for CCl_2 additions to the same alkenes, all at $25°C$.[86,87] Furthermore, m_{CXY} can be estimated from the dual substituent parameter equation (Eq. 12)[86] if appropriate resonance and inductive σ values are available for the carbenic substituents.

$$m_{CXY} = -1.10\Sigma_{X,Y}\sigma_R^+ + 0.53\Sigma_{X,Y}\sigma_I - 0.31 \qquad (12)$$

Equation 12 readily identifies CH_3OCCl ($m^{calcd} = 1.59$), CH_3OCF ($m^{calcd} = 1.85$), and $(CH_3O)_2C$ ($m^{calcd} = 2.22$) as carbenes that should be more selective toward simple alkenes than CF_2 ($m^{obsd} = 1.48$).[86,87] Because $(CH_3O)_2C$ is *nucleophilic* toward alkenes (adding to dimethylfumarate, but not to cyclohexene[88]), whereas CF_2 is electrophilic,[42] intuition alone suggests interesting properties for the "bracketed" carbenes, CH_3OCCl and CH_3OCF. The former species has been subjected to close scrutiny.

CH_3OCCl is readily generated by pyrolysis of the corresponding diazirine (Eq. 13).[89] Activation parameters ($E_a^{30°C, \text{ hexane}} = 23.9 \pm 0.6$ kcal/mol, $\log A = 13.2$), solvent rate effects (minimal), and decomposition products (in ether: N_2,

$$\text{(13)}$$

CO, CH_3Cl, $HCOOCH_3$, $CH_3OCCl=CClOCH_3$, and $CH_3OCCl=N-N=CClOCH_3$) have been carefully examined. Cogent analysis leads to the conclusion that Eq. 13 involves a concerted, two-bond cleavage of the diazirine, leading directly to CH_3OCCl *without* intervention of a diazoalkane. The latter complication is known for dialkyldiazirines but is suppressed when the product carbene is highly stabilized by heteroatomic substituents.[89]

Intermolecular additions of CH_3OCCl are of the greatest interest.[90] A wide range of alkenes is cyclopropanated, including $Me_2C=CMe_2$, $Me_2C=CH_2$, t-$MeCH=CHMe$, $CH_2=CHCOOMe$, $CH_2=CHCN$, and t-$MeOOCCH=CHCOOMe$.[90] Moreover, the rate constants for the decomposition of the diazirine and for the appearance of nitrogen are comparable and independent of the electronic character of the alkene, indicating that carbene generation is the rate-determining step in these cyclopropanation reactions. There is no evidence for pyrazoline intermediacy with the electron-poor substrates. Qualitatively, at least, CH_3OCCl appears to behave as an *ambiphile,* a carbene that displays electrophilic selectivity toward electron-rich substrates and nucleophilic selectivity toward electron-poor substrates.

The ambiphilicity of CH_3OCCl was quantitatively demonstrated by the relative reactivities summarized in Table 3.5.[91] Whereas the typical electrophiles CH_3CCl ($m^{obsd} = 0.50$) and CCl_2 ($m = 1.00$) exhibit steadily decreasing reactivi-

TABLE 3.5. Relative Reactivities of CXY

Alkene	CH_3CCl (25°C)	CCl_2 (80°C)	CH_3OCCl (25°C)
$Me_2C=CMe$	7.44	78.4	12.6
$Me_2C=CH_2$	1.92	4.89	5.43
t-$MeCH=CHMe$	1.00	1.00	1.00
$CH_2=CHCOOMe$	0.078	0.060	29.7
$CH_2=CHCN$	0.074	0.047	54.6

ties as alkene nucleophilicity declines (i.e., as alkene π-electron ionization potential increases), ambiphilic CH_3OCCl describes a *parabolic* relation between selectivity and alkene π-electronic character. This carbene is reactive toward either electron-rich or electron-poor carbenes, but exhibits minimal reactivity toward alkenes of intermediate nature.[92-94] It is particularly gratifying that this behavior is both predictable and (at least) semiquantitatively understandable from application of frontier molecular orbital theory to the carbene–alkene cycloadditions.[79,85,95]

The addition of a singlet carbene to an alkene involves simultaneous interactions of the vacant carbenic p-orbital (LUMO) with the filled alkene π-orbital (HOMO), and of the filled carbenic σ-orbital (HOMO) with the vacant alkene π^*-orbital (LUMO).[95] A singlet carbene is inherently *both* an electrophile and a nucleophile. Behaviorally decisive, however, is whether in the transition state of a given addition, the $LUMO_{CXY}$–$HOMO_{C=C}$ or $HOMO_{CXY}$–$LUMO_{C=C}$ orbital interaction dominates and determines the electronic distribution in the transition state. The dominant orbital interaction is determined both by the differential energies of the "competitive" interactions (the smaller differential energy makes the larger contribution to the stabilization energy afforded by interactions of the reactant FMOs at the transition state) and by the comparative extents of orbital overlaps.

From calculated 4-31G energies (electron volts)[79] of CXY LUMOs (CCl_2, 0.31; CF_2, 1.89; CH_3OCCl, 2.46) and HOMOs (-11.44, -13.38, -10.82, respectively), and from experimental values of alkene π^*- ($Me_2C=CMe_2$, 2.27; $CH_2=CHCN$, 0.21)[79,85] and π-orbitals (-8.27, -10.92, respectively),[79,85] we derive the differential energies ($E_{CXY}^{LU} - E_{C=C}^{HO}$) and ($E_{C=C}^{LU} - E_{CXY}^{HO}$). For reactions of CCl_2 or CF_2 with *either* $Me_2C=CMe_2$ or $CH_2=CHCN$, the former difference is smaller than the latter, indicating that these are electrophilic additions. On the other hand, although ($E_{CXY}^{LU} - E_{C=C}^{HO}$) is also smaller than ($E_{C=C}^{LU} - E_{CXY}^{HO}$) for reaction of CH_3OCCl with $Me_2C=CMe_2$ (10.73 versus 13.09 eV), the *reverse* is true for reaction of CH_3OCCl and $CH_2=CHCN$ (13.38 versus 11.03 eV). Therefore, the addition of CH_3OCCl to $Me_2C=CMe_2$ can still be characterized as electrophilic, but the analogous reaction with $CH_2=CHCN$ should be considered nucleophilic. The ambiphilicity of CH_3OCCl is thus consistent with FMO theory on a semiquantitative level. For a fuller discussion, involving the consideration of comparative orbital overlaps and of differential FMO energies for all the alkenes included in Table 3.5 with CCl_2, CF_2, CH_3OCCl, and $(CH_3O)_2C$, the reader is referred to the original publications.[79,85,91]

Not only do m_{CXY} values (Eq. 12) provide a convenient measure of carbenic selectivity, but they correlate linearly with *ab initio* values of E_{CXY}^{LU}. Electron-donating substituents on the carbene raise its LUMO energy and increase its selectivity, whereas electron-withdrawing groups lower E_{CXY}^{LU} and decrease selectivity. Both m_{CXY} and E_{CXY}^{LU} are related to the exothermicity of carbene-alkene addition, correlating with ΔE_{Stab}, the carbene stabilization energy. The latter is defined as the negative of the 4-31G energy of the isodesmic reaction (Eq. 14). ΔE_{Stab} is inversely related to the exothermicity of carbene-alkene

$$CH_2 + CH_3X + CH_3Y \longrightarrow CXY + 2CH_4 - \Delta E_{Stab} \qquad (14)$$

addition.[79] The most stable carbenes, those that react least exothermically, exhibit the greatest selectivity. (These considerations prove useful in understanding the selectivity of cyclopropylchlorocarbene.[96]) Calculations (4-31G) carried out on ground states, transition states, and products of CXY additions to ethylene gave the following values of E_a and ΔE_{rx} (kcal/mol): CCl_2, 8, -70; CF_2, 27, -46; FCOH, 37, -31; $C(OH)_2$, 47, -18.[79] As selectivity (m_{CXY}^{calcd}) increased, E_a increased but ΔH_{rx} decreased.

FMO considerations have also been applied to Hammett studies of the additions of CF_2, CFCl, CFBr, CBr_2, and CCl_2 to a constant series of styrene derivatives.[97] The observed ρ values (σ^+, 0°C) were -0.57, -0.65, -0.55, -0.44, and -0.69, respectively. Carbenic relative reactivity depended on substrate type: CF_2 and CFCl were more selective than CCl_2 with alkylethylenes,[86] but the inverse order held with styrenes,[97,98] a result that could be rationalized by FMO theory. Additionally, m_{CXY} was also related to FMO parameters.[97]

It is appropriate here to consider the report of Giese and Meister, who studied the selectivities of CH_2, CCl_2, and CBr_2 toward the $Me_2C=CHMe/Me_2C=CH_2$ alkene pair from -40 to 160°C.[99] Below ~42°C, the selectivity order was $CF_2 >$ $CCl_2 > CBr_2$, but above ~72°C the order was reversed. In the temperature range 57 ± 15°C, the selectivities were similar. Clearly, it is dangerous to draw conclusions from the temperature dependence of k_{rel} for CXY toward a single pair of alkenes. CCl_2, for example, shows almost *no* temperature dependence with this pair over the entire -40 to 160°C range, although its behavior toward the $Me_2C=CMe_2/Me_2C=CH_2$ pair is temperature dependent.[42a] It is, however, fair to state that "conclusions concerning differences in reactivities or· stabilities of . . . carbenes from experimentally available selectivities are possible only if the influence of temperature is observed."[99] It will be noted that all m_{CXY} data[86] refer to 25°C, below the reported isoselective temperature (~57°C),[99] and that the m_{CXY} values correlate linearly with various *ab initio* parameters.[79]

It was also suggested that the electrophilic character of CCl_2 and CBr_2 "is due to entropic effects."[99] This conclusion was based on an Arrhenius analysis of the temperature dependence of $(k_{rel})_{CXY}$ for the $Me_2C=CHMe/Me_2C=CH_2$ pair. The argument ignores any k_{rel} and Hammett relation data that are more in keeping with selectivities based on differential activation enthalpies.[42] The differential entropy suggestion was considered in detail over a decade ago, when it was warned that "this activation parameter-reactivity correlation, although appealing, is contingent on the ability of the Arrhenius equation to separate effectively $\Delta H^{\ddagger}$ and $\Delta S^{\ddagger}$ terms . . . of small magnitude . . . . should the Arrhenius equation prove to be too crude a tool to effect this separation, meaningful detailed interpretation of the $\Delta \Delta H^{\ddagger}$ and $\Delta \Delta S^{\ddagger}$ values is obviously impossible."[100] On the whole, it seems best to discuss "electrophilicity" and other "philicities" in terms of differential *free energies* of the reactions.

b. Linear Free Energy Relations

Several reports deserve brief mention. Watanabe et al. have examined carbenic insertions into aryl and alkylsilanes, determining k_{rel} for Si—H insertion reactions as a function of substrate structure.[101] Data were analyzed by Hammett (σ) or Taft ($\Sigma\sigma^*$) methods, and ρ or ρ^* was obtained for appropriate substrate series. Typical results included ($\rho^* =$) -0.36 ($CHCOOCH_3$ from $N_2CHCOOCH_3/$ Cu, 90°C), -1.07 (CCl_2 from CCl_3COONa, DME, 100°C), -1.88 ("CH_2" from $Hg(CH_2Br)_2$, C_6H_6, 80°C), -1.56 ("CH_2" from $CH_2I_2/ZnEt_2$, ether, 24°C), and +2.13 ($Me_3SiOCPh$ from $Me_3SiC(=O)Ph + h\nu$, 22°C).[101] Note that all the species behaved as electrophiles ($\rho^* < O$) except for $Me_3SiOCPh$, which appeared to be strongly nucleophilic ($\rho^* > O$). The ρ or ρ^* values for Si—H insertions of $CHCOOCH_3$ and CCl_2 were roughly proportional to the sums of the σ^* values for the substituents of these carbenes.

In a related study ρ^* was determined from relative addition rates (18°C) of CXY to $CH_2=CHCH_2Z$ as a function of $\sigma^*(Z)$.[94] The data included ($\rho^* =$) -0.05 (PhCH), -0.18 ($CHCOOCH_3$), -0.25 (cyclopentadienylidene), and -0.36 $[C(COOCH_3)_2]$, with all carbenes generated by direct diazoalkane photolysis. A correlation of ρ^* with $\Sigma\sigma_I$ for the substituents of the various carbenes (including CCl_2) showed that ρ^* became more negative as $\Sigma\sigma_I$ became more positive, indicating that CXY became more electrophilically selective toward $CH_2=CHCH_2Z$ with more inductively withdrawing X and Y.

Mention should also be made of Nefedov's study of CCl_2 and CBr_2 additions to **75-77**.[102] For CCl_2 additions, log (k_{rel}) correlated with appropriate substituent constants: $\rho = -0.58$ (σ^+) or -0.94 (σ) for **75** (temperature *variable*, -12 and 82°C); $\rho^* = -0.46$ for **76** (minimum value, -12°C); and $\rho^* = -1.10$ for **77** (-12°C). Note that $|\rho|$ appears to decrease with increasing reactivity (nucleophilicity) of the substrate set.

75   76   77

Finally, CCl_2 relative reactivities were reported for additions to a variety of cyclopropylethylenes.[103] Opposing steric and electronic effects attend cyclopropyl substitution on the alkenic sp^2 carbon atoms, making it difficult to discern general trends. For example, toward competitive pair **78**, CCl_2 showed a kinetic selectivity of 100, but its selectivity toward pair **79** was only 0.96.

78   **79**

c. 1,2,2-Trifluoroethylidene

Absolute reactivity parameters for intermolecular reactions are quite rare in carbene chemistry, so that we note reports that difluoromethylfluorocarbene underwent competitive 1,2-hydride migration and addition to alkenes.[104] At 140–200°C (X = F) or ~300–370°C (X = CH_3), F_2CHCF was pyrolytically generated in the gas phase, apparently as a singlet (stereospecific addition to cis-2-butene). The relative rate of intermolecular addition versus intramolecular rearrangement was determined as a function of temperature (167–350°C), per-

mitting derivations of the relative Arrhenius parameters $(E_a^{\sim H} - E_a^{addn}) = 11.2$ kcal/mol and log $A^{\sim H}$/log $A^{addn} = 2.1$. Various approximations were then applied to the isomerization reaction, leading to estimated values of $E_a^{\sim H}$ 22.9 kcal/mol and $A^{\sim H}$ $10^{13.4}$ sec^{-1}. Thus absolute values for the F_2CHCF addition reactions would be $E_a = 11.7$ kcal/mol and $A \sim 10^{11.3}$ sec^{-1}. For comparison, E_a for addition of CF_2, to $CF_2=CF_2$ (gas phase) was reported to be 6 kcal/mol.[105] These E_a values, the latter particularly, seem rather low.

C. Addition Reactions of Halogenated Alkylcarbenes

A persistent problem in carbene chemistry is the difficulty of achieving intermolecular reactions of alkylcarbenes. Intramolecular additions and insertions are typically too fast to allow for the intermolecular reactions to compete favorably. One solution to this problem is to generate the alkylcarbene in a triplet state, thus rendering intramolecular reaction relatively unfavorable.[106] This approach is often thwarted, however, by a rapid interconversion of singlet and triplet states combined with the very rapid reaction of the singlet. Even though one first generates the triplet, the reactions of the carbene come from the related singlet state. Moss and his co-workers have achieved another solution to this problem that appears to have very few limitations. In a group of five communications they described the intermolecular additions of chloroethylcarbene, tert-butyl-chlorocarbene, chlorocyclopropylcarbene, and chlorocyclobutylcarbene.[96,107] In a typical example[107b] it was found that chlorocyclopropylcarbene gave yields of 33–78% of cycloaddition products to various olefins with only 5–10% of the ring expansion product, 1-chlorocyclobutene (see Eq. 15). By contrast, singlet cyclopropylcarbenes have not proved easy to trap in intermolecular reactions.[106,107b]

However, chlorocyclopropylcarbene turns out to be substantially less selective than predicted by the empirically developed equation of Moss.[108] *Ab initio*

$$\text{33–78\%} \qquad \text{5–10\%} \tag{15}$$

calculations support the intuitive feeling that chlorocyclopropylcarbene should prefer a bisected conformation (**80a**).[96] However, reaction of such a conformation is retarded by a steric interaction between a pair of hydrogens (H_a) encountering the substrate olefin (see Eq. 16). Twisting about the appropriate carbon-

$$\tag{16}$$

80b   **80a**

carbon bond to give **80b** relieves these interactions at the calculated maximum cost of some 9.5 kcal/mol.[96] A new m value for twisted chlorocyclopropylcarbene can be calculated ($m = 0.48$) and this value compares very well with the experimental number ($m = 0.41$).[109]

D. Triplet Halocarbenes (?) and Spinoffs[110]

Although halocarbenes have been regarded as archetypal singlets, several theoretical studies of the last 3 years suggest that, for some halocarbenes at least,

TABLE 3.6.   Energy Differences Between
             Triplet and Singlet Halocar-
             benes

Carbene	$E_{\text{triplet}} - E_{\text{singlet}}$ (kcal/mol)
CF_2	-44.5
CCl_2	-13.5
CHF	-9.2
CHCl	-1.6
CHBr	$+1.1$

the triplets may be accessible intermediates.[10] Thus the *ab initio* calculations of Bauschlicher, Schaefer, and Bagus gave the singlet–triplet energy differences shown in Table 3.6.[10] From these calculations one would estimate that CBr_2 and the iodocarbenes should have accessible triplet states. This does not mean, of course, that *reactions* of these triplets are observable, as most of these carbenes are born in the singlet state. If the rate of reaction of the singlet is much greater than that of the triplet, it may well be that singlet reactions are all that will be observed regardless of the ground state of the molecule.

In 1978 Lambert, Kobayashi, and Mueller made the remarkable discovery that reaction of dibromocarbene, derived from phenyltribromomethylmercury, with 1,2-dichloro- or 1,2-dibromoethylene, led not only to the expected dihalocyclopropane, but also to the rearranged compounds shown in Eq. 17.[111] Note

$$:CBr_2 + \overset{Cl}{\underset{}{\diagdown}}\!\!\diagup^{Cl} \rightarrow \quad + \quad \diagup\!\!-CHCl_2 \qquad (17)$$

that the cyclopropane is formed in a stereospecific fashion, thus implicating the singlet state of the carbene. Lambert suggested that the triplet dihalocarbene was in equilibrium with the singlet state. The rearranged compound formed by the rapid migration of halogen in the diradical formed by addition of the triplet. This interesting and sensible suggestion was picked up by others,[112] who noted the calculations mentioned previously[10] and pointed out that other seemingly anomalous reactions could be explained using triplet halocarbenes. Thus the curious reaction (Eq. 18) of Yang and Marolewski,[113] in which vinylcyclopropanes are formed from cyclobutenes, was originally[113] rationalized using a dipolar intermediate. This intermediate could be replaced with the diradical produced by addition of a triplet carbene. It could also be demonstrated that the halogen shift was indeed a reaction induced by triplet addition.[112] Singlet dicar-

$$\text{(18)}$$

X = I, Cl

bomethoxycarbene reacted with 1,2-dichloroethylene to give the stereospecifically formed cyclopropane, whereas the triplet gave rearrangement in exactly the anticipated sense (see Eq. 19). This work has spawned further research, and it has now been shown that diphenylcarbene also adds to 1,2-dichloroethylene

singlet $(ROOC)_2C$:

triplet $(ROOC)_2C$:

$$\text{(19)}$$

to give a mixture of stereospecifically formed cyclopropane and rearranged compound.[114]

Ironically, the original suggestion by Lambert that triplet dibromocarbene was responsible for the rearrangements he noted[111] has turned out to be wrong. Further kinetic studies by Lambert and co-workers have shown that although two intermediates *are* involved in this reaction, the one leading to rearranged product is not the triplet, but rather a carbene metal complex.[115] We think it is important to note the repercussions of this "mistake." First of all, attention has been focused on an important and neglected area and others have been provoked into doing experiments that have led to the discovery of the very rearrangement of halogen initially suggested. Moreover, if the surmise that a carbene metal complex is producing a trimethylene intermediate is correct, then this phenomenon itself is certainly worthy of further study.

It really should be possible to find reactions of triplet halocarbenes, at least for bromo- and iodocarbenes. Here the triplets cannot be far from the singlets in energy, and the whole trick seems to be to find a way to decrease the rate of singlet addition while maintaining or increasing, if possible, the rate of triplet addition. In the haloethylene studied by Lambert et al.,[111] the singlet reaction is hampered by the electron-withdrawing nature of the halogens and the triplet reaction is enhanced by the stabilization afforded one end of the trimethylene by the halogen. There must be other tricks one could use to accomplish the same end, and it seems to us that there is a good chance that the dihalo-ethylenes would be effective in "siphoning off" triplets from equilibrating pairs of spin states. One would expect increasing work on such systems.

Although the calculations mentioned above,[10] as well as earlier work,[116] placed triplet CF_2 at least 45 kcal/mol higher than the ground-state singlet, it is this unlikely species that has received the most attention among the dihalocar-benes. In the latest of a series of works dealing with the reaction of oxygen and ozone with tetrafluoroethylene, Toby and Toby have observed both singlet and triplet emission from difluorocarbene.[117] Similarly, Koda[118,119] has found that the reaction of (^3P) oxygen atoms with tetrafluoroethylene gave both singlet and triplet CF_2 emission. This work echoed a result of Hsu and Lin,[120] who also claimed that triplet CF_2 was formed in the oxygen/tetrafluoroethylene system. The only quantitative data to emerge from this work is Koda's estimate[118] of a singlet–triplet gap for CF_2 of 57 kcal/mol. This value is considerably higher than theoretical estimates[10,116] and led Koda to postulate that a singlet state excited by some 7 kcal/mol was produced by triplet–triplet annihilation.

The report by Levi, Taft, and Hehre[121] that the heat of formation of di-chlorocarbene was 53.1 ± 2 kcal/mol has been confirmed by Ausloos and Lias.[122] The former workers used the bracketing proton-transfer technique and the latter workers removed the possibility that reactions other than proton transfer between CCl_2H^+ and bases were actually being measured. In the work of Ausloos and Lias[122] a slightly lower value of 47.8 ± 2 kcal/mol was obtained, but this does not seem to us to be significantly different from that obtained by Levi et al.[121]

There was also a report in 1977 of a new reaction of dichlorocarbene.[123] Considering the rather long history of this intermediate, this new reaction qual-ifies as something of a mechanistic triumph, especially when the nature of the product is considered. Applequist and Wheeler not only corrected the earlier report[124] that dichlorocarbene reacted with methyl 3-methylbicyclobutane-1-carboxylate to give a 3% yield of a bicyclo[1.1.1]pentane, but also isolated a 27% yield of the real bicyclopentane (see Eq. 20a).[1] The compound previously

$$(20a)$$

$$(20b)$$

assigned that structure is the diene **81**. As shown in Eq. 20b, other bicyclo-butanes give variable yields of bicyclopentanes. Neither Doering and Coburn,[125] who allowed methylene to react with 1,3-dimethylbicyclobutane, nor Wiberg et al.,[126] who examined the reaction of dichlorocarbene with methyl bicyclo-butane-1-carboxylate, noticed any bicyclopentanes. Instead, only products of insertion reactions along with the mysterious dienes appeared.

Although Applequist and Wheeler[123] ventured no mechanistic speculation—perhaps wisely—both Doering and Coburn[125] and Wiberg et al.[126] wrote a step-wise process in which the carbene reacts with the central bond of the bicyclo-butane to give an intermediate that suffers bond cleavage to give the diene (Eq. 21). The problem with this mechanism is that these reactions are almost

$$(21)$$

certainly singlet-state processes and we know of no precedent for a stepwise reaction of a singlet carbene with a hydrocarbon. It is very tempting to us to speculate that the reaction occurs in a single step in which the attacking carbene plucks one of the bicyclobutane carbons, leaving behind an olefin as shown in Eq. 22a. This reaction even has precedent, at least in its intramolecular version.

$$(22a)$$

$$(22b)$$

Cyclopropylcarbenes are known to undergo fragmentation along with ring expansion (Eq. 22b). This fragmentation reaction has been shown to be a con-certed process[106] and is related in an obvious way to the "plucking" postulated above. We repeat here[112] our feeling that these reactions are extraordinarily curious and urge further study.

E. Stereochemistry of the Intramolecular Carbon–Hydrogen Insertion Reaction

A long-standing problem in carbene chemistry has involved the preferred arrangement for carbon–hydrogen insertion. This problem has been attacked theoretically many times[127] and the generally reasonable conclusion has been reached that a process in which the LUMO of the carbene—the empty p-orbital—interacts with the HOMO of the carbon–hydrogen bond is the best description of the reaction. The first-experimental attack on the problem was that by Nickon and co-workers,[128] who used the carbene derived from brexan-4-one (**82**) to determine whether the *exo-* or *endo-* hydrogen migrated preferentially in the insertion reaction (Eq. 23). The purpose of the "extra" bridge in the brexyl

$$(23)$$

82

systems was to maximize the overlap of the *exo*-hydrogen with the empty p-orbital and minimize the overlap of the *endo*-hydrogen. An *exo/endo* ratio of 138 was observed and this was taken as evidence that the previously described theoretical description was accurate. That is, a process in which the carbene empty orbital overlapped efficiently with the bond undergoing the insertion reaction was greatly favored. It is worth noting, relative to our previous discussion on page 63 of the possible role of excited carbenes, that the photochemical generation of the brexylcarbene led to an *exo/endo* ratio of only 4:8! Thus the thermal process was far more selective than the photochemical.

It was soon pointed out, however, by Kyba and Hudson[129] and by Freeman and his co-workers[130] that there is a natural preference in norbornyl systems for *exo* migration to an empty orbital and that Nickon's factor of 138 might be misleadingly high. Thus some portion of that preference must be due to factors present in cationic systems as well as carbenic systems. In Nickon's

$$(24)$$

defense, however, it must be noted that not only are the reasons for this so-called "natural" preference poorly understood, but the carbenic and cationic processes are clearly closely related. However, it is certainly true that this modified norbornyl system is not an *ideal* framework on which to make this study. Two other systems studied are shown in Eq. 24, and the *exo/endo* ratios found were between 13 and 20.[129,130] If this represents the natural *exo/endo* preference, then the extra bridge in the brexyl system accounts for roughly a factor of 10 in increasing the efficiency of the *exo* carbon-hydrogen insertion reaction.

Another way to fix stereochemistry and thus to study stereochemical preferences is to use a cyclohexyl system anchored by a tertiary butyl group. Seghers and Shechter have used such a system[131] and have determined that axial phenyl migrated approximately five time as fast as equatorial phenyl (Eq. 25). This may

$$\text{(25)}$$

have very little to do with the carbon-hydrogen insertion process, as mechanisms are available for phenyl migration that look much more like an addition process than an insertion reaction.[131] Seghers and Shechter did find that axial hydrogen migrated faster than equatorial hydrogen, but only by a factor of 1.4.[131]

A strikingly similar ratio of 1.5 was determined by Kyba and John[132] in an anchored cyclohexyl system (Eq. 26). At first glance this number seems astonishingly small in view of the reasonable theoretical ideas mentioned before.

$$\text{(26)}$$

Kyba has used both MINDO/3 and MNDO semiempirical calculations in an attempt to rationalize this low ratio.[133] When the calculations are carried out without geometrical restrictions, it is always the empty p-orbital to which the migrating hydrogen goes. The transition states for migration of axial (83) and equatorial (84) hydrogen in the cyclohexyl system both involve overlap with the carbenic p-orbital. The molecule is flexible enough to permit such overlap and, strikingly, the two transition states reached are of comparable energies.

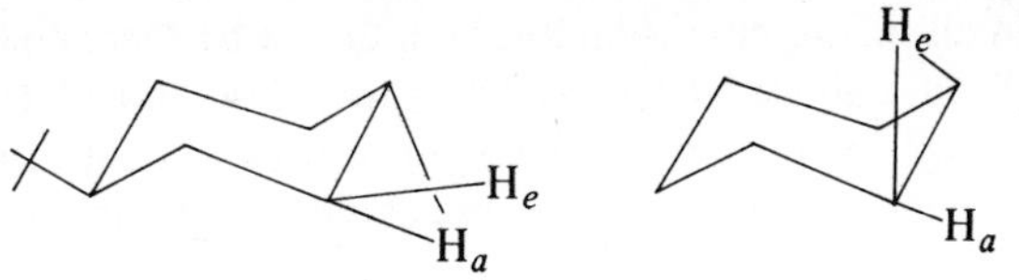

83   84

Kyba promised a further rationalization of the larger exo/endo ratios obtained in the rigid bicyclic systems,[133] but this has not yet appeared.

In apparent sharp contradiction comes a report from Press and Shechter that deals mainly with migratory aptitudes in 2-methoxy-4-tert-butylcyclohex-anylidenes.[134] On the assumption that migration of hydrogen from C_6 (the adjacent methylene group to the carbene) would be equally probable in both carbenes of Eq. 27, it was found that H_a migrated faster than H_e by a factor

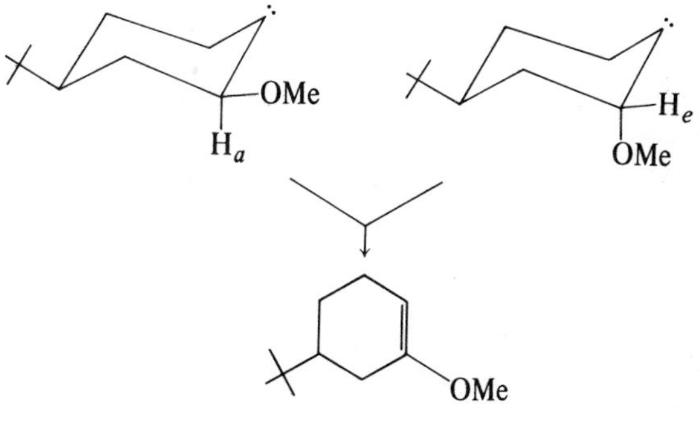

(27)

of 35–46. These ratios were determined by normalization relative to migration from the adjacent methylene. In the justification for this normalization procedure, the authors noted that the methoxy group has only a small A value and anticipated that electronic effects on the remote methylene would be negligible. We do not see anything wrong with these assumptions, but they do remain assumptions, and the axial/equatorial ratio here is much greater than that determined by Kyba and John.[132]

We are not able to resolve this conflict, but can repeat the suggestion[134] that as the methoxy group is known to promote migrations of adjacent groups,[134,135] the migrations in this work[134] are more exothermic than those in the unsubstituted case studied by Kyba and John[132,133] and thus their transistion states will come earlier. Hydrogen bridging should be less and *perhaps* the orbital interactions in the starting materials—here[134] resembling the transition states more than they do in the earlier case[132,133] —are determining. This is a feeble straw to grasp, but it is all we have at the moment.

Of course, in all of this discussion it is tempting to become too one-sided—to focus too much on a single HOMO-LUMO interaction. Su and Thornton have measured isotope effects for the 1,2-hydrogen shift to give substituted styrenes (Eq. 28) and have come to the reasonable conclusion that a "pull-push"mechan-

$$Ar-CHD-\ddot{C}-CH_3 \longrightarrow ArCD=CHCH_3 + ArCH=CDCH_3 + ArCHDCH=CH_2$$

(28)

ism applies in which electrophilic attack on the carbon–hydrogen bond by the empty p-orbital is accompanied by backside nucleophilic attack by the carbene-filled orbital.[136] This second half of the reaction represents the interaction of the carbene HOMO with the carbon–hydrogen bond LUMO.[136]

F. Phenycarbenes, Cycloheptatrienylidenes, and Their Interconversion

a. Reactions of Phenylcarbenes

In a series of four papers Crow and McNab have explored the intramolecular insertion reactions of o-substituted arylcarbenes in the gas phase.[137] When possible, five-membered ring formation is the favored reaction, competing favorably even with 1,2 carbon–hydrogen insertion when R = methyl (Eq. 29). Six-

(29)

membered ring formation is less favored and, as one would expect, the presence of a heteroatom at the 4-position increases five-membered ring formation. Sulfur is less effective than either oxygen or nitrogen in favoring the five-membered ring.

Two other fascinating reactions are described by Crow and McNab.[137] One of these, a 1,2 *vicinal* abstraction reaction to give alkenes (see Eq. 30a), was pre-viously mentioned by Baer and Gutsche,[138] but the other, a *geminate* abstraction to give substituted benzaldehydes, is new. Labeling experiments were used to

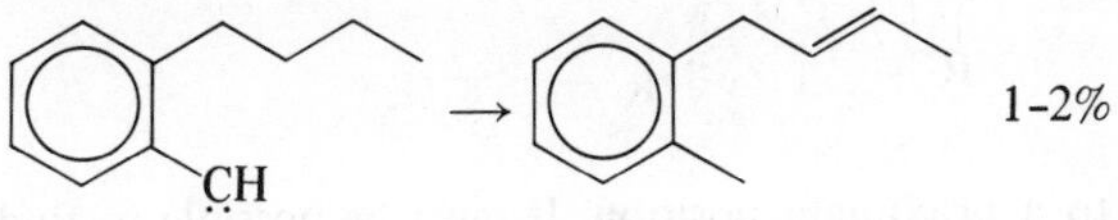

$$1-2\% \qquad\qquad (30a)$$

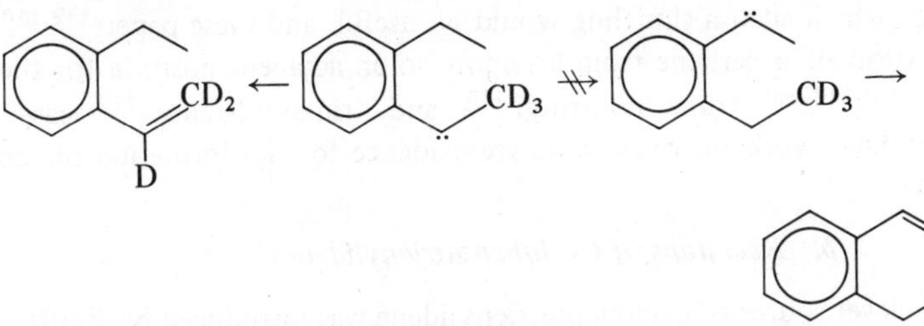

$$(30b)$$

show that the vicinal abstraction reaction did not proceed by a geminate abstrac-tion process followed by hydrogen shift (Eq. 30b). In a similar fashion, deuterium labels were used to demonstrate that the benzaldehydes were formed by an irreversible sequence of reactions (Eq. 31). Thus *no* H appears in the aldehydic

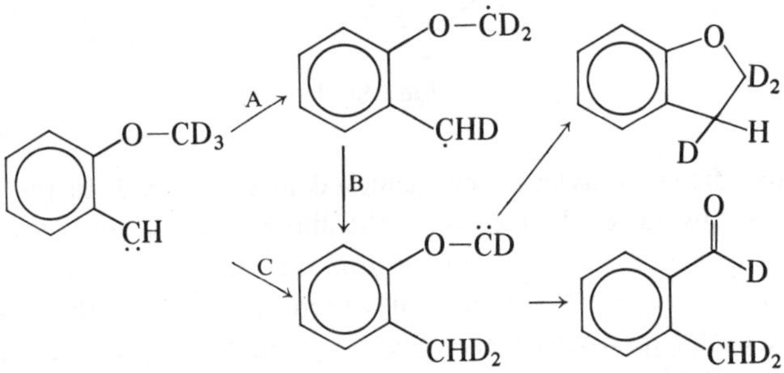

$$(31)$$

85

position. Path C could not be differentiated from Path A + B. Note that in the diagram on page 114 of the Crow and McNab series,[137] there is an important typographical error. A deuterium is omitted from the aldehydic position of compound 85.

Two papers[139,140] have emphasized the potential for using the gas-phase phenylcarbene rearrangement in a practical way. In the phenylcarbene rearrange-ment all the parts of the original molecule are still present at the end of the reaction. Of course, there has been a substantial shuffling of positions, but as Eq. 32 shows, the benzene ring functions as a delivery system for the carbene,

$$\text{(32)}$$

moving it from a remote to a proximate position. It must be possible to find many cases for which such a shuffling would be useful, and these papers[139,140] describe migration of a carbene from a remote to an adjacent position for the R = trimethylsilyl,[139,140] trimethylgermyl,[140] and trimethylstannyl[140] cases. This last-mentioned work provides tentative evidence for the formation of the first stannirane.

b. Reactions of Cycloheptatrienylidenes

A new and clever source of cycloheptatrienylidene was introduced by Reiffen and R. W. Hoffmann,[141] who treated trimethylsilylcycloheptatrienylium fluoroborate with fluoride ion (Eq. 33). The familiar cycloadditions to dimethyl-

$$\text{(33)}$$

fumarate and dicyanoethylene were achieved in good yield. In the absence of good traps for this nucleophilic carbene, the dimer heptafulvalene was isolated.

The additions noted in Eq. 33 confirm the nucleophilic character of this carbene. A ρ value of +1.05 had been measured earlier[142] and this has been confirmed for a cycloheptatrienylidene generated from a nitrogen-free precursor.[143] Thus a ρ value of +1.02 was determined by measuring the addition to substituted styrenes of the species generated from chlorocycloheptatriene and potassium t-butoxide.[143] This is important because one must be certain that addition of a diazo compound (which would give a positive ρ) to the styrenes was not occurring in the earlier work.[142]

The relationship between cycloheptatrienylidene and the related allene cycloheptatetraene (86) has been explored in various ways. The question of the mechanism of formation of cycloheptatrienylidenes from chlorocycloheptatrienes has been pursued by C. Mayor and W. M. Jones.[144] Treatment of 2-chlorocycloheptatriene with base could lead either to cycloheptatrienylidene or cycloheptatetraene and the two species are in any event closely related (Eq. 34). Indeed, it was been argued based on MINDO/3 calculations, that the tetraene is

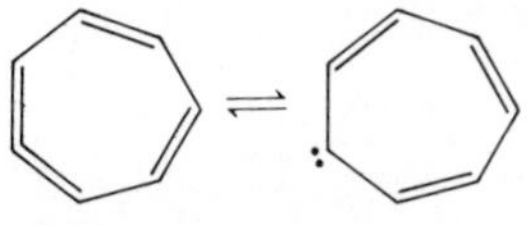

86

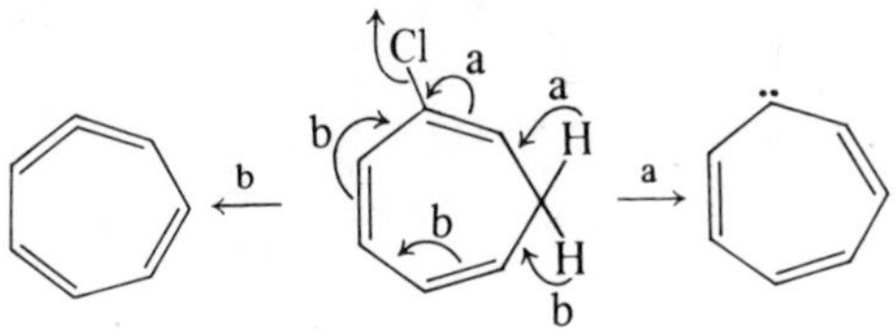

(34)

the better description of the two.[145] Should this turn out to be the case, it would not of course affect arguments concerning the *reacting* state of the molecule.

The products isolated from the base-catalyzed generation of cycloheptatriene in styrenes are the spirononatrienes formed by "classical" addition of the carbene to the double bond. These can be rearranged thermally as shown in Eq. 35 to give the hypothetical adducts of the allene. That these 2 + 2 cycloaddition products are not formed *directly* in the reaction would seem to argue that elimination

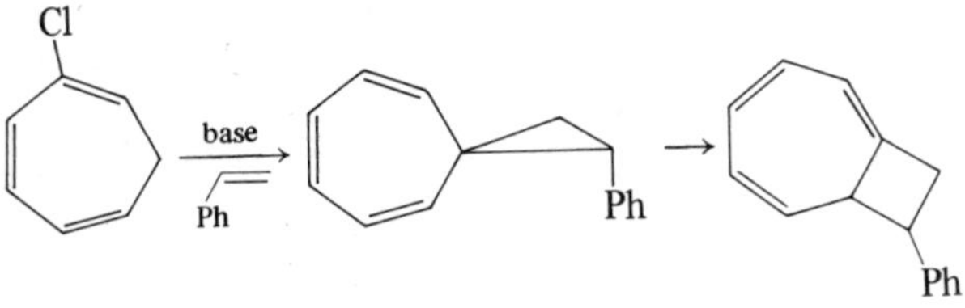

(35)

takes place to give a carbene and not a cyclic allene. However, this ignores possible equilibration between the two species, with the carbene being the more reactive. Evidence from benzo systems[144] argues that it is in fact the allene that is first formed (Eq. 36). Thus treatment of 87 with potassium butoxide at 0°C leads to products from the cyclic allene. Here direct carbene formation would have to disrupt aromaticity in the benzene ring. The related compound 88, however, could only give the allene by a process through which the resonance energy was disrupted, although the carbene could be formed without such problems. The fact that 88 is stable under conditions far more severe than those sufficing to convert 87 to product argues that, in these cases at least, allene formation precedes carbene generation.

c. Phenylcarbene–cycloheptatrienylidene Interconversions

This area has been reviewed twice[146, 147] and we focus here on events not touched on or only briefly mentioned in the reviews.

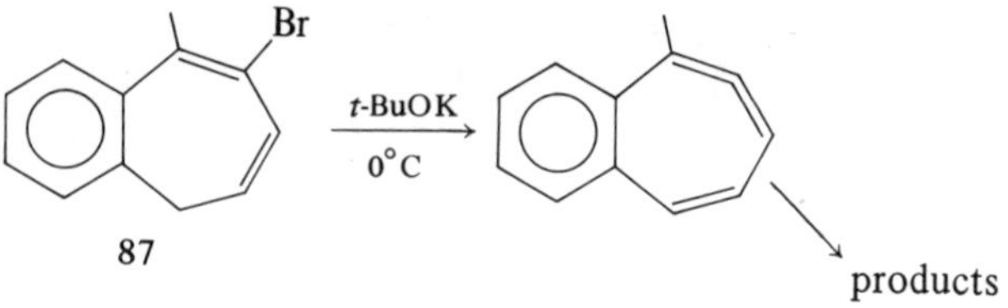

(36)

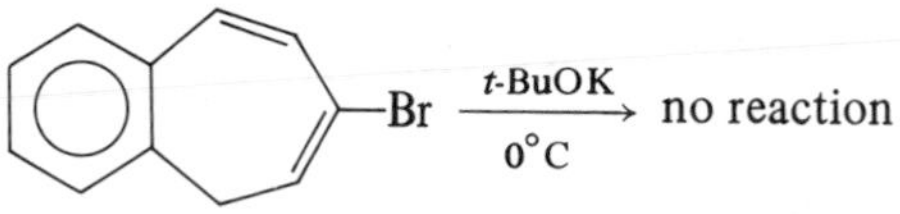

A number of attempts have been made to achieve the phenylcarbene rearrangement in solution. In principle, one might imagine at least three ways of achieving this gas-phase reaction in solution. The first would be dilution to minimize intermolecular reactions, and the second would be introduction of groups producing a steric barrier to dimerization, the favored pathway for cycloheptatrienylidenes in solution. Finally, one could imagine the introduction of benzene rings to reduce loss of aromaticity in the rearrangement process. The first of these three techniques fails, but the other two succeed. Thus introduction of 2,7-disubstitution in cycloheptatrienylidene leads to ring contraction, presumably by the mechanism shown in Eq. 37.[148] Curiously, no benzocyclobutene (89) is formed, and we wonder if this lack is mechanistically important, or whether it simply reflects the fact that methyl-o-tolylcarbene must be the first-formed phenylcarbene.

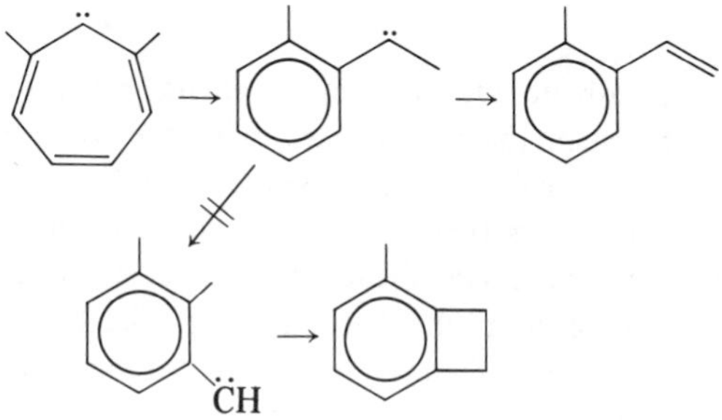

(37)

Although variants are possible, the most economical and widely supported mechanism for the interconversion of phenylcarbene and cycloheptatrienylidene involves the intermediacy of a bicyclo[4.1.0]heptatriene.[145-147] In a simple aromatic system, formation of such an intermediate will involve rather considerable loss of resonance stabilization in the precursor benzene. When this energy cost is attenuated, however, by introduction of other benzene rings, one might hope that the reaction could be observed, as it indeed appears to be. Thus Billups and Reed[149] isolated compounds 90 and 91 from treatment of 92 with potassium t-butoxide (Eq. 38). As the authors note, it is difficult to avoid

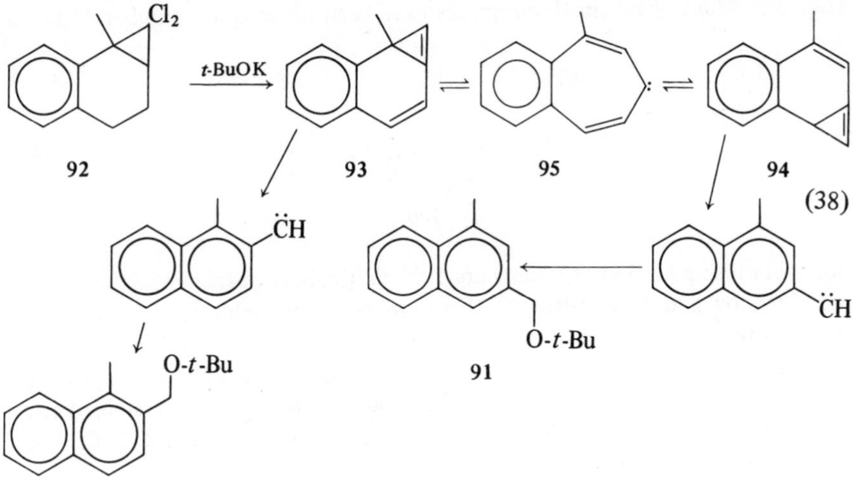

Cl$_2$

t-BuOK

92   93   95   94

(38)

ĊH

O-t-Bu   O-t-Bu

91

O-t-Bu

90

the conclusion that an equilibration of compounds containing the bicyclo[4.1.0] structure (93, 94) has resulted in the generation of two naphthylcarbenes. This work does not require that the naphthylcarbenes themselves revert to 93 and 94, but it does show that the cycloheptatrienylidene 95 can isomerize in solution.

Another method of attenuating aromaticity loss is to make the precursor less aromatic from the start. This appears to be the case with the annulenylcarbenes 96 and 97, which have been shown to generate the annulenylidenes 98 and 99

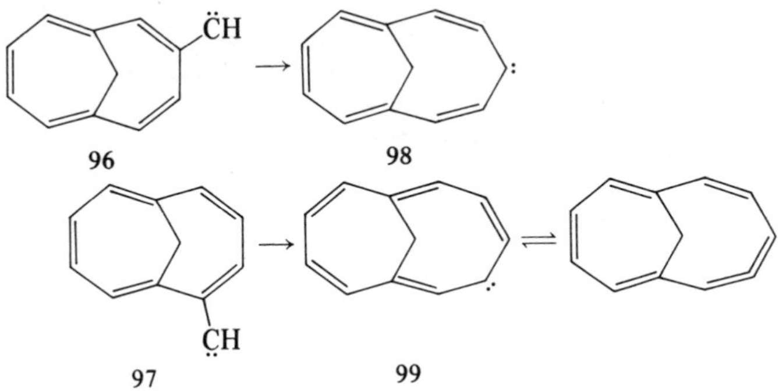

ĊH

→   :

96   98

(39)

→   ⇌

ĊH

97   99

(Eq. 39). Carbene **98** behaves as a normal cycloheptatrienylidene, but **99** yields products consistent either with reaction as the cyclic allene form or stepwise processes and rearrangements of initially formed products.[150]

The mechanism of rearrangement of phenylcarbene itself has also received attention. In the absence of intra- or intermolecular traps, phenylcarbene ring-contracts at high temperature to fulvenallene (**100**) and ethynylcyclopentadienes (Eq. 40). Many mechanistic proposals have been made and a number of some-

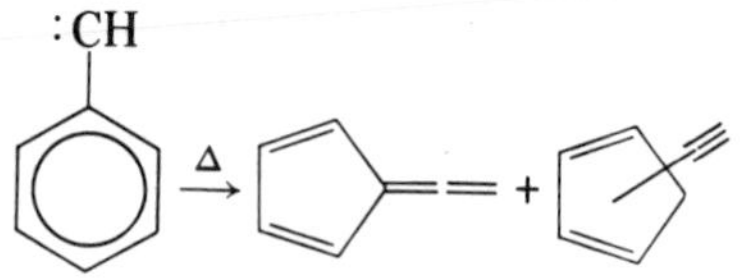

$$\begin{array}{c} :CH \\ \bigcirc \end{array} \xrightarrow{\Delta} \quad + \qquad \qquad \text{(40)}$$

100

times confusing label experiments done.[151] Particularly puzzling was the observation by Crow and Paddon-Row[152] that pyrolysis of phenyldiazomethane, labeled at C_1 with ^{13}C, produced uniformly labeled fulvenallene (Eq. 41). Wentrup and co-workers[153] have reexamined this pyrolysis, carrying it out at 590°C, which is the lowest temperature sufficient to give **100**. The label is still scrambled to all positions but not *equally*. The labeling order is $C_5 > C_1, C_4 > C_2, C_3 > C_6 > C_7$.

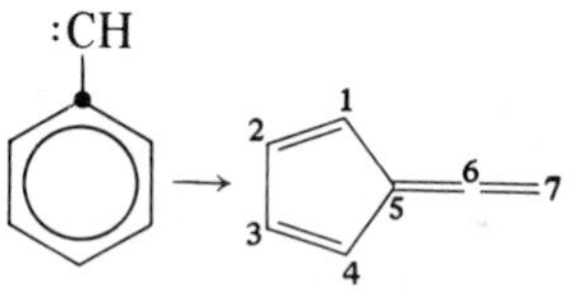

$$\begin{array}{c} :CH \\ \bigcirc \end{array} \rightarrow \qquad \qquad \text{(41)}$$

100
uniformly labeled

The mechanism shown in Eq. 42, though complex and *ad hoc*, fits this labeling

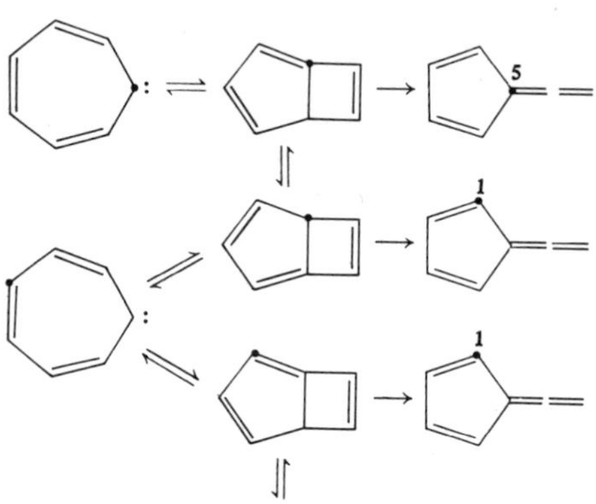

$$\text{(42)}$$

pattern brilliantly. There is even recent evidence that the new intermediate postulated in this mechanism, bicyclo[3.2.0]heptatriene, undergoes the required reversion to cycloheptatrienylidene.[154] Generation of 7-norbornadienylidene at relatively low temperature leads to a dimer of the bicyclo[3.2.0] compound that can be shown *not* to lead to heptafulvalene. At higher temperatures heptafulvalene does appear in pyrolyses of norbornadienylidene. Thus it seems that the [3.2.0] compound can equilibrate with the seven-membered ring (Eq. 43).

$$\text{(43)}$$

d. Extensions of the Reaction

A reasonable question to ask is whether the phenylcarbene rearrangement occurs in aromatic systems other than simple benzenes. At least for the boron cluster compound, *o*-carborane, the answer is yes.[155] Generation of the carboranylcarbene **101** in the gas phase leads to the benzocyclobutene analogue **102** and vinyl-*o*-carborane. A labeling experiment has demonstrated that **102** is not a precursor of vinylcarborane and thus the "phenylcarbene rearrangement" shown in Eq. 44 is implicated.

G. Nucleophilic Carbenes

New reports concern carbenes **103–106**.[156, 157] *N*-Methylbenzothiazolenylidene (**103**)[156a] and 1,3-diphenylimidazolidenylidene (**104**)[156a] are generated by

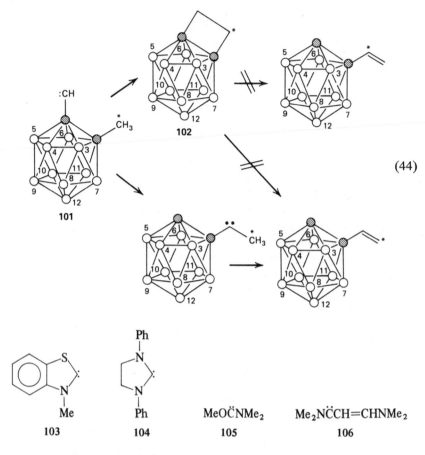

(44)

$$103 \qquad 104 \qquad MeO\overset{..}{C}NMe_2 \qquad Me_2N\overset{..}{C}CH=CHNMe_2$$
$$\qquad\qquad\qquad\qquad\qquad\qquad 105 \qquad\qquad 106$$

thermolytic elimination (140°C) of piperidine and methanol, respectively, from their formal N−H or O−H insertion products. The carbenes are readily trapped by ArNCO or ArNCS (2 equiv), with formation of heterocycles such as **107** from **103** and PhNCO. The carbenes are nucleophilic (see **104a**) in their selectivities toward ArNCO; Hammett studies afford $\rho = +1.7 \pm 0.5$ (**103**) and $\rho = +1.9 \pm 0.5$ (**104**), versus $\rho = +2.0 \pm 0.5$ for $(MeO)_2C$, all at 140°C. Although the ρ values are

$$107 \qquad\qquad 104a \qquad\qquad (MeO)_2CHNMe_2$$
$$\qquad\qquad\qquad\qquad\qquad\qquad\qquad 108$$

equal, in terms of magnitudes of relative reactivity, the selectivity order appears to be **104** > **103** > $(MeO)_2C$.[156a]

Methoxydimethylaminocarbene (**105**) is apparently generated by pyrolytic elimination of methanol from amide acetal **108** (80°C).[156b] This species can also be intercepted by ArNCO or PhNCS, with the formation of products analogous to **107**. Carbene **105** is especially interesting in that $m^{calcd} = 2.91$,[85, 86] an extremely high m value, which points toward strong nucleophilicity. Species such as **103–105** are probably best regarded as C_2-deprotonated-1,3-diheteroatomic allylic cations (see **109**). Thus **105** can be generated by NaH deprotonation of **110**,[156b] and pyrolysis of **108** is suggested to occur by initial reversible

109	**110**	**111**

ionization to **110** and methoxide, followed by removal of H$^+$. As might be expected, $(MeO)_3CH$ also eliminates methanol (145°C), forming $(MeO)_2C$, which can be trapped by PhNCO to give a hydantoin analogous to **107**.[156c]

Ethoxide-promoted cleavage of **111** is believed to afford carbene **106**, a vinylogous bisdimethylaminocarbene. Protonation of **106** affords stable salt **112**.[157] The carbene can also be generated by decarboxylation (160°C) of zwitterion **113**, with subsequent nucleophliic addition to various electrophiles.

112	**113**

Of interest is the suggested nucleophilicity of thioethylvinylcarbene (**114**), generated by photolytic elimination of nitrogen from pyrazoline **115**.[158] Cyclopropanations of $Me_2C{=}CMe_2$ or $EtOCH{=}CH_2$ proceed in only low yield (~20–25%), whereas additions to $CH_2{=}CHCN$ or $CH_2{=}CHCOOMe$ occur in 90–95%

114	**115**	**116**	**117**

yield. Additions also proceed with dimethyl fumarate (80%) or maleate (60%), affording the *trans*-dicarbomethoxycyclopropane in each case.

An alternative mechanism that must be considered is photolytic reversion of 115 to diazoalkane 116, which could add to electrophilic alkenes by 1,3-dipolar addition, followed by loss of nitrogen. However, the rate of photolytically driven nitrogen evolution from 115 is reported to be identical in both inert solvents and dimethyl maleate.[158] Despite the fact that 114 appears to be nucleophilic (see 117), its electrophilic properties are not fully suppressed. It can be trapped by allylic sulfides (90%) via initial S-ylide formation followed by [2,3]-sigmatropic rearrangement.

Nucleophilic properties also have been reported for thioxanthenylidene, 118.[159, 160] Generated photolytically from the corresponding diazothioxanthene, 118 added to dimethyl fumarate (32%),[159] but not to cyclohexene,[159] tetra-

118   119

methylethylene,[160] or trimethylethylene.[160] Semiempirical (extended Hückel, CNDO/2) calculations indicated high electron density on the carbenic C atom of 118.[160] As an alternative to addition of 118, we must consider 1,3-dipolar addition of 9-diazothioxanthene, followed by decomposition of the derived pyrazoline. However, the pyrazoline derived from dimethyl maleate was reportedly quite stable to heat or light.[159, 160] Also, the photochemical reaction of 9-diazothioxanthene with dimethyl fumarate in THF (which afforded the adduct of 118), produced nitrogen at a rate that was independent of olefin concentration.[159] A more detailed study of 118 would be welcome, especially one involving styrene substrates and a Hammett correlation.

Such a study has been reported for 9-xanthenylidene (119) which, despite predicted nucleophilicity,[160] afforded an unusual reactivity pattern.[161] Generated by photolysis of 9-xanthone tosylhydrazone sodium salt, k_{rel} (-18 to -36°C) for additions to styrenes were p-OCH$_3$, 1.56; p-CH$_3$, 1.33; H, 1.00; p-Cl, 1.52; p-Br, 2.63; m-Br, 3.48. The logs of these relative reactivities gave a parabolic relation with σ or σ^+. Among proferred explanations[161] were additions proceeding via triplet 119, or competitive additions of either singlet and triplet 119, or of 119 and photoexcited 9-diazoxanthene. An intriguing possibility to be added to the list is that singlet 119 behaves as an ambiphile toward styrene substrates.

H. 7-Norbornenylidene and Related Species

7-Norbornenylidene (120) offers the potential for chemically dominant, homoallylic $p-\pi$ interaction.[162] Attempts to detect this have focused on the

120   syn-121   anti-121

intermolecular selectivity of 120 and on its intramolecular rearrangement, which can be usefully formulated as a "foiled methylene" addition (Eq. 45).[163]

$$(45)$$

It was suggested that $p-\pi$ interaction in 120 would cause the carbenic bridge to bend $\sim 20°$ away from the vertical toward the π-bond, so that sterically bulky substituents on an addend would prefer the *anti* side of 120.[162] However, reaction of 120 with *t*-butylethylene (220°C) gave a 7:1 ratio of *syn*-121/*anti*-121, indicating approach of the alkene across the π side of 120.[164] This contradictory result may be rationalized on steric grounds: The π side of 120 offers less hindrance to attack of a nucleophile at C_7 than does the ethano side; the same situation obtains in nucleophilic attack on 7-norbornenone. Whatever the extent of bridge-bending in 120, it does not control the stereochemistry of alkene addition.

A related carbene, 2-methyl-7-norbornenylidene (122) has two available $p-\pi$ rearrangement paths: $122 \rightarrow A \rightarrow 123$, and $122 \rightarrow B \rightarrow 124$ (see Eq. 46).

$$(46)$$

124   B   122   A   123

The former would seem electronically better, and is indeed preferred by a factor of 1.7-2.0 at ~180°C. (Photolytic generation of **122** at 25°C gave **123/124** = 1.9.[165]) The regioselectivity expressed in the rearrangement of **122** is modest, although in keeping with Eq. 45, and the result cannot be regarded as mechanically definitive.

The related 7-norbornadienylidene (**125**) was generated via pyrolysis of 7-diazonorbornadiene at 250°C.[154] In a gas-phase reaction, **125** rearranged to **126**, which then gave 2 + 2 dimers in 14–34% yield. The **125** → **126** rearrange-

125   **126**   **127**   **128**

ment may be formulated analogously to Eq. 45. As is mentioned in Section V.F, at 350°C heptafulvalene is formed from **125**, suggestive of a **126** → cycloheptatrienylidene → dimer pathway at high temperature.

The C_7-π interaction in **120** is known to be stronger than the alternative interaction of C_7 with the saturated ethano side.[163] Murahashi has examined carbenes **127** and **128**, in which the p-π interaction typical of **120** is allowed to complete with p-cyclopropyl-σ or alternative p-π interactions.[166] Mechanistically significant among the products derived from pyrolytically generated (180°C) **127** was triene **129**. Its origin can be formulated as in Eq. 47 by analogy to Eq. 45. No rearrangement products attributable to p-π interaction were detected,

127   **129**   (47)

suggesting that with carbene **127**, as with the corresponding carbonium ion, p-σ interaction dominates over p-π interaction.

Carbene **128** was also pyrolytically generated and afforded rearrangement products from competitive p-π_a and p-π_b interactions. The preference (1.7-fold) for the former was judged from molecular models to stem from better overlap of the carbenic p-orbital with π_a than with π_b. However, the carbonium ion corresponding to **128** appears to prefer p-π_b interaction.[166]

I. The Skattebøl Rearrangement

The Skattebøl rearrangement (SR) is formally a vinylcyclopropane rearrangement of a vinylcyclopropylidene; it can occur in competition with the well-known cyclopropylidene-allene ring opening (see Eq. 48).[167] Examples

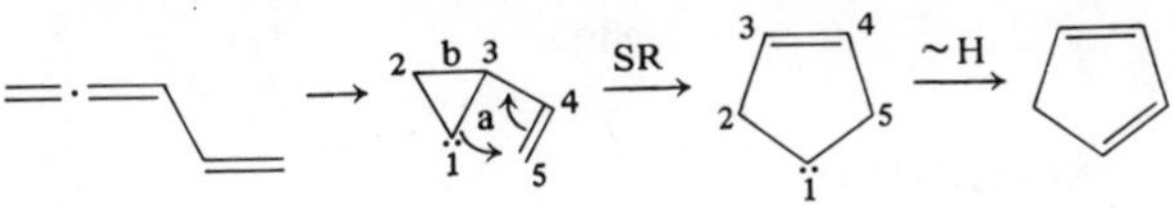

$$
\text{(48)}
$$

of the SR have been reported under carbenoid-generative conditions, (Eqs. 49–52),[167–169] as well as in metal halide-free reactions (Eqs. 53 and 54).[170, 171]

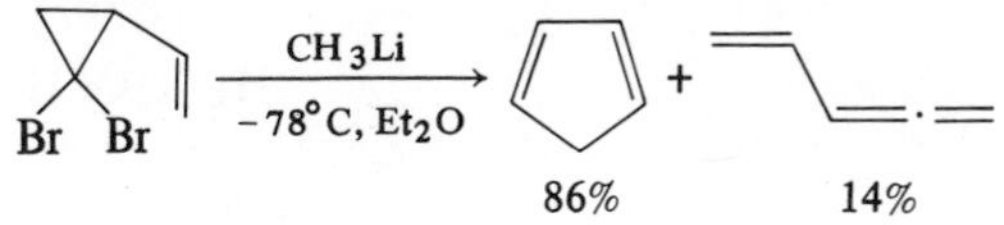

$$
\text{(49)}^{167}
$$

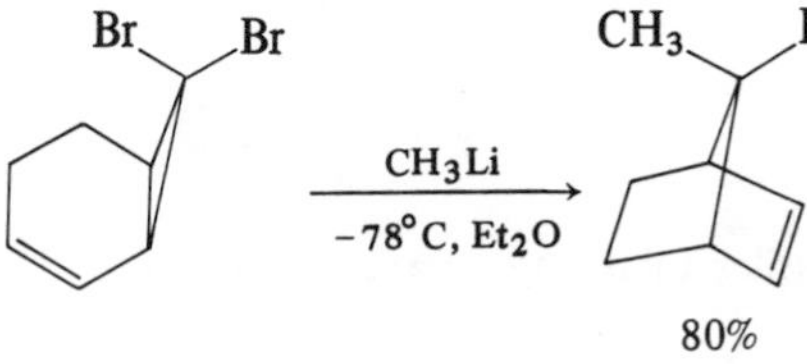

86%   14%

$$
\text{(50)}^{167}
$$

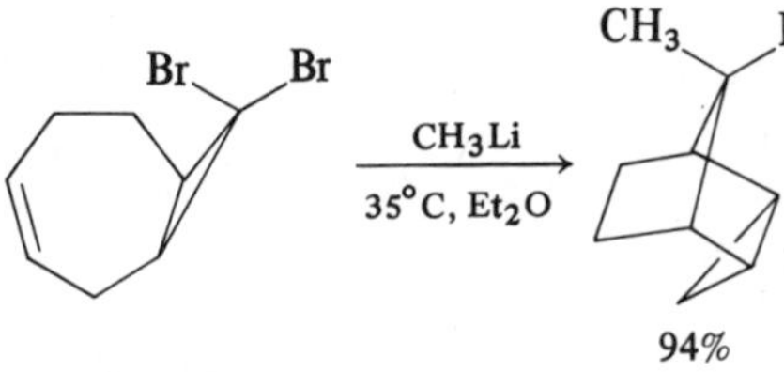

80%

$$
\text{(51)}^{168}
$$

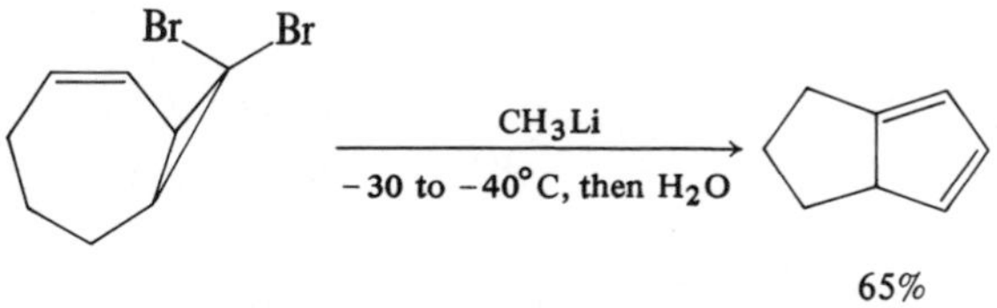

94%

$$
\text{(52)}^{169}
$$

65%

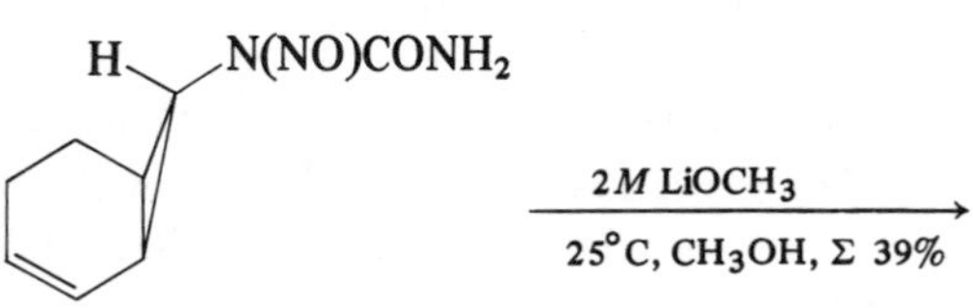

$$
\text{(53)}^{170}
$$

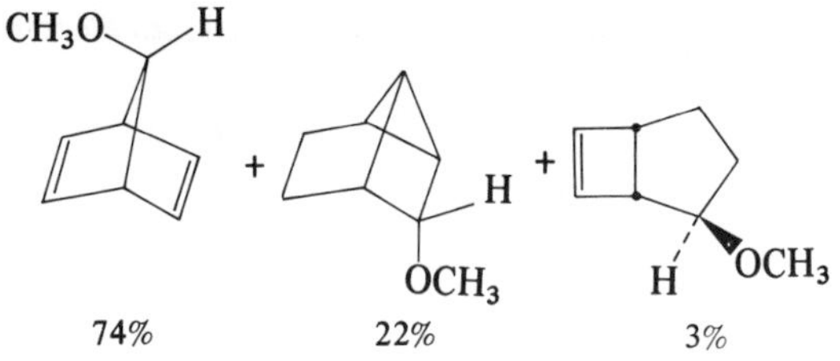

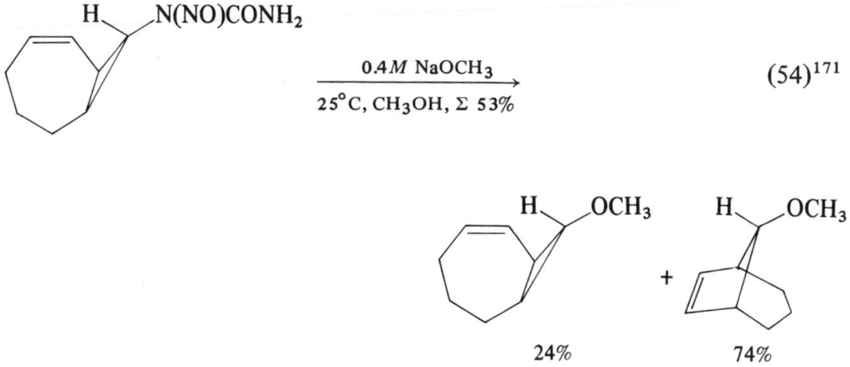

74%   22%   3%

$$\xrightarrow[25°C, CH_3OH, \Sigma\ 53\%]{0.4M\ NaOCH_3}$$

$(54)^{171}$

24%   74%

Several additional examples of reactions related to this rearrangement have been recorded.[172, 173]

Outstanding questions include: (a) What is the mechanism of the SR? (b) What are the scope and limitations of the SR? (c) How does the mechanism vary with reaction conditions?

MINDO/3 calculations on the prototypal vinylcyclopropylidene → 3-cyclopentenylidene rearrangement (Eq. 48), indicate that the reaction is initiated by (singlet) carbene p-π interaction (130), which strengthens to afford structure 131, featuring a three-center, two-electron "nonclassical" interaction.[174] In this

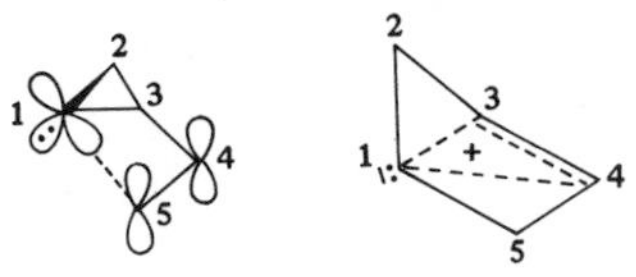

130   131

view the SR is related to the "foiled methylene" additions of 7-norbornenylidene, formulated as in Eq. 45. The activation energy calculated for the SR of Eq. 48 is 13.8 kcal/mol (believed to be a maximum), consistent with the mild conditions under which many SR's occur.

Note that Eq. 48 indicates SR by bond a cleavage; that is, the carbenic carbon (C_1) "migrates" to C_5, the terminal sp^2-carbon of the π-bond.[167] (This feature has been incorporated into the calculations.[174]) *A priori*, one might also formulate Eq. 48) with cleavage of bond b and migration of C_2 to C_5. However, bond a cleavage has been definitely established in a reaction analogous to Eq. 49 using ^{12}C labeling (see Eq. 55).[175] Bond b cleavage must be restricted to allene formation. A similar experiment with identical results (carbenic carbon migration) has

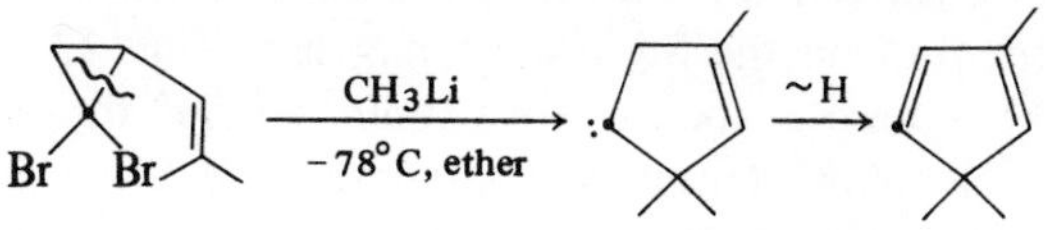

$$(55)$$

47% as maleic
anhydride adduct

been reported for the rearrangement of Eq. 52.[175] Further mechanistic discussion appears below.

Regarding the scope and limitations of the SR, note the apparent vinylogous (1,5) SR outlined in Eq. 56.[176] For convenience, the process is formulated via an intermediate 3,5-cycloheptadienylidene, which affords cycloheptatriene by hydride migration. Other products (not shown) stem from competing cyclopropylidene → allene reaction.[176] The *trans*-isomer of **132** is precluded from

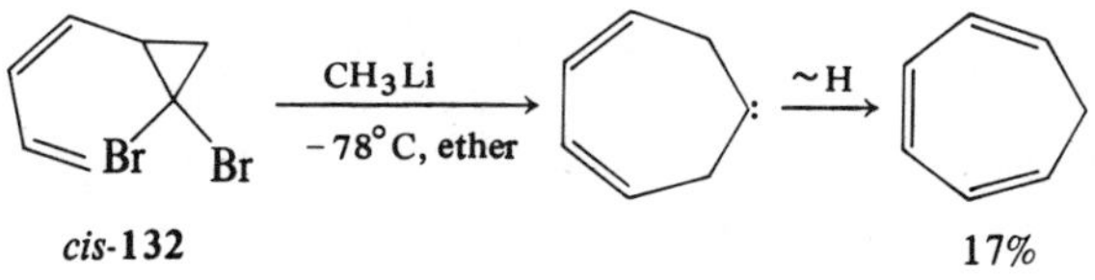

cis-**132**   17%

$$(56)$$

1,5-SR and affords only 1,3-SR products, as well as an allene (Eq. 57).[176]

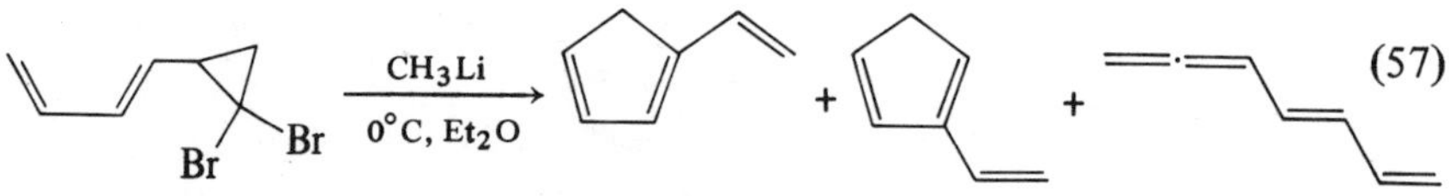

$$(57)$$

trans-**132**

Even the "normal" 1,3-SR is geometry dependent because the distal carbon of the vinyl group must be within bonding distance of the carbenic carbon. The distal carbon must therefore be either freely rotatory (see Eq. 49) or fixed in the "cis" arrangement (see Eq. 50 or **133**)[172] If the distal carbon is locked into an unfavorable geometry, as in "*trans*"-**134**, SR does not occur and only

133   **134**   **135**   **136**

allene formation is observed.[172] (Species **133** and **134** were generated from *gem*-dibromides and CH_3Li; free carbenes are drawn for convenience.)

Further restrictions on the SR involve ring size. Thus **135** does *not* afford **136**, presumably because of the excessive strain required for carbenic attack on the methylenecyclobutane terminus. The appropriate allene forms instead.[172] Note that no such problem attends the SR of **133**. Finally, observe that the vinyl*cyclobutylidene* SR does not occur; **137** (from tosylhydrazone salt pyrolysis at 200°C) affords no products from carbene **138**.[177] Instead, **139** is formed via cyclobutylidene–methylenecyclopropane contraction.

137   **138**   **139**

Turning to speculative, detailed SR mechanisms, it is helpful to consider separately SRs that involve dibromide/CH_3Li initiation and SRs that involve nitrogen loss. The key mechanistic point for dibromide/CH_3Li SR is that free carbenes are not involved; stereospecific reactions of intermediate lithium carbenoids control product structures. The stereospecific product formation in Eqs. 50 and 51 has long been recognized as making free carbene intermediacy

140   **141**   **142**   (58)

143   **144**

unlikely,[167,168] but consistent detailed mechanisms have not been offered. Consider Eq. 50, which is paradigmatic for this reaction class (see Eq. 58).

In analogy to reactions of 7,7-dibromonorcarane, the initial dibromide of Eq. 58 is first converted to *exo*-Li carbenoid **140**, which is unstable relative to the epimeric carbenoid **141**.[72,178] Cyclopropylidene carbenoid **141** is here represented with the partly ionic C—Br bond that follows from the work of Seebach et al.,[71b] (see **65a**). Rear-side π-assisted displacement of bromide then converts **141** to ion pair **142**, which, after bond reorganization, is best represented as a 7-lithio-7-norbornenyl cation–bromide ion pair (**143**). We must now account for the stereospecific conversion of **143** to **144**. It is tempting to postulate conversion of **143** to **145**, followed by S_N2 attack of **145** on CH_3Br (formed in the initial dibromide/CH_3Li reaction) leading to **144**. However, we must note the experiments of Warner and Chang, summarized in Eq. 59.[179]

145

(59)

146   **147, 49%**   **148**
(as diphenyliso-
benzofuran adduct)

When dibromide **146** was subjected to SR, a typical norbornenyl product (**147**) was obtained, along with a small quantity of **148**. Formation of **147** from **146** can be accounted for by steps analogous to those of Eq. 58, shown in Eq. 60. A crucial observation is that sequestering Li by addition of 12-crown-4 *prevents* formation of **147** in favor of **148**. Hence a lithium carbenoid is the precursor of **147**. Moreover, if the reaction is carried out in CD_3I as solvent, **147** (I instead of Br) carries mainly CH_3 (93%) rather than CD_3 groups.[179a,179c] This would seem to exclude formation of **147** from **149** by subsequent S_N2 reaction of an alkyllithium derivative with methyl halide. (By analogy, such a conversion of **143** to **144** also seems inadmissible.) Instead, it is suggested that electrophilic **149** is trapped by CH_3Li, forming **150**, which then affords **147** by Li–halide

$$(60)$$

exchange with CH_3X.[179a] An analogous final step is perforce included in Eq. 58.
The CD_3I experiment[179a] appears compelling, but it is initially surprising that rapid exchange of CH_3Li with CD_3I solvent does not competitively afford CD_3Li and ultimately introduce CD_3 groups into product 147. Either the entire sequence of Eq. 60 is very fast or the exchange reaction must be slow. Indeed, the exchange reaction may be bypassed; at room temperature CH_3Li and CD_3I appear to give CH_3CD_3 and LiI rapidly and exothermically.[179c]

The mechanism in Eq. 58 is readily extendable to the transformations in Eqs. 51[168] and 52.[169] Key elements of the former are outlined in Eq. 61 and involve conversion of carbenoid 151 to ion pair 152 by internal π-displacement, followed

$$(61)$$

by collapse of 152 to delocalized lithiocation 153. Presumably, 153 is converted to the observed product by CH_3Li methylation, followed by Li/Br exchange.

The conversion in Eq. 52 can be formulated as in Eq. 62. Note that delocalized cation 154 naturally accounts for exclusive formation of homoannular diene 155; delocalization forces 1,2-migration of the saturated ethano bridge. This rearrangement is related to the 7-norbornenylidene type (see above), but if free carbene 156 were involved in place of 154, one would expect vinyl "migration" to 157.[163, 180]

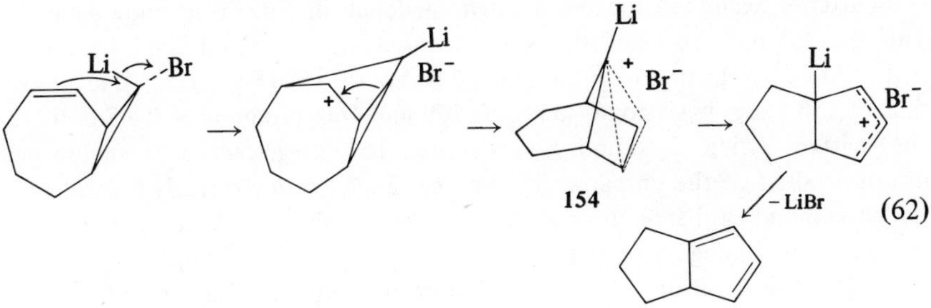

(62)

155

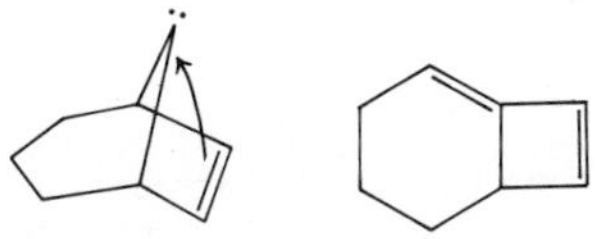

156   157

Note that although **154** completes a norbornenylidene-style rearrangement (Eq. 62), intermediates **143**, **149**, and **153** do not advance along this pathway, but are trapped by CH_3Li. Presumably the strain in the products to which they would rearrange is kinetically prohibitive relative to intermolecular trapping. The added strain should be least in **155** so that rearrangement of **154** proceeds.

The *gem*-dibromide/CH_3Li-generated 7-norbornenylidene intermediates of the SR are usually trapped by CH_3Li,[179a] precluding subsequent rearrangement. Similarly, generation of the corresponding intermediates by deaminative procedures in protic solvents (Eqs. 53 and 54 normally leads to interception by protonation, again precluding further rearrangement.[170, 171] Indeed, one must wonder whether SR reactions initiated from diazo or diazonium ion precursors in protic solvents may entirely bypass the rearranged carbenes by direct diazonium to carbonium ion conversion.

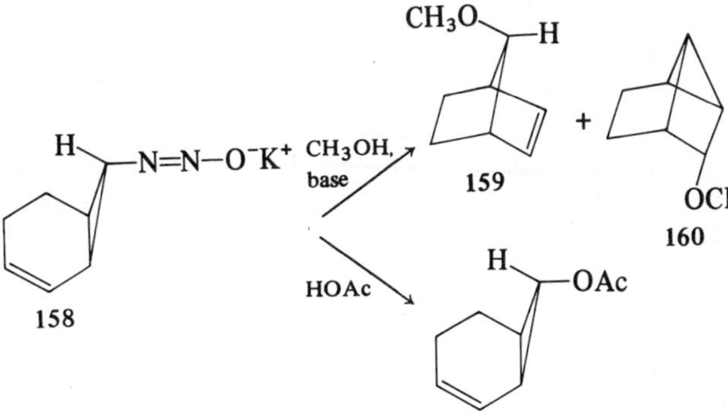

(63)

Skattebøl examined the deaminative analogue of Eq. 50 in some detail.[181] This reaction has also been studied by Kirmse (Eq. 53),[170] but Skattebøl simplified matters by starting with the isolated diazotate 158 (Eq. 63).[182] Decomposition of 158 under basic conditions gave 159 and 160, products of the 7-norbornenyl cation. Acidic decomposition afforded 161, suggested to be an ion pair return product of the unrearranged 7-norcar-2-enyl cation directly formed from the corresponding diazonium ion. Acidic decomposition under other conditions gave products analogous to 161.

Basic decomposition was suggested to occur via the mechanism in Eq. 64.[181]

$$(64)$$

Basic conditions favor conversion of 158 to the unstable diazonorcarene, from which 7-norcarenylidene 162 arises by nitrogen loss. SR converts 162 to 7-norbornenylidene, which is rapidly protonated by methanol, affording the 7-norbornenyl cation and thence ethers 159 and 160. As anticipated, the reaction carried out in CH_3OD affords 159 with > 97% *syn-7-d*. Protonation not only precludes further rearrangement of 7-norbornenylidene,[163] but also intermolecular trapping by *cis*-2-butene, diethyl fumarate, or fumaronitrile (although the latter does intercept the diazo compound as a pyrazoline). In aprotic solution olefinic capture of 7-norbornenylidene is well known.[163,164]

The major question concerning Eqs. 63 and 64 is whether the 162 → 7-norbornenylidene SR is firmly established or whether it is bypassed by a cationic rearrangement; that is, does the 7-norbornenyl cation arise by protonation of 7-norbornenylidene or by rearrangement of a precursor carbonium ion (162-H⁺) or diazonium ion? On balance, we believe that the evidence favors the SR mechanism. The principal support for this carbenic mechanism is the consistently noted increase in SR products that accompanies increasingly basic reaction conditions.[170,171,181] Were these cationic rather than carbenic rearrangements, they would be most strongly expressed under the least nucleophilic (i.e., acidic) reaction conditions. In fact, however, the rearrangements (e.g., 158 → 159) are *absent* under acidic conditions,[170,181] and most dominant under strongly basic conditions.

It remains to generate carbenes such as 162 under aprotic, nonorganometallic conditions so that the 7-norbornenylidene SR product can fulfill its ultimate destiny of rearrangement as in Eq. 45.[163]

J. Reactions of Carbenes in Matrices

As early as 1971 it was evident that fascinating chemistry would emerge from studies of reactions of carbenes in frozen matrices.[183] In their early paper[183] Moss and Dolling revealed that in contrast to reactions at ambient temperatures, which lead to nearly stereospecific cyclopropanation, along with less than 10% olefins, reaction of phenylcarbene in a matrix of *cis*-2-butene gave as much as 52% olefins along with the still mostly stereospecifically formed cyclopropanes (Eq. 65). It was logically concluded that the olefins derived from reaction of

$$Ph-\ddot{C}H \xrightarrow{-196°C} \begin{array}{c} \\ \triangledown \\ Ph \end{array} + \begin{array}{c} \\ \triangledown \\ Ph \end{array} + \begin{array}{c} \\ \triangledown \\ Ph \end{array} + Ph- + \quad + \quad + \quad (65)$$

3.2   43.9   18.5   29.0   4.9

triplet phenylcarbene by an abstraction–recombination process. This work has been expanded upon in recent years by the groups of both Moss and Tomioka and, although we are not able to "wrap up" the area here, it seems well worth our time to highlight these increasingly fascinating reactions.

A key question, of course, is, "Why does triplet reactivity appear at low temperature?" Given that spin states of arylcarbenes are generally in equilibrium, one must wonder why the normally slower triplet reactions appear at low temperature and why triplet abstraction prevails over triplet (and singlet) addition. Moss and Joyce extended the earlier[183] work in a study of methylphenylcarbene.[184] They made the suggestion that in a matrix where motion is obviously restricted, singlets may be hampered by an inability to obtain a reasonable geometry for formation of two bonds. Triplet carbenes that can react in stepwise fashion are relatively unhindered. This seems reasonable to us. Calculations suggest that abstraction by triplet occurs in a more or less linear fashion,[185] whereas in singlet insertion—or addition—the carbene must penetrate more substantially to the inner parts of the reacting molecule. Moss and Joyce[184] noted that methylphenylcarbene produced large amounts of styrene—presumably by intramolecular singlet insertion—even at -196°C. Clearly, the intramolecular singlet reaction would be relatively unhindered, at least compared to intermolecular reactions.

The ability of methylphenylcarbene to produce styrene, even at matrix temperatures, also has been noted by Tomioka and Izawa[186] in their study of the reactions of arylcarbenes in alcoholic matrices. Tomioka and Izawa found that as the temperature decreased, insertion into the α-carbon-hydrogen bond increased at the expense of "insertion" into the oxygen-hydrogen bond, a known singlet reaction (Eq. 66). Further, at low temperatures tertiary or secondary

$$Ph-CHN_2 \xrightarrow[R-CH_2-OH]{h\nu} R-CH_2-O-CH_2-Ph + R-\underset{\underset{CH_2-Ph}{|}}{CH}-OH \qquad (66)$$

carbon-hydrogen bonds were preferred to primary, a selectivity reasonably associated with a triplet species. Large amounts of styrene were formed from methylphenylcarbene, even at low temperatures. Tomioka and Izawa seemed disturbed by this observation, although they ultimately reached a conclusion very similar to that of Moss ("thus matrix might affect the kinetics of the competitive intermolecular singlet and triplet carbene processes occurring within it,"), and went on to worry about the diminution of O—H insertion under conditions where the singlet methylphenylcarbene was producing styrene. We see no real cause for worry and are happy with the proposition that the mixture of intermolecular reactions available to an equilibrating singlet–triplet pair in a matrix may well be different from that in solution.

In a paper examining the insertion reactions of arylcarbenes into carbon-hydrogen bonds in hydrocarbons at various temperatures, Tomioka has provided more evidence for extreme steric selectivity in matrix reactions.[187] For instance, at $0°C$ the statistically corrected tertiary/primary insertion ratio for phenylcarbene and isobutane is 118. This drops at $-196°C$ to a remarkable 4.1. Thus insertion into the primary position becomes *more* favorable. This seems to us in accord with the idea that reactivity in a matrix is controlled to a great extent by what the carbene can reach. Tomioka notes that this is not a temperature effect. That is, initially, as the temperature drops, the carbene gets more selective and it is only after the matrix forms that selectivity in favor of the more substituted position declines. Tomioka reasonably concludes that matrix effects may be—indeed it looks as though they must be—important and cautions the reader to be on the alert for such effects in other systems.

Phenylcarbene was shown to be sensitive to its *precursor* in its selectivity for carbon-hydrogen insertion.[188] Although there was no effect detectable in solution, in the solid phase the ratios of products depended on the source (diazo compound or oxirane) of the carbene (Eq. 67). The same effect appeared for diphenylcarbene, but here, even in the liquid phase, the precursor influenced the product ratio. The authors go on to caution against deriving singlet/triplet carbene ratios from observed product distributions, thus implying that site preferences may determine all.

$$PhCHN_2 \xrightarrow[\text{(CH}_3)_2\text{CHOH}]{hv, -196°C} PhCH_2O-CH(CH_3)_2 + PhCH_2\underset{\underset{CH_3}{|}}{\overset{\overset{CH_3}{|}}{C}}OH + PhCH_2CH_2\underset{\underset{OH}{|}}{\overset{\overset{CH_3}{|}}{C}}H$$

	15.7	68.5	15.8

(67)

$$\xrightarrow[\text{(CH}_3)_2\text{CHOH}]{hv, -196°C}$$

	14.3	79.5	6.2

$$\xrightarrow[\text{(CH}_3)_2\text{CHOH}]{hv, -196°C}$$

	15.9	76.4	7.7

Of course, a precursor dependence may mean that carbenes are not involved in all these reactions. We and, especially, others[26,27] have repeatedly cautioned that excited precursors may be involved in some "carbene" chemistry. Indeed, there is new evidence supplied by Zupancic and Schuster[189] that electronically excited diazofluorene may be involved in insertion reactions. This has obvious implications for the most recently mentioned work[188] and indeed for all of this low-temperature insertion chemistry. Zupanicic and Schuster have managed a direct observation of two transients generated by irradiation of diazofluorene. The short-lived species, with a half-life of 17 ± 2 nsec, is reasonably identified with the singlet carbene (although triplet diazofluorene remains a remote possibility) and the longer-lived species with the triplet. By monitoring the yield of the triplet signal, a number of remarkably similar rate constants for singlet additions have been measured. In contrast to photolyses in acetonitrile, which generate a signal for singlet fluorenylidene, irradiation of diazofluorene in pentane or hexane leads only to the 9-fluorenyl radical. By 10 nsec after the initial flash, no singlet or triplet fluorenylidene appears. Irradiation of diazofluorene in acetonitrile doped with toluene shows that neither singlet nor triplet fluorenylidene is sufficiently reactive as a hydrogen atom abstractor to produce the fluorenyl radical as rapidly as it appears in the hydrocarbon solvents. The authors note that there are two possibilities. Either fluorenylidene abstracts hydrogen much more rapidly from a pure hydrocarbon than from toluene in acetonitrile or abstraction is occurring from the electronically excited state of the diazo compound. Although it is *possible* that a large solvent effect is being observed, these results do strongly raise the possibility that excited diazo compound chemistry may be observable. Whether such chemistry intrudes on the reactions of the Moss and Tomioka groups remains to be seen.

Moss and Huselton[190] have investigated the reaction of diphenylcarbene with isobutylene at various temperatures. They did not investigate various carbon-hydrogen insertions, but rather the competition between "insertion" and addition to isobutylene. At O°C and above addition is dominant, but by the time the temperature has reached -196°C, the "insertion" product, which can be shown by a labeling experiment to be in reality a hydrogen abstraction-recombination product, is nearly the exclusive compound formed (Eq. 68). However, there

$$Ph_2C: + \quad \diagup\!\!\!\diagdown_* \quad \rightarrow \left(Ph_2\dot{C}H + \diagdown\!\!\!\diagup(:)_* \right) \xrightarrow[matrix]{-196°C} \begin{array}{l} Ph_2CH \diagdown\!\!\!\diagup_* \quad 72\% \\[2em] Ph_2CH \diagdown\!\!\!\diagup_* \quad 28\% \end{array} \qquad (68)$$

seems to be no matrix effect on the ratio of adduct to insertion product. That is, as the medium changes from liquid to solid, no dramatic effect is observed. A plot of ln (abstraction product/adduct) versus $1/T$ is linear over the entire range investigated. This does *not* mean that there cannot be a matrix effect on carbon-hydrogen abstraction, that is, it seems to us that the results of Moss and Tomioka *results* are not incompatible. Indeed, Moss and Huselton show that there is a matrix effect—or as they call it a "memory" effect—on the radical abstraction-recombination process. Labeled isobutylene does not produce a symmetrical intermediate in the matrix, although it does in solution at -77°C. Rather, most of the label remains where it originates, in the vinyl position (Eq. 68).

A kinetic analysis shows that the triplet abstraction reaction wins out over the singlet addition process because as the temperature decreases, the increased favoring of the lower energy, ground-state triplet overcomes the higher activation energy needed for triplet abstraction.

Not all diarylcarbenes behave in the same way, however. Moss and Joyce have investigated the reactions at various temperatures of fluorenylidene.[191] Significant differences appear between the reactions of this carbene and its less restricted relative diphenylcarbene. The general outlook is the same; as the temperature is lowered, products of hydrogen abstraction-recombination appear at the expense of cyclopropanes. However, this time the plot of ln (abstraction/addition) versus $1/T$ is not linear. There is not the smooth relationship between these two quantities that exists for diphenylcarbene. However, the "break" occurs at -77°C (i.e., in solution), which indicates that the points at -100 (solution) and -160 and -196°C (matrix) are linear. Thus here there is no simple matrix effect. One must not only worry about the cause of this behavior, but also about the fact that fluorenylidene changes its preference—*even in the triplet state*—from addition at relatively high temperature to abstraction at low temperature. Finally, we note that there is no "memory" effect operating here in the reaction with labeled isobutylene. That is, equal amounts of isomerically labeled products are produced even at $-196°C$ (Eq. 69).

(69)

50%   50%

In a sense, the answer to the second problem is obvious. There are two different reactions—triplet addition and triplet abstraction—and they must have different temperature dependencies. As Moss points out, in a quantitative analysis, the activation enthalpies must be close enough so that a difference in the $T \Delta S^{\ddagger}$ term determines the course of the reaction.

Problems remain and new work is revealing properties of carbenes in matrices that lead to somewhat different explanations for the reactions observed and that raise questions for which we do not have complete answers. Thus Lin and Gaspar[192] have shown that diphenylcarbene decays only 8% over 1 hr at $-196°C$ in isobutylene. Thus at these temperatures essentially no reaction occurs. What are we to make of this in the light of the observation of products by Moss and Huselton at these temperatures? Lin and Gaspar might be observing in their ESR experiment only those carbenes remaining in unreactive matrix sites. Carbenes may also be located in more reactive sites, thus leading to the products observed by Moss. However, the signal intensity observed by Lin and Gaspar is different by but a factor of 2 in isobutylene and fluorolube, and this does not seem a likely explanation to us. Perhaps some of Moss's products are formed on warm-up, but this explanation does not seem reconcilable with the kinetic analysis of Moss.[190]

There certainly are important effects of matrix structure. New work by Senthilnathan and Platz[193] shows, among other things, that the observed rates of decay of triplet arylcarbenes are dependent critically on the composition of the matrix. Thus if a matrix is generated in the usual way and the diazo compound photolyzed, the decay of a triplet ESR signal can be measured. If the frozen sample is simply reirradiated, the results are different than those obtained when the sample is first thawed and then reirradiated. The latter procedure reproduces the original data. That is, a new matrix identical to the initial one must be regenerated. However, simple rephotolysis leads to a decrease in the observed pseudo first-order rate constants.

Although Senthilnathan and Platz find the previously observed predominance of products of hydrogen abstraction-recombination, their kinetic analysis and measured isotope effects lead to a rather different explanation. The predominant carbene decay path occurs by hydrogen atom tunneling. It is the availability of this tunneling pathway at low temperatures that leads to the hydrogen abstraction products, not the emergence of "new" triplet behavior.

VI. REFERENCES

1. E. Wasserman and R. S. Hutton, *Acc. Chem. Res.*, **10**, 27 (1977).

2. H. D. Roth, *Acc. Chem. Res.*, **10**, 85 (1977).

3. D. J. Burton and J. L. Hahnfield, *Fluorine Chem. Rev.*, **8**, 119 (1977).

4. P. J. Stang, *Acc. Chem. Res.*, **11**, 107 (1978).

5. P. J. Stang, *Chem. Rev.*, **78**, 383 (1978).

6. A. Padwa, *Acc. Chem. Res.*, **12**, 310 (1979).

7. H. F. Schaeffer III, *Acc. Chem. Res.*, **12**, 288 (1979).

8. P. F. Zittel, G. B. Ellison, S. V. ONeill, E. Herbst, W. C. Lineberger, and W. P. Reinhardt, *J. Am. Chem. Soc.*, **98**, 3731 (1976).

9. L. B. Harding and W. A. Goddard III, *J. Chem. Phys.*, **67**, 1777 (1977).

10. C. W. Bauschlicher, Jr., H. F. Schaefer III, and B. J. Bagus, *J. Am. Chem. Soc.*, **99**, 7106 (1977).

11. R. R. Lucchese and H. F. Schaefer III, *J. Am. Chem. Soc.*, **99**, 6765 (1977).

12. B. O. Roos and D. M. Siegbahn, *J. Am. Chem. Soc.*, **99**, 7716 (1977).

13. D. Gervey and G. Verhaegen, *Int. J. Quantum Chem.*, **12**, 115 (1977).

14. C. W. Bauschlicher, Jr., and I. Shavitt, *J. Am. Chem. Soc.*, **100**, 739 (1978).

15. (a) J. W. Kenney III, J. Simons, G. D. Purvis, and R. J. Bartlett, *J. Am. Chem. Soc.*, **100**, 6930 (1978);
     (b) J. H. Davis, W. A. Goddard III, and L. B. Harding, *J. Am. Chem. Soc.*, **99**, 2919 (1977);
     (c) Y. Apeloig and R. Schreiber, *Tetrahedron Lett.*, 4555 (1978).

16. S.-K. Shih, S. D. Peyerimhoff, R. J. Buenker, and M. Perić, *Chem. Phys. Lett.*, **55**, 206 (1978).

17. L. B. Harding and W. A. Goddard III, *Chem. Phys. Lett.*, **55**, 217 (1978).

18. R. Noyori, M. Yamakawa, and W. Ando, *Bull. Chem. Soc. Japan*, **51**, 811 (1978).

19. N. C. Baird and K. F. Taylor, *J. Am. Chem. Soc.*, **100**, 1333 (1978).

20. H. L. Hase, G. Lauer, K.-W. Schutte, A. Schweig, and W. Thiel, *Chem. Phys. Lett.*, **54**, 494 (1978).

21. M. J. S. Dewar and H. S. Rzepa, *J. Am. Chem. Soc.*, **100**, 784 (1978).

22.* (a) R. K. Lengel and R. N. Zare, *J. Am. Chem. Soc.*, **100**, 7495 (1978).
  * (b) J. Danon, S. V. Filseth, D. Feldmann, H. Zacharias, C. H. Dugan, and K. H. Welge, *Chem. Phys.*, **29**, 345 (1978).

(c) H. M. Frey and G. J. Kennedy, *J. Chem. Soc., Faraday Trans. I*, **73**, 164 (1977);

(d) E. A. Halevi, R. Pauncz, I. Schek, and H. Weinstein, *Jerusalem Symp. Quant. Chem. Biochem.*, **VI**, 167 (1974);

* (e) C. D. Duncan and C. Trimble, *Tetrahedron Lett.*, 2251 (1977).

23. G. B. Ellison and W. C. Lineberger, private communication.

24.* R. S. Hutton and H. D. Roth, *J. Am. Chem. Soc.*, **100**, 4324 (1978);

* (a) J. H. Davis, W. A. Goddard III, and R. G. Bergman, *J. Am. Chem. Soc.*, **99**, 2427 (1977);

(b) R. Hoffmann, G. D. Zeiss, and G. W. Van Dine, *J. Am. Chem. Soc.*, **90**, 1485 (1968); E. Wasserman, L. Barash, and W. A. Yager, *J. Am. Chem. Soc.*, **87**, 2075 (1965); W. H. Smith and G. E. Leroi, *Spectrochim. Acta, Part A*, **25**, 1917 (1969);

* (c) R. R. Lucchese and H. F. Schaefer III, *J. Am. Chem. Soc.*, **99**, 13 (1977).

25. R. R. Rando, *J. Am. Chem. Soc.*, **92**, 6706 (1970); **94**, 1629 (1972).

26.* H. Tomioka, H. Kitagawa, and Y. Izawa, *J. Org. Chem.*, **44**, 3072 (1979).

27. D. S. Wulfman, B. Poling, and R. S. McDaniel, *Tetrahedron Lett.*, 4519 (1975).

28. W. L. Hase and P. M. Kelley, *J. Chem. Phys.*, **66**, 5093 (1977).

29. T. H. Richardson and J. W. Simons, *J. Am. Chem. Soc.*, **100**, 1002 (1978).

30.* K.-T. Chang and H. Shechter, *J. Am. Chem. Soc.*, **101**, 5082 (1979).

31. H. M. Frey and I. D. R. Stevens, *J. Am. Chem. Soc.*, **84**, 2647 (1962); H. M. Frey, *J. Chem. Soc.*, 2293 (1962); H. M. Frey and I. D. R. Stevens, *J. Chem. Soc.*, 3514 (1963); *J. Chem. Soc.*, 4700 (1964); *J. Chem. Soc.*, 3101 (1965); H. M. Frey, *Pure Appl. Chem.*, **9**, 527 (1964); A. M. Mansoor and I. D. R. Stevens, *Tetrahedron Lett.*, 1733 (1966).

32.* G. R. Chambers and M. Jones, Jr., *J. Am. Chem. Soc.*, **102**, 4516 (1980).

33.* N. Kawabata, I. Kamemura, and M. Naka, *J. Am. Chem. Soc.*, **101**, 2139 (1979).

34. H. E. Simmons, T. L. Cairns, S. A. Vladuchick, and C. M. Hoiness, *Org. React.*, **20**, 1 (1973).

35. J. Nishimura, J. Furukawa, N. Kawabata, and M. Kitayama, *Tetrahedron*, **27**, 1799 (1971).

36. N. Kawabata, M. Tanimoto, and S. Fujiwara, *Tetrahedron*, **35**, 1919 (1979).

37. N. Kawabata, N. Yamagishi, and S. Yamashita, *Bull. Chem. Soc. Japan*, **50**, 466 (1977).

38.* S. Brandt and P. Helquist, *J. Am. Chem. Soc.*, **101**, 6473 (1979).

39. P. W. Jolly and R. Pettit, *J. Am. Chem. Soc.*, **88**, 5044 (1966).

40.* C. P. Casey, S. W. Polichnowski, A. J. Shusterman, and C. R. Jones, *J. Am. Chem. Soc.*, **101**, 7283 (1979).

41. G. L. Closs and R. A. Moss, *J. Am. Chem. Soc.*, **86**, 4042 (1964).

42. R. A. Moss, in *Carbenes*, Vol. I, M. Jones, Jr., and R. A. Moss, Eds., Wiley-Interscience, New York, 1973: (a) pp. 153ff; (b) pp. 223–225.

43. R. A. Moss, *J. Org. Chem.*, **30**, 3261 (1965).

44. Such a mechanism was favored initially: C. P. Casey and S. W. Polichnowski, *J. Am. Chem. Soc.*, **99**, 6097 (1977).

45. P. J. Stang and M. G. Mangum, *J. Am. Chem. Soc.*, **97**, 6478 (1975).

46. M. S. Newman and T. B. Patrick, *J. Am. Chem. Soc.*, **91**, 6461 (1969).

47.* P. J. Stang, J. R. Madsen, M. G. Mangum, and D. P. Fox, *J. Org. Chem.*, **42**, 1802 (1977).

48.   D. Seyferth, J. Y.-P. Mui, and R. Damrauer, *J. Am. Chem. Soc.*, **90**, 6182 (1968).

49.   T. B. Patrick, E. C. Haynie, and W. J. Probst, *J. Org. Chem.*, **37**, 1553 (1972).

50.*   J. C. Gilbert, U. Weerasooriya, and D. Giamalva, *Tetrahedron Lett.*, 4619 (1979).

51.   P. J. Stang and D. P. Fox, *J. Org. Chem.*, **42**, 1667 (1977).

52.   R. F. Cunico and Y.-K. Han, *J. Organomet. Chem.*, **105**, C29 (1976).

53.   Review: H. D. Hartzler, in *Carbenes*, Vol. II, R. A. Moss and M. Jones, Jr., Eds., Wiley-Interscience, New York, 1975, pp. 73ff.

54.*   T. B. Patrick and D. J. Schmidt, *J. Org. Chem.*, **42**, 3354 (1977).

55.*   W. J. le Noble, C.-M. Chiou, and Y. Okaya, *J. Am. Chem. Soc.*, **101**, 3244 (1979).

56.   R. A. Moss and F. G. Pilkiewicz, *J. Am. Chem. Soc.*, **96**, 5632 (1974).

57.   (a) W. E. Keller, *Compendium of Phase-Transfer Reactions and Related Synthetic Methods*, Fluka AG, Buchs, Switzerland, 1979.
    (b) C. M. Starks and C. Liotta, *Phase Transfer Catalysis: Principles and Techniques*, Academic, New York, 1978.
    (c) W. P. Weber and G. W. Gokel, *Phase Transfer Catalysis in Organic Synthesis*, Springer, New York, 1977.
    (d) E. V. Dehmlow, *Angew. Chem., Int. Ed. Engl.*, **16**, 493 (1977).
    (e) M. Makosza, *Russ. Chem. Rev.*, **46**, 1151 (1977).

58.   E. V. Dehmlow and M. Lissel, *J. Chem. Res. (S)*, 310 (1978); *J. Chem. Res. (M)*, 4163 (1978).

59.   E. V. Dehmlow, M. Lissel, and J. Heider, *Tetrahedron*, **33**, 363 (1977).

60.*   M. Fedorynski, *Synthesis*, 783 (1977).

61.*   E. V. Dehmlow and M. Slopianka, *Justus Liebigs Ann. Chem.*, 1465 (1979).

62.   K. Steinbeck, *Tetrahedron Lett.*, 1103 (1978); *Chem. Ber.*, **112**, 2402 (1979).

63.   K. Steinbeck and J. Klein, *J. Chem. Res. (S)*, 396 (1978); *J. Chem. Res. (M)*, 4771 (1978).

64.   D. Seyferth and Y. M. Cheng, *J. Am. Chem. Soc.*, **95**, 6763 (1973).

65.*   Y. Kimura, K. Isagawa, and Y. Otsuji, *Chem. Lett.*, 951 (1977).

66.   M. Makosza, A. Kacprowicz, and M. Fedorynski, *Tetrahedron Lett.*, 2119 (1975).

67.*   A. Gambacorta, R. Nicoletti, S. Cerrini, W. Fedeli, and E. Gavuzzo, *Tetrahedron Lett.*, 2439 (1978).

68.*   T. Aratani, Y. Yoneyoshi, and T. Nagase, *Tetrahedron Lett.*, 2599 (1977).

69.*   (a) A. Nakamura, A. Konishi, Y. Tatsuno, and S. Otsuka, *J. Am. Chem. Soc.*, **100**, 3443 (1978).
    (b) A. Nakamura, A. Konishi, R. Tsujitani, M-a. Kudo, and S. Otsuka, *J. Am. Chem. Soc.*, **100**, 3449 (1978).

70.*   (a) T. Clark and P. v. R. Schleyer, *J. Chem. Soc., Chem. Commun.*, 883 (1979).
    (b) T. Clark and P. v. R. Schleyer, *Tetrahedron Lett.*, 4963 (1979).
    (c) T. Clark and P. v. R. Schleyer, *J. Am. Chem. Soc.*, **101**, 7747 (1979).

71.*   (a) H. Siegel, K. Hiltbrunner, and D. Seebach, *Angew. Chem., Int. Ed. Engl.*, **18**, 785 (1979);
    (b) D. Seebach, H. Siegel, K. Müller, and K. Hiltbrunner, *Angew. Chem., Int. Ed. Engl.*, **18**, 784 (1979).

72.   M. Jones, Jr., and R. A. Moss, Eds., *Reactive Intermediates*, Vol. 1, Wiley-Interscience, New York, 1978, pp. 84–89.

73. D. A. Hatzenbühler, L. Andrews, and F. A. Carey, *J. Am. Chem. Soc.*, 97, 187 (1975).

74. K. Kitatani, T. Hiyama, and H. Nozaki, *Bull. Chem. Soc. Jap.*, 50, 1600, 3288 (1977); K. Kitatani, H. Yamamoto, T. Hiyama, and H. Nozaki, *Bull. Chem. Soc. Jap.*, 50, 2158 (1977).

75. W. W. Schoeller and E. Yurtsever, *J. Am. Chem. Soc.*, 100, 7549 (1978).

76. M. Jones, Jr., and R. A. Moss, Eds., *Reactive Intermediates*, Vol. 1, Wiley-Interscience, New York, 1978, pp. 92–95.

77.* (a) C. W. Jefford and P. T. Huy, *Tetrahedron Lett.*, 755 (1980). For applications of these ideas to the reactions of CF_2 and CCl_2 with quadricyclane, see C. W. Jefford, J.-C. E. Gehret, and V. de los Heros, *Bull. Soc. Chim. Belg.*, 88, 973 (1979).
  (b) C. W. Jefford, J. Mareda, J.-C. E. Gehret, nT. Kabengele, W. D. Graham, and U. Burger, *J. Am. Chem. Soc.*, 98, 2585 (1976).

78. Y. Jean, *Tetrahedron Lett.*, 2689 (1977).

79.* N. G. Rondan, K. N. Houk, and R. A. Moss, *J. Am. Chem. Soc.*, 102, 1770 (1980).

80. G. W. Klumpp and P. M. Kwantes, submitted for publication. We thank Prof. Klumpp for permission to cite these results.

81.* U. Burger and G. Gandillon, *Tetrahedron Lettt.*, 4281 (1979).

82. A. G. Anastassiou and R. P. Cellura, *Tetrahedron Lett.*, 5267 (1970).

83. M. Jones, Jr., W. Ando, M. E. Hendrick, A. Kulczycki, Jr., P. M. Howley, K. F. Hummel, and D. S. Malament, *J. Am. Chem. Soc.*, 94, 7469 (1972).

84. M. E. Hendrick, private communication, 1971.

85.* R. A. Moss, *Acc. Chem. Res.*, 13, 58 (1980).

86.* R. A. Moss, C. B. Mallon, and C.-T. Ho, *J. Am. Chem. Soc.*, 99, 4105 (1977).

87. M. Jones, Jr., and R. A. Moss., Eds., *Reactive Intermediates*, Vol. 1, Wiley-Interscience, New York, 1978, pp. 95–97.

88. R. W. Hoffmann, W. Lilienblum, and B. Dittrich, *Chem. Ber.*, 107, 3395 (1974).

89.* N. P. Smith and I. D. R. Stevens, *J. Chem. Soc., Perkin Trans. II*, 213 (1979).

90.* (a) N. P. Smith and I. D. R. Stevens, *Tetrahedron Lett.*, 1931 (1978);
  (b) R. A. Moss and W.-C. Shieh, *Tetrahedron Lett.*, 1935 (1978).
  (c) N. P. Smith and I. D. R. Stevens, *J. Chem. Soc., Perkin Trans. II*, 1298 (1979);

91.* R. A. Moss, M. Fedorynski, and W.-C. Shieh, *J. Am. Chem. Soc.*, 101, 4736 (1979); R. A. Moss and R. C. Munjal, *Tetrahedron Lett.*, 4721 (1979).

92. Thermolytic decomposition of N_2=CHCOOMe in arylnitriles to form oxazoles has been claimed to involve ambiphilic additions of triplet CHCOOMe.[93] The range of relative reactivities was small: p-$CH_3OC_6H_4CN/C_6H_5CN/p$-ClC_6H_4CN = 1.46 : 1.00 : 1.32. The addition of "singlet" $C(COOMe)_2$ to CH_2=CHCH$_2$X was also reported to display an ambiphilic reactivity pattern, but reactivity differences were again small: (X =) $SiMe_3$/OMe/CN = 2.00 : 1.00 : 1.22.[94]

93. M. I. Komendantov, R. R. Bekmukhametov, and R. R. Kostikov, *J. Org. Chem. USSR*, 14, 1448 (1978).

94. T. Migita, K. Kurino, and W. Ando, *J. Chem. Soc., Perkin Trans. II*, 1094 (1977).

95. W. M. Jones, R. A. LaBar, U. H. Brinker, and P. H. Gebert, *J. Am. Chem. Soc.*, 99, 6379 (1977), Footnote 27; W. M. Jones and U. H. Brinker, A. P. Marchand and R. E. Lehr, Eds., *Pericyclic Reactions*, Vol. 1, Academic, New York, 1977, pp. 110–117.

96.   R. A. Moss, M. Vezza, W. Guo, R. C. Munjal, K. N. Houk, and N. G. Rondan, *J. Am. Chem. Soc.*, **101**, 5088 (1979).

97.   R. R. Kostikov, A. P. Molchanov, G. V. Golovanova, and I. G. Zenkevich, *J. Org. Chem. USSR*, **13**, 1846 (1977); see also R. R. Kositkov, A. F. Khlebnikov, and K. A. Ogloblin, *J. Org. Chem. USSR*, **13**, 1857 (1977).

98.   *Cf.* R. A. Moss and C. B. Mallon, *J. Am. Chem. Soc.*, **97**, 344 (1975).

99.*   B. Giese and J. Meister, *Angew. Chem., Int. Ed. Engl.*, **17**, 595 (1978).

*100.   P. S. Skell and M. S. Cholod, *J. Am. Chem. Soc.*, **91**, 7131 (1969).

101.   H. Watanabe, N. Ohsawa, T. Sudo, K. Hirakata, and Y. Nagai, *J. Organomet. Chem.*, **128**, 27 (1977).

102.   O. M. Nefedov, R. N. Shafran, and A. I. Ioffe, *Izv. Akad. Nauk SSSR, Ser. Khim.*, 2292 (1977).

103.   E. V. Dehmlow and A. Eulenberger, *Angew. Chem., Int. Ed. Engl.*, **17**, 674 (1978).

104.*   R. N. Haszeldine, C. Parkinson, P. J. Robinson, and W. J. Williams, *J. Chem. Soc., Perkin Trans. II*, 954 (1979); see also R. N. Haszeldine, R. Rowland, J. G. Speight, and R. E. Tipping, *J. Chem. Soc., Perkin Trans. I*, 1943 (1979).

105.   W. J. Tyerman, *Trans. Faraday Soc.*, **65**, 1188 (1969).

106.   R. R. Gallucci and M. Jones, Jr., *J. Am. Chem. Soc.*, **98**, 7704 (1976), and references therein.

107.*   (a) R. A. Moss and R. C. Munjal, *J. Chem. Soc., Chem. Commun.*, 775 (1978).

        (b) R. A. Moss and M. E. Fantina, *J. Am. Chem. Soc.*, **100**, 6788 (1978).

        (c) R. A. Moss, M. E. Fantina, and R. C. Munjal, *Tetrahedron Lett.*, 1277 (1979).

        (d) R. A. Moss and R. C. Munjal, *Tetrahedron Lett.*, 2037 (1980).

108.   See Ref. 85, 86, and 96.

109.   One of us (MJ Jr) feels bound to point out that the quote with which Ref. 107b begins is not complete and suggests that the reader buy the book and look it up.

110.   It is a pleasure to acknowledge the assistance of Prof. S. Toby in preparing part of this section.

111.*   J. B. Lambert, K. Kobayashi, and P. H. Mueller, *Tetrahedron Lett.*, 4253 (1978).

112.*   M. Jones, Jr., V. J. Tortorelli, P. P. Gaspar, and J. B. Lambert, *Tetrahedron Lett.*, 4257 (1978).

113.   N. C. Yang and T. A. Marolewski, *J. Am. Chem. Soc.*, **90**, 5644 (1968).

114.*   P. P. Gaspar, B. L. Whitsel, M. Jones, Jr., and J. B. Lambert, *J. Am. Chem. Soc.*, **102**, 6108 (1980).

115.   J. B. Lambert, P. H. Mueller, and P. P. Gaspar, *J. Am. Chem. Soc.*, **102**, 6615 (1980).

116.   V. Staemmler, *Theor. Chim. Acta*, **35**, 309 (1974).

117.   S. Toby and F. S. Toby, *J. Phys. Chem.*, **84**, 206 (1980).

118.   S. Koda, *Chem. Phys. Lett.*, **55**, 353 (1978).

119.   S. Koda, *J. Phys. Chem.*, **83**, 2065 (1979).

120.   D. S. Y. Hsu and M. C. Lin, *Chem. Phys.*, **21**, 235 (1977).

121.   B. A. Levi, R. W. Taft, and W. J. Hehre, *J. Am. Chem. Soc.*, **99**, 8454 (1977).

122.   P. Ausloos and S. G. Lias, *J. Am. Chem. Soc.*, **100**, 4594 (1977).

123.*   D. E. Applequist and J. W. Wheeler, *Tetrahedron Lett.*, 3411 (1977).

124.   H. K. Hall, Jr., C. D. Smith, E. P. Blaushard, Jr., S. C. Cherkofsky, and J. B. Sieja, *J. Am. Chem. Soc.*, **93**, 121 (1971).

Carbenes   131

125.   W. von E. Doering and J. F. Coburn, Jr., *Tetrahedron Lett.*, 991 (1965).
126.   K. B. Wiberg, G. M. Lampman, R. P. Ciula, D. S. Connor, P. Schertlen, and J. Lavanish, *Tetrahedron*, 21, 2749 (1965).
127.   For representative earlier calculations see: R. Hoffmann, G. D. Zeiss, and G. W. Van Dine, *J. Am. Chem. Soc.*, 90, 1485 (1968); H. E. Zimmerman, *Acc. Chem. Res.*, 5, 393 (1972); O. S. Tee and K. Yates, *J. Am. Chem. Soc.*, 94, 3074 (1972); N. Boda and M. J. S. Dewar, *J. Am. Chem. Soc.*, 94, 9103 (1972); J. A. Altmann, I. G. Csizmadia, and K. Yates, *J. Am. Chem. Soc.*, 96, 4196 (1974); J. A. Altmann, I. G. Csizmadia, and K. Yates, *J. Am. Chem. Soc.*, 97, 5217 (1975); J. A. Altmann, O. S. Tee, and K. Yates, *J. Am. Chem. Soc.*, 98, 7132 (1976).
128.   A. Nickon, F. Huang, R. Weglein, K. Matsuo, and H. Yagi, *J. Am. Chem. Soc.*, 96, 5264 (1974).
129.* E. P. Kyba and C. W. Hudson, *J. Org. Chem.*, 42, 1935 (1977).
130.* P. K. Freeman, T. A. Hardy, J. R. Balyeat, and L. D. Wescott, Jr., *J. Org. Chem.*, 42, 3356 (1977).
131.   L. Seghers and H. Shechter, *Tetrahedron Lett.*, 1943 (1976).
132.* E. P. Kyba and A. M. John, *J. Am. Chem. Soc.*, 99, 8329 (1977).
133.* E. P. Kyba, *J. Am. Chem. Soc.*, 99, 8330 (1977).
134.* L. S. Press and H. Shechter, *J. Am. Chem. Soc.*, 101, 509 (1979).
135.   W. Kirmse and M. Buschoff, *Chem. Ber.*, 100, 1491 (1967).
136.   D. T. T. Su and E. R. Thornton, *J. Am. Chem. Soc.*, 100, 1872 (1978).
137.* W. D. Crow and H. McNab, *Aust. J. Chem.*, 32, 89, 99, 111, 123 (1979).
138.   T. A. Baer and C. D. Gutsche, *J. Am. Chem. Soc.*, 93, 5180 (1971).
139.   A. Sekiguchi and W. Ando, *Bull. Chem. Soc. Japan*, 50, 3007 (1977).
140.   G. R. Chambers and M. Jones, Jr., *Tetrahedron Lett.*, 5193 (1978).
141.   M. Reiffen and R. W. Hoffmann, *Tetrahedron Lett.*, 1107 (1978).
142.   L. W. Christianson, E. E. Waali, and W. M. Jones, *J. Am. Chem. Soc.*, 94, 2118 (1972).
143.   B. L. Duell and W. M. Jones, *J. Org. Chem.*, 43, 4901 (1978).
144.* C. Mayor and W. M. Jones, *Tetrahedron Lett.*, 3855 (1977).
145.   M. J. S. Dewar and D. Landman, *J. Am. Chem. Soc.*, 99, 6179 (1977).
146.   W. M. Jones, *Acc. Chem. Res.*, 10, 353 (1977).
147.   U. H. Brinker and W. M. Jones, "Pericyclic Reactions of Carbenes," in *Pericyclic Reactions*, Vol. 1, R. Lehr and A. Marchand, Eds., Academic, New York, 1977.
148.   C. Mayor and W. M. Jones, *J. Org. Chem.*, 43, 4498 (1978).
149.* W. E. Billups and L. E. Reed, *Tetrahedron Lett.*, 2239 (1977).
150.* W. M. Jones, R. A. LaBar, U. H. Brinker, and P. H. Gebert, *J. Am. Chem. Soc.*, 99, 6379 (1977).
151.   C. Wentrup, *Top. Curr. Chem.*, 62, 173 (1976).
152.   W. D. Crow and M. N. Paddon-Row, *Aust. J. Chem.*, 26, 1705 (1973).
153.* C. Wentrup, E. Wentrup-Byrne, and P. Müller, *J. Chem. Soc., Chem. Commun.*, 210 (1977).
154.* W. T. Brown and W. M. Jones, *J. Org. Chem.*, 44, 3090 (1979).
155.* S. Chari, G. K. Agopian, and M. Jones, Jr., *J. Am. Chem. Soc.*, 101, 6125 (1979).

156.* (a) R. W. Hoffmann, B. Hagenbruch, and D. M. Smith, *Chem. Ber.*, **110**, 23 (1977); (b) M. Reiffen and R. W. Hoffmann, *Chem. Ber.*, **110**, 23 (1977); (c) R. W. Hoffmann and M. Reiffen, *Chem. Ber.*, **110**, 49 (1977).

157. R. Gompper and R. Sobatta, *Angew. Chem., Int. Ed. Engl.*, **10**, 762 (1978).

158.* M. Franck-Neumann and J. J. Lohmann, *Tetrahedron Lett.*, 3729 (1978); 2075 (1979).

159.* T. B. Patrick, M. A. Dorton, and J. G. Dolan, *J. Org. Chem.*, **43**, 3303 (1978).

160. H. Dürr, S. Fröhlich, and M. Kausch, *Tetrahedron Lett.*, 1767 (1977).

161. G. W. Jones, K. T. Chang, and H. Shechter, *J. Am. Chem. Soc.*, **101**, 3906 (1979).

162. R. Gleiter and R. Hoffmann, *J. Am. Chem. Soc.*, **90**, 5457 (1968).

163. R. A. Moss and U.-H. Dolling, *Tetrahedron Lett.*, 5117 (1972); R. A. Moss, U.-H. Dolling, and J. R. Whittle, *Tetrahedron Lett.*, 931 (1971).

164. R. A. Moss and C.-T. Ho, *Tetrahedron Lett.*, 1651 (1976).

165. R. A. Moss and C.-T. Ho, *Tetrahedron Lett.*, 3397 (1976).

166.* K. Okamura and S.-I. Murahashi, *Tetrahedron Lett.*, 3281 (1977).

167. L. Skattebøl, *Tetrahedron*, **23**, 1107 (1967). For a brief review see Ref. 95 (Jones and Brinker), pp. 159–165.

168.* M. S. Baird and C. B. Reese, *J. Chem. Soc., Chem. Commun.*, 523 (1972).

169.* M. S. Baird and C. B. Reese, *Tetrahedron Lett.*, 2895 (1976).

170.* W. Kirmse and H. Jendralla, *Chem. Ber.*, **111**, 1873 (1978).

171. W. Kirmse and U. Richarz, *Chem. Ber.*, **111**, 1883, 1895 (1978).

172. R. B. Reinarz and G. J. Fonken, *Tetrahedron Lett.*, 4591 (1973).

173. D. N. Butler and I. Gupta, *Can. J. Chem.*, **56**, 80 (1978).

174.* W. W. Schoeller and U. H. Brinker, *J. Am. Chem. Soc.*, **100**, 6012 (1978).

175.* K. H. Holm and L. Skattebøl, *Tetrahedron Lett.*, 2347 (1977).

176.* U. H. Brinker and I. Fleischhauer, *Angew. Chem., Int. Ed. Engl.*, **18**, 396 (1979).

177. U. H. Brinker and L. König, *J. Am. Chem. Soc.*, **101**, 4738 (1979).

178. D. Seyferth, R. L. Lambert, Jr., and M. Massal, *J. Organomet. Chem.*, **88**, 255 (1975).

179.* (a) P. Warner and S.-C. Chang, *Tetrahedron Lett.*, 3981 (1978). (b) P. Warner and S.-C. Chang, *Tetrahedron Lett.*, 4141 (1979). (c) P. Warner, private communication.

180. M. H. Fisch and H. D. Pierce, Jr., *J. Chem. Soc., Chem. Commun.*, 503 (1970).

181.* K. J. Holm and L. Skattebøl, *J. Am. Chem. Soc.*, **99**, 5480 (1977).

182. For a review of alkane diazotate chemistry, see R. A. Moss, *Acc. Chem. Res.*, **7**, 421 (1974).

183. R. A. Moss and U. H. Dolling, *J. Am. Chem. Soc.*, **93**, 954 (1971).

184. R. A. Moss and M. A. Joyce, *J. Am. Chem. Soc.*, **99**, 1262 (1977). Be certain to see also the important correction to this paper: *J. Am. Chem. Soc.*, **99**, 7399 (1977).

185. C. W. Bauschlicher, Jr., C. F. Bender, and H. F. Schaefer III, *J. Am. Chem. Soc.*, **98**, 3072 (1976); S. Nagase and T. Fueno, *Theor. Chim. Acta*, **41**, 59 (1976).

186. H. Tomioka and Y. Izawa, *J. Am. Chem. Soc.*, **99**, 6128 (1977).

187. H. Tomioka, *J. Am. Chem. Soc.*, **101**, 256 (1979).

188. H. Tomioka, G. W. Griffin, and K. Nishiyama, *J. Am. Chem. Soc.*, **101**, 6009 (1979).

189. J. J. Zupancic and G. B. Schuster, *J. Am. Chem. Soc.*, **102**, 5958 (1980). We thank Prof. Schuster for a preprint of this work.

190. R. A. Moss and J. K. Huselton, *J. Am. Chem. Soc.*, **100**, 1314 (1978).

191. R. A. Moss and M. A. Joyce, *J. Am. Chem. Soc.*, **100**, 4475 (1978).

192. C. T. Lin and P. P. Gaspar, *Tetrahedron Lett.*, 3553 (1980). We thank Prof. Gaspar for a preprint of this work.

193. U. P. Senthilnathan and M. S. Platz, *J. Am. Chem. Soc.*, **102**, 7637 (1980). We thank Prof. Platz for a preprint of this work.

4

METAL-CARBENE COMPLEXES

CHARLES P. CASEY

Department of Chemistry, University of Wisconsin, Madison, Wisconsin 53706

I. INTRODUCTION

This chapter covers reports on the synthesis and reactions of metal–carbene complexes for the years 1977–1979 with some references to work in early 1980. The review is by no means comprehensive and concentrates on the new types of reactions of metal–carbene complexes observed in the last several years. A number of earlier reviews of metal–carbene complexes are available.[1-11]

II. ELECTROPHILIC CARBENE COMPLEXES

Fischer prepared the first metal–carbene complexes in 1964 from the reaction of organolithium reagents with metal carbonyl compounds followed by alkylation of the resulting acyl metal anions. The carbene complexes are strongly polarized and have an electrophilic carbene carbon atom. The use of metal–carbene complexes in organic chemistry has been reviewed.[8-11]

$$Cr(CO)_6 \xrightarrow{CH_3Li} \xrightarrow{(CH_3)_3O^+} (CO)_5Cr=C\big<^{OCH_3}_{CH_3} \longleftrightarrow (CO)_5\bar{Cr}-C\big<^{+OCH_3}_{CH_3}$$

One approach to using metal–carbene complexes in organic synthesis has been to exploit the acidity of protons attached to the carbon atom α to the carbene carbon atom. Reaction of $(CO)_5CrC(OCH_3)CH_3$ with base yields carbene anion 1, which reacts with carbon electrophiles to give carbon–carbon bond formation.

$$(CO)_5Cr=\big<^{OCH_3}_{CH_3} \xrightarrow{base} (CO)_5\overset{\ominus}{Cr}-\big<^{OCH_3}_{CH_2} \xrightarrow{\diagup\diagup\diagdown Br} (CO)_5Cr=\big<^{OCH_3}$$

1

The conjugate addition of carbene anions to α,β-unsaturated esters and ketones can be controlled to give either mono- or dialkylation.[12] The reaction of the stoichiometrically generated anion of 2 with methyl vinyl ketone gave a 41% yield of the monoalkylated conjugate addition product 3 and no dialkylated material. When the reaction mixture was quenched with DCl, monodeuterated

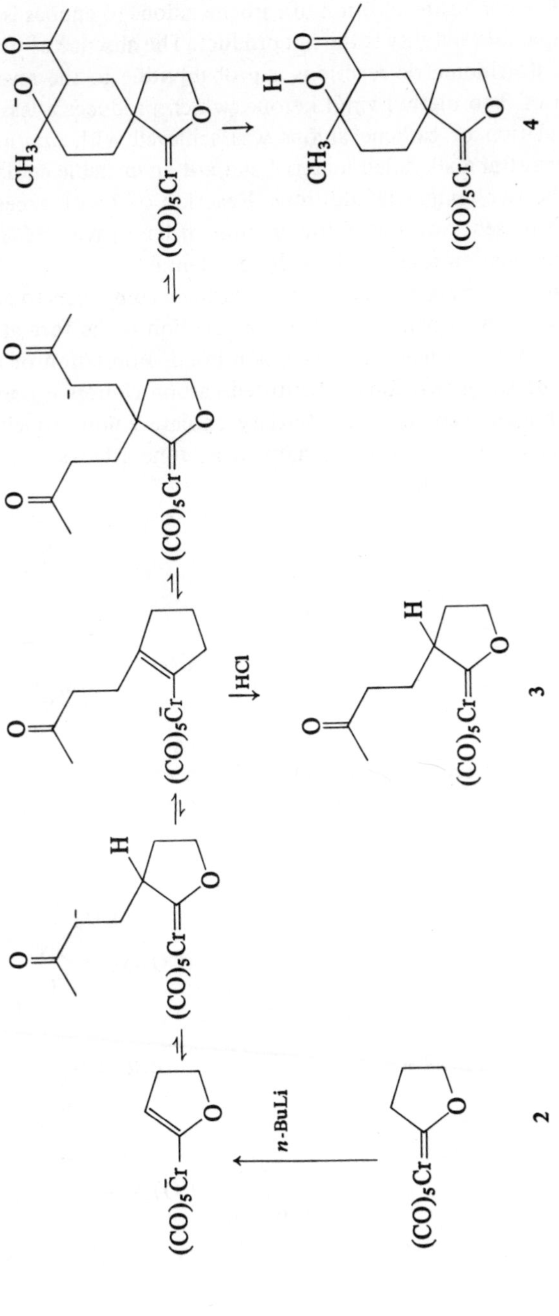

$(CO)_5Cr$
$(CO)_5\bar{C}r$
n-BuLi
HCl
CH_3
O
H
2
3
4

3 was obtained, indicating that the predominant species in solution was the anion of **3**. The conjugate addition of carbene anions to enones is favorable if an anion of comparable stability is the net product. The absence of dialkylated products in these stoichiometric reactions is probably due to the reversible addition of the anion of **3** to methyl vinyl ketone, which produces a less stable enolate anion. Dialkylation of carbene anions was achieved with *catalytic amounts of base* since a neutral dialkylated material and not an unstable enolate anion is the product of the two conjugate additions. Reaction of **2** with excess methyl vinyl ketone in the presence of a catalytic amount of base gave a 36% yield of **4**, the product of dialkylation followed by aldol condensation.

Enolate anions conjugately add to vinylcarbene complexes to give stable addition products.[13] The driving force for this reaction is the formation of a stable carbene anion and of a new carbon–carbon bond. Formation of a new carbon–carbon bond between two fully substituted carbon centers is possible using this reaction. Although secondary and tertiary enolate anions attack at the vinylic carbon, the less hindered enolate anion of acetone attacks vinyl carbene complexes at the carbene carbon atom.

Fischer has reported that conjugate addition to acetylenic carbene complex **5** is favored at $-20°C$, but that attack at the carbene carbon atom is favored at $-60°C$.[14]

III. REACTIONS OF ELECTROPHILIC CARBENE COMPLEXES WITH ALKENES

The electrophilic carbene complex $(CO)_5WC(C_6H_5)_2$ reacts with alkenes to give mixtures of cyclopropanes and alkene scission products.[15] The reaction was suggested to proceed by CO dissociation, formation of a metal–carbene–alkene complex, and isomerization to a metallacyclobutane. The mechanism proposed is similar to the mechanism suggested for the olefin metathesis reaction. Katz has shown that $(CO)_5WC(C_6H_5)_2$ and $(CO)_5WC(OCH_3)C_6H_5$ can act as olefin metathesis catalysts.[16]

The reaction of 2,3-dimethylbutadiene with $(CO)_5WC(C_6H_5)_2$ gives 48% 1,1-diphenylethylene and 15% isoprene.[17] The isoprene could be formed by rearrangement of an intermediate vinylcarbene complex.

Reaction of cyclic enol ethers with $(CO)_5WC(C_6H_5)_2$ gives ring-opened products.[18]

In contrast to $(CO)_5WC(C_6H_5)_2$, the much more electrophilic carbene complex $(CO)_5WCH(C_6H_5)$ (6) reacts with alkenes to give only cyclopropanes and no alkene scission products. Reaction of $(CO)_5WC(OCH_3)C_6H_5$ with $K^+ HB[OCH(CH_3)_2]_3^-$ followed by cation exchange given $N(CH_2CH_3)_4^+ (CO)_5WCH(OCH_3)C_6H_5^-$ (7) which is a stable crystalline solid.[19] Protonation of 7 with CF_3CO_2H in CH_2Cl_2 at $-78°C$ gives 6, which was characterized by low-temperature ^{1}H NMR and by reaction which $P(n$-$Bu)_3$, which gave $(CO)_5WCH[P(n$-$Bu)_3]C_6H_5$.[20] Thermal decomposition of 6 proceeded with a half-life of 24 min at $-56°C$. In spite of its great kinetic instability, 6 reacts with alkenes to given high yields of cyclopropanes; no methathesis-like products have been observed. Both *cis*- and *trans*-2-butene react with 6 to give cyclopropanes with retention of the alkene geometry. In the case of *cis*-2-butene, the less stable *syn*-isomer is formed with a 41:1 preference over the *anti*-isomer.

For terminal alkenes, $RCH=CH_2$, the ratio of *cis/trans* phenylcyclopropanes decreased from 1.8 (R = CH_3) to 0.01 (R = t-Bu) as the size of the alkyl group increased. Complex 6 is 310 times as reactive toward $CH_2=C(CH_3)_2$ as toward $CH_2=CHCH_3$; ethylene is the only alkene that does not react with 6. This reactivity order is interpreted in terms of electrophilic attack of the electron-deficient carbene carbon of 6 on the least substituted end of the alkene.

The reactions of 6 with propene, styrene, and *cis*-2-butene all led to the predominant formation of the thermodynamically less stable *syn*-cyclopropanes. This result was initially interpreted in terms of formation of the more stable puckered metallacyclobutane intermediate.[21] However, the reaction of 2-ethyl-2-butene with **6** gave a 94:1 excess of the *syn*-isomer, which is inconsistent with control of stereochemistry by preferential formation of the sterically preferred metallacyclobutane, since the metallacyclobutane that leads to *syn* product has an axial methyl group on the central carbon atom.

Casey and Polichnowski have now advanced a totally different explanation for the preferential formation of *syn*-cyclopropanes from **6**.[20] The reaction is viewed as proceeding through closed transition state **8**, in which initial bond formation between the carbene carbon atom and the less substituted alkene carbon has taken place and in which the positive charge developing on the more substituted carbon atom is stabilized by donation of π-electrons from the benzene ring. The preference for formation of *syn*-cyclopropane from 2-methyl-2-butene is determined by the approach of the alkene to the carbene carbon atom so that steric interactions between $W(CO)_5$ and the methyl group are minimized.

8

The stereochemistry of cyclopropanes formed from reaction of **6** with monosubstituted alkenes requires consideration of a second transition state (**9**) for reaction with alkenes bearing very bulky substituents. In the reaction of **6** with *tert*-butylethylene, a transition state involving phenyl participation similar to that which explains the selective formation of *cis*-cyclopropane from 2-methyl-2-butene would be destabilized by a steric interaction between the phenyl group and the *tert*-butyl group of the olefin. Little reaction appears to go via this closed transition state. Instead, the reaction proceeds via "open" transition state **9**, in which *tert*-butylethylene approaches the carbene carbon atom with the bulky *tert*-butyl group away from the $W(CO)_5$ fragment.

A number of very electron-deficient carbene complexes have recently been reported. Brookhart reported the isolation of the cationic phenylcarbene iron

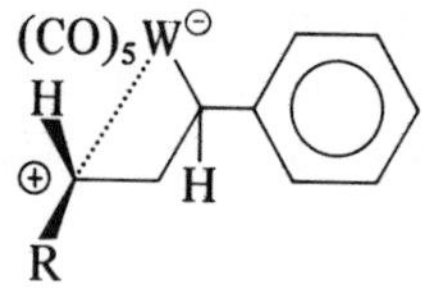

9

complex **10**, which decomposed in less than 1 hr at 25°C. The derivative **11**, in which an electron-donating phosphine ligand replaced CO, had a half-life of 60 hrs at 25°C. By using a chelating diphosphine, Brookhart and Flood[22] were able to observe cationic methylene iron complex **12**. The H—C—H plane of the carbene ligand of **12** lies perpendicular to the plane of the cyclopentadienyl ring. Below -60°C, the $Fe=CH_2$ protons are nonequivalent in the ^1H NMR spectrum. The barrier to rotation about the $Fe=CH_2$ bond was found to be 10.4 kcal/mol.

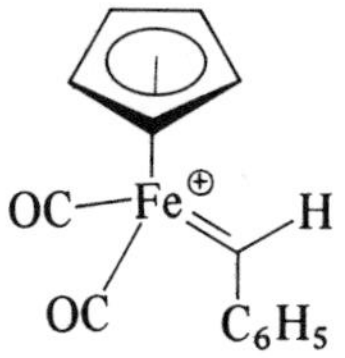

    10 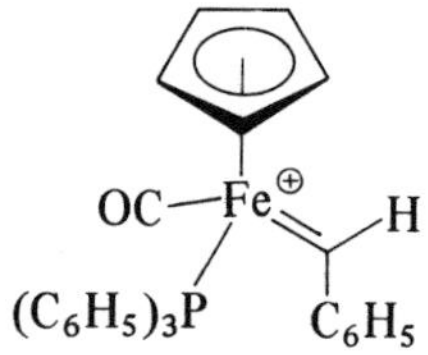   11 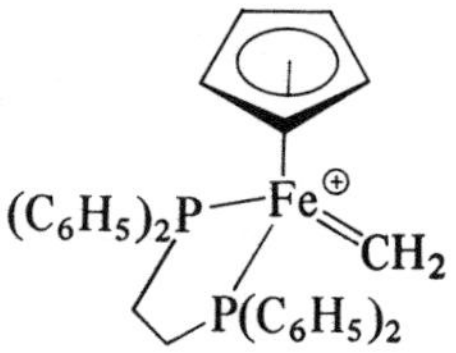   12

Brookhart has reported cyclopropane formation from reaction of cationic iron carbene complexes with alkenes.[21,22] In related work, Helquist has found that sulfonium salt **13** is a stable, isolable cyclopropanating agent that reacts with alkenes under neutral conditions in refluxing dioxane to give high yields of cyclopropanes.[23] Presumably, $(C_5H_5)Fe(CO)_2CH_2^+$ (**14**) is generated as a reactive intermediate under the reaction conditions. Beauchamp has observed **14** in the

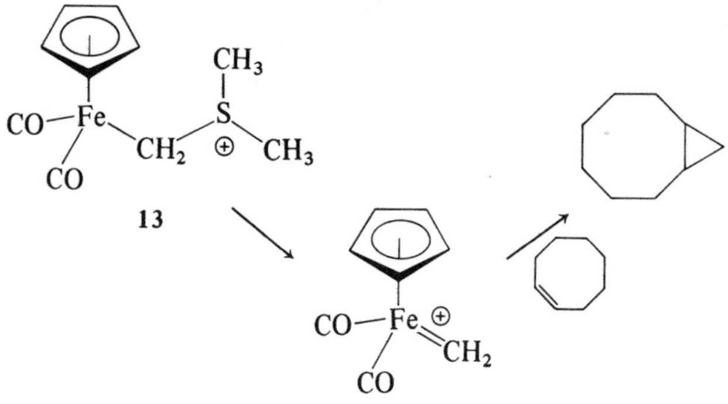

    13

    14

gas phase by ion cyclotron resonance and noted its transfer of methylene to cyclohexene.[24]

Cutler has reported the preparation of the secondary alkoxycarbene complex 15 by hydride abstraction from the corresponding alkoxymethyl complex.[25]

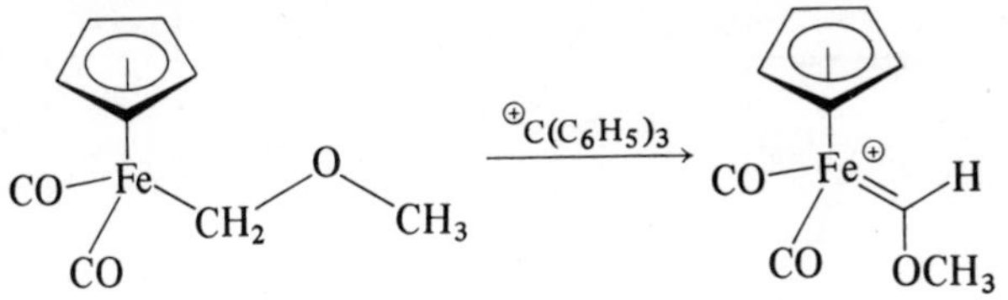

15

Gladysz has reported the observation by NMR spectroscopy of a number of cationic rhenium–carbene complexes formed by hydride abstraction with trityl cation.[26] The formation of ethylidene complex 16 is particularly remarkable since α-hydride abstraction was preferred over β-hydride abstraction.

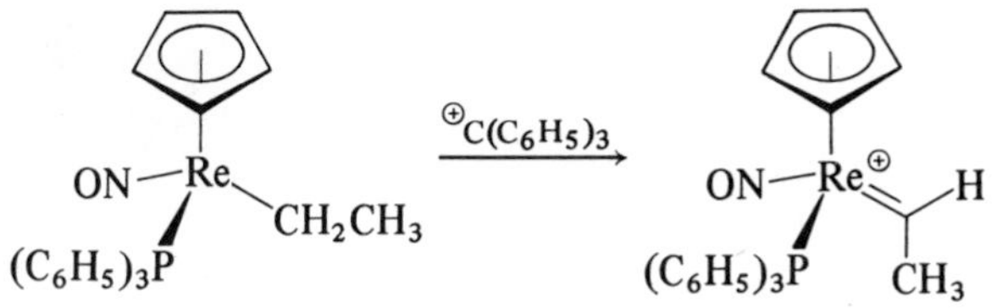

16

IV. NUCLEOPHILIC ALKYLIDENE COMPLEXES

A large number of alkylidene complexes of niobium and tantalum have been prepared by Schrock's group in the past several years.[27] These early transition metal complexes are polarized $\delta + M = CR_2 \ \delta -$, in contrast to Fischer's carbene complexes, such as $(CO)_5 WC(OCH_3)C_6H_5$, which are polarized $\delta - M = CR_2 \delta +$. The first tantalum–alkylidene complex was prepared in 1974 by Schrock from the reaction of $Ta[CH_2C(CH_3)_3]Cl_2$ with $LiCH_2C(CH_3)_3$.[28] Instead of obtaining an $R_5 Ta$ compound as observed in the preparation of $Ta(CH_3)_5$, Schrock found that an α-elimination took place to give tantalum–neopentylidene complex 17. One of the important requirements for an α-elimination process noted by Schrock is that the transition metal alkyl precursor be a crowded molecule. Similar steric accelerations of α-elimination reactions are seen in the syntheses of cyclopentadienyl tantalum alkylidene complexes such as 18[30] and in the phosphine-induced α-eliminations to form compounds such as 19.[31]

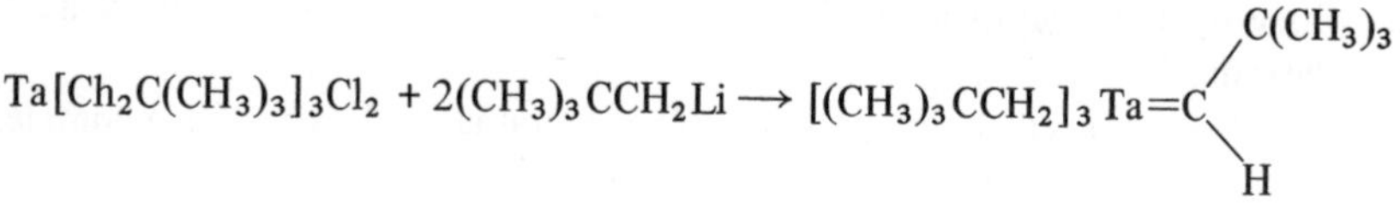

$$Ta[Ch_2C(CH_3)_3]_3Cl_2 + 2(CH_3)_3CCH_2Li \longrightarrow [(CH_3)_3CCH_2]_3Ta=C\begin{matrix} C(CH_3)_3 \\ \\ H \end{matrix}$$

17

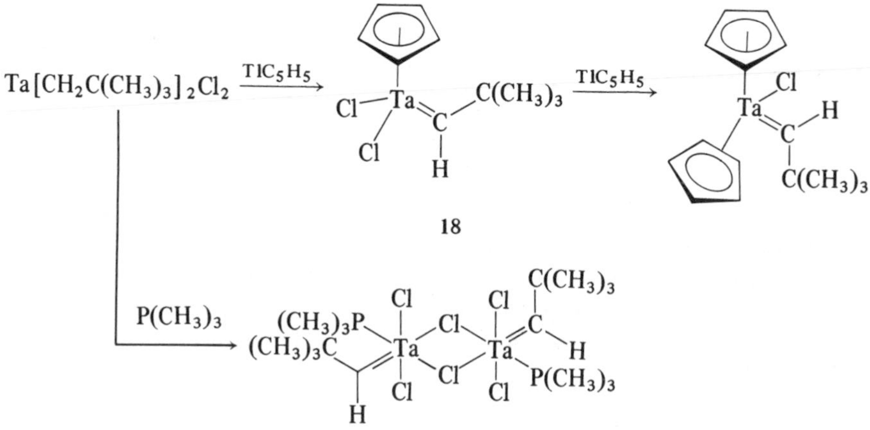

18

19

The α-elimination route to metal alkylidene complexes fails with organometallic alkylating reagents possessing β-hydrogen atoms because of competing alkene elimination. A route to alkylidene complexes that circumvents this limitation involves the reaction of phosphoranes with Ta(III) complexes.[32] Since $(C_5H_5)_2Ta[P(CH_3)_3]CH_3$ undergoes rapid ligand exchange, its reaction with $(CH_3CH_2)_3P=CHCH_3$ was proposed to proceed via phosphorane attack on a coordinatively unsaturated intermediate followed by loss of $P(CH_2CH_3)_3$.

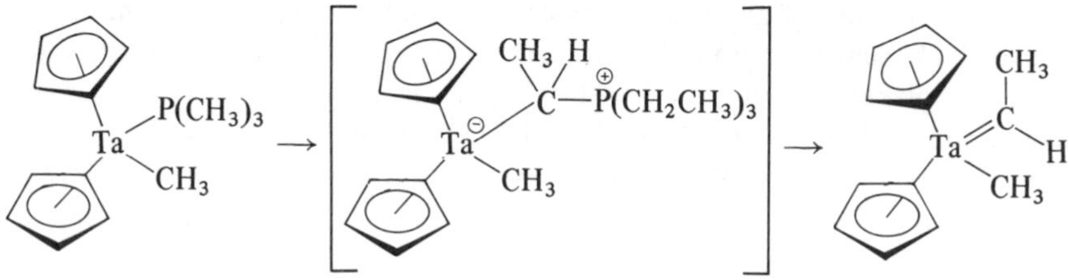

Schwartz[33] has reported the related reaction of $(C_5H_5)_2Zr(PR_3)_2$ with $CH_2 = P\text{-}(C_6H_5)_3$, which gives $(C_5H_5)_2Zr(PR_3)(CH_2)$. The crystal structures of a number of tantalum-alkylidene complexes have been reported.[34-36] Tantalum-carbon multiple bonding is apparent in all the structures: the Ta=C bonds are ~0.22 Å shorter than the Ta—C bonds. For biscyclopentadienyl compounds, the plane of the alkylidene ligand is nearly perpendicular to the plane passing between the two cyclopentadienyl rings. This orientation allows extensive π-bonding between tantalum and carbon using a metal d-orbital lying in the plane between the two cyclopentadienyl rings. Consequently, there is a substantial barrier to rotation about the Ta=C bond.[35] The barrier is > 21 kcal/mol for $(C_5H_5)(C_5H_4CH_3)Ta(CH_2)(CH_3)$.

The alkylidene carbon of these tantalum complexes is electron rich, as demonstrated by reactions with electrophiles. Reaction of $[(CH_3)_3CCH_2]_3$-Ta=CHC$(CH_3)_3$ with DCl gave a mixture of neopentane-d_2, neopentane-d_1, and neopentane-d_0. The formation of d_2 material indicates that the alkylidene carbon is protonated by DCl.[29] $(C_5H_5)_2Ta(CH_2)(CH_3)$ reacts with Al$(CH_3)_3$ to give the adduct **20**.[37] Early transition metal–alkylidene compounds react with ketones and with esters to give alkenes and vinyl ethers.[38,39]

20

The reactions of tantalum alkylidene compounds with alkenes give chain-extended alkenes derived from an intermediate metallacyclobutane. In the reaction with propene, the more substituted end of the alkene becomes bonded to the alkylidene carbon[40]. This contrasts with the regiochemistry observed in the reactions of $(CO)_5WC(C_6H_5)_2$ with alkenes. The tantalum product is a metallacyclopentane formed by dimerization of two alkenes in the coordination sphere of tantalum. These metallacyclopentanes are a new class of alkene dimerization catalysts.

The mechanism of α-elimination from $(C_5H_5)TaX_2[CH_2C(CH_3)_3]_2$ (21) has been studied in detail.[41] Both the *cis*- and *trans*-configurations of 21 are observable by NMR. The chloro compound exists predominantly in the *trans* form, while the less stable bromo compound has a higher proportion of the *cis* form. For the pentamethylcyclopentadienyl derivative, the *trans*-configuration is the only one observed and the compound decomposes thousands of times more slowly than the related C_5H_5 compound. The rate of decomposition is first order in tantalum and does not depend on the concentration of added chloride ion. Crossover experiments demonstrated the intramolecular nature of the α-elimination process. A primary isotope effect k_H/k_D of 5.4 ± 0.5 was found. The reaction therefore involves rate-determining internal hydrogen abstraction from the *cis*-isomer.

The decomposition of $(C_5H_5)_2Ta(CH_2)(CH_3)$ in the presence of $P(CH_3)_3$ has been studied in detail.[37] The products of the reaction are ethylene complex 22 and phosphine complex 23. The rate of decomposition is second order in tantalum and does not depend on the concentration of added $P(CH_3)_3$. The reaction was suggested to proceed via a doubly bridged intermediate, which then decomposes to **22** and a coordinatively unsaturated species, which reacts rapidly with $P(CH_3)_3$.

V. METAL–CARBENE COMPLEXES AND OLEFIN METATHESIS

The mechanism of the olefin metathesis reaction is now generally accepted to involve interconversion of isomeric metal–carbene–alkene complexes via a metallacyclobutane intermediate. Since several excellent reviews of olefin methathesis have recently appeared,[42–45] only a few comments are made here.

One of the most exciting developments was the discovery by Tebbe and Parshall of an isolable titanium methylene compound that catalyzes olefin methathesis. Reaction of 2 equiv of $Al(CH_3)_3$ with $(C_5H_5)TiCl_2$ at room temperature gave bridging methylene complex **24**.[46] The reaction of **24** with cyclohexanone at $-15°C$ gave methylenecyclohexane, indicating the **24** possesses a nucleophilic methylene unit. The reaction of **24** with ethylene and propene gave propene and isobutylene respectively. These reactions with alkenes were proposed to proceed via a metallacyclobutane in which the less substituted carbon of the alkene becomes bonded to titanium. Alkene homologation then takes place by β-hydride elimination and reductive elimination of alkene.

The titanium methylene complex 24 was stable in the presence of 1,1-disubstituted alkenes.[47] However, ^{13}C labeling studies demonstrated that degenerate metathesis was occurring. Methylene exchange between isobutylene-1-C_{13} and methylenecyclohexane was catalyzed by 24 at 51 °C. Significantly, there was little decomposition of 24 and a $^{13}CH_2$ was incorporated into 24. These results demonstrate that the metallacyclobutane is stable if it has no β-hydrogen atoms and that it is an intermediate in olefin metathesis.

Platinacyclobutanes are stable compounds that have been suggested as models for the reactive intermediate in the olefin metathesis reaction. Puddephatt has reported that phenylcyclopropane reacts with platinum complexes to selectively give an α-phenylplatinacyclobutane 25. Upon heating, the initially formed α-isomer 25 rearranges to an equilibrium mixture of 25 and β-phenylplatinacyclobutane 26.[48]

Mechanisms involving reductive elimination of phenylcyclopropane or cleavage to give styrene and a platinum–methylene complex were ruled out by Casey et al.[49] and by Johnson[50] on the basis of crossover experiments. A mechanism involving independent rotation of styrene in a styrene–methylene–platinum complex was ruled out on the basis of the retention of stereochemistry observed in the rearrangement of cis-deuterated 25 to cis-deuterated 26.[49,51] Retention of stereochemistry was explained by Casey et al. in terms of a concerted rotation of the methylene and styrene units.[49]

Green et al. have shown that tungsten and molybdenum metallacyclobutanes can be prepared by addition of $NaBH_4$ or allylmagnesium chloride to cationic π-allyl metal complexes.[52] Thermal decomposition of **27** gives predominantly butenes (26%) and very little of the metallacycle fragmentation products propene (0.4%) and ethylene (0.7%). In contrast, photolysis of **27** gave 12% propene and 1% ethylene as major products formed via fragmentation processes related to olefin metathesis.[53] Green proposed that photolysis of **27** gives the slipped η^3-cyclopentadienyl intermediate **28**, which has a vacant coordination site and can act as the immediate precursor of a metal–carbene–alkene complex.

Green speculated that the initiation of metathesis may proceed by conversion of a metal–alkene complex to a metallacyclobutane by a 1,2-hydride shift.[53]

Green has noted the similarity between olefin metathesis catalysts and Ziegler-Natta olefin polymerization catalysts and has speculated that the reac-

tions may be mechanistically related.[54] Experiments designed to distinguish between Green's mechanism and the insertion of alkenes into metal–carbon bonds have not been reported.

The migration of a hydrogen atom from the α-carbon of a metal-alkyl to the metal to produce a metal-carbene-hydride complex had been proposed several years ago by Green.[54] Shaw has proposed such a pathway for the dehydrogenation of an alkyl iridium hydride to an iridium carbene complex.[109] More recently Shaw has reported NMR evidence for the rapid reversible migration of hydride from rhodium to a carbene ligand.[110]

Casey et al. reported the observation by NMR spectroscopy of a metastable metal-carbene-alkene complex of tungsten **29**.[55] The complex decomposes to give quantitative yields of cyclopropane. The factors that determine whether a metal-carbene-alkene complex or metallacyclobutane complex will decompose to give cyclopropane or olefin metathesislike products are still not understood.

For olefin metathesis catalysts generated from transistion metal complexes and alkyl tin, aluminum, or lithium compounds, the initial carbene complex could be generated by an α-elimination process. However, the generation of carbene ligands in metathesis catalyst systems that do not employ an alkylating agent has been puzzling. In a study of the mechanism of alkene dimerization catalyzed by tantalum complexes, Schrock has found that metallacyclopentanes can undergo ring contraction to give metallacyclobutanes.[56] Since the reaction

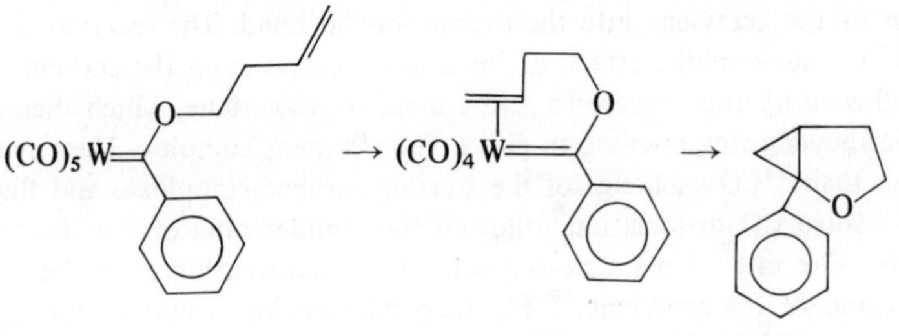

29

of alkenes with metal complexes has been shown to give metallacyclopentanes, this new discovery indicates that routes from alkene to metal–alkylidene complexes may be available.

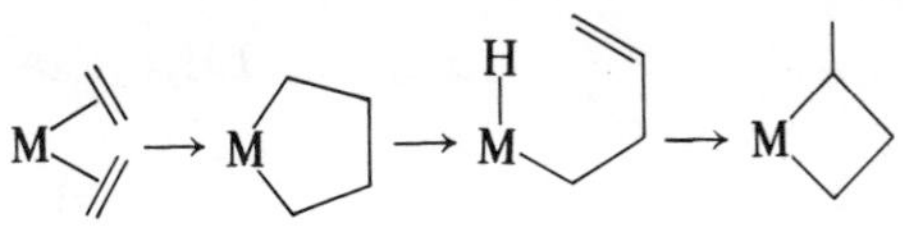

Casey and Tuinstra have shown that the degenerate methathesis of labeled terminal alkenes occurs with substantial retention of stereochemistry.[57] This result is consistent with a chain-carrying M=CHR catalyst in which stereochemistry is preserved as a result of preferential formation of a diequatorial puckered metallacyclobutane **30**. However, retention of stereochemistry can also be explained in terms of a M=CH$_2$ chain-carrying intermediate (**31**) if the catalyst is asymmetric and if there is a substantial barrier to rotation about the M=CHD bond.

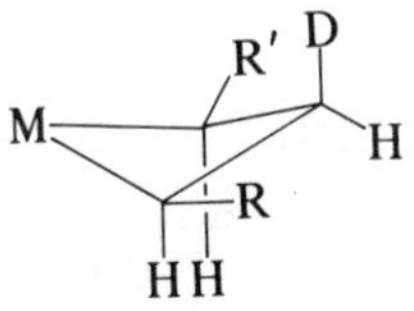

30 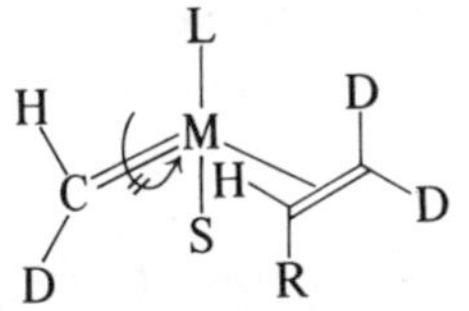   **31**

VI. REACTIONS OF CARBENE COMPLEXES WITH ACETYLENES

A. Insertion

Dötz and Kreiter[58] were the first to report that electron-rich acetylenes react with carbene complexes to give vinylcarbene complexes, the product of a formal insertion of the acetylene into the carbene–metal bond. The reaction probably proceeds by nucleophilic attack of the amino acetylene on the carbene carbon atom followed by ring closure to give a metallacyclobutene, which then undergoes electrocyclic ring opening to give a vinyl carbene complex. These reactions are faster than [13]CO exchange of the starting carbene complexes and therefore do not involve CO dissociation followed by complexation of the electron-rich acetylene. The rate of reaction depends on the concentration of the carbene complex and of the acetylene.[59] Electron-withdrawing groups on the aryl ring of $(CO)_5WC(OCH_3)C_6H_4$-p-x accelerate the reaction with acetylenes.

The very reactive electrophilic phenylcarbene complex 6 reacts with ethoxyacetylene to produce a styrylethoxycarbene complex 32.[20] Reaction of diphenyl-

carbene–manganese complex 33 with an amino acetylene led to acetylene insertion.[60]

Schrock et al.[41] have found that the nucleophilic alkylidene–tantalum complex 18 reacts with diphenylacetylene to give an insertion product 34. The reaction could well involve prior coordination of acetylene to coordinatively unsaturated tantalum complex 18.

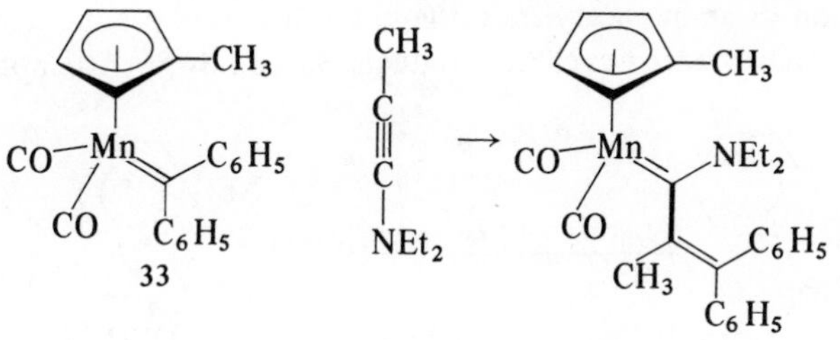

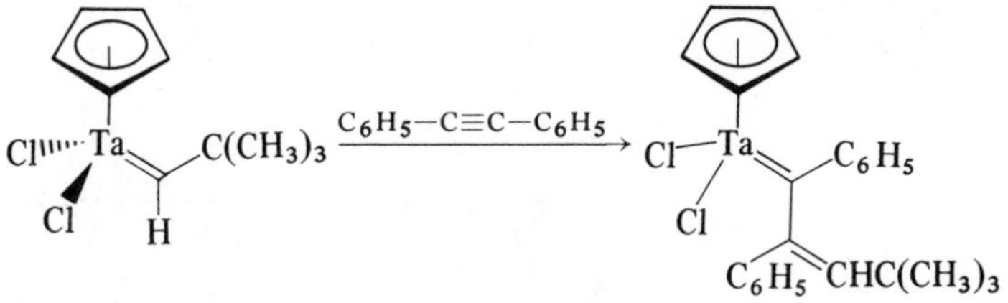

18   34

Tebbe has reported that reaction of diphenylacetylene with the bridging titanium–carbene complex **24** leads to the formation of an isolable metallacyclobutene[47]

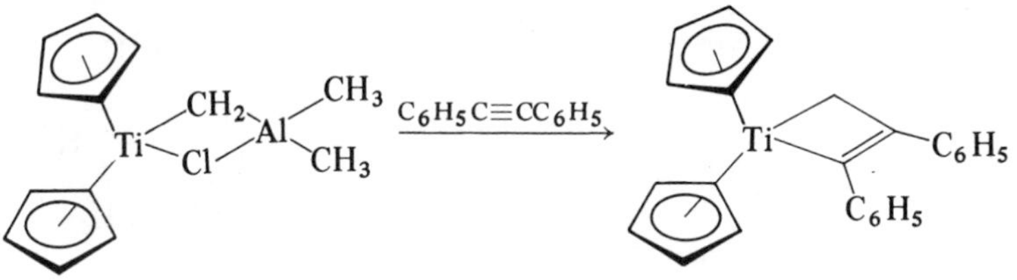

24

B. Formation of Naphthols and Indenes

The reaction of arylmethoxycarbene complexes with acetylenes gives mixtures of naphthols and indenes.[61]

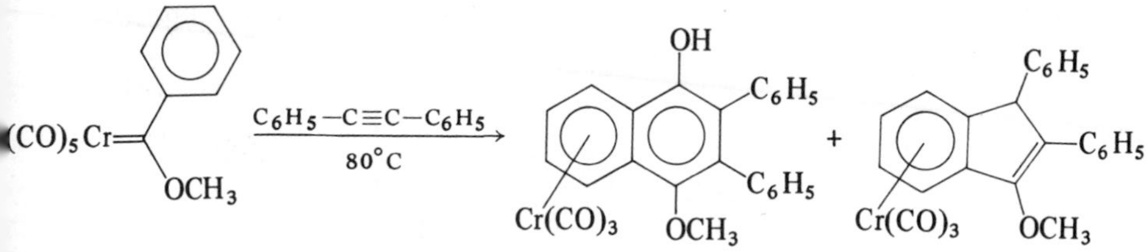

The reaction probably proceeds via formation of a simple insertion product, which then undergoes an intramolecular attack on the aryl ring. In the case of the reaction of amino acetylenes the initial insertion product has been isolated and then converted to cyclized products **35** and **36**.[62] Attempts to block the

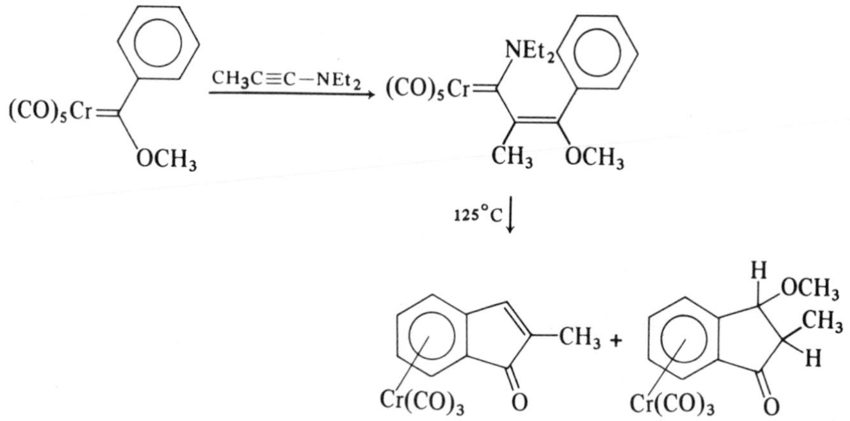

cyclization reaction with *ortho* methyl substitution have failed.[63]

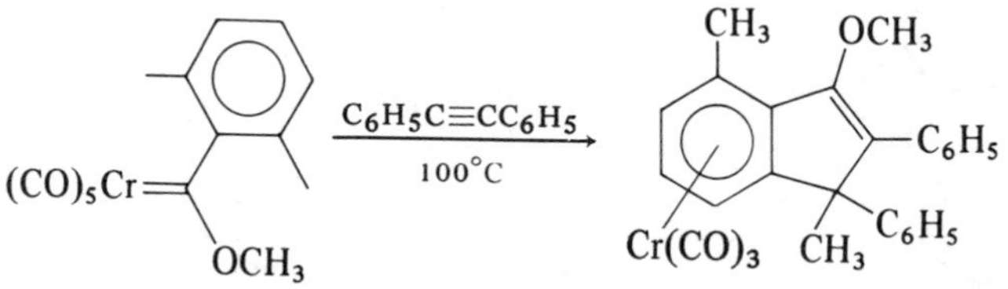

In the formation of naphthols, the metal is initially bonded to the hydroxy-substituted ring but then shifts to the unsubstituted ring.[64]

The formation of cyclized products from reactions of phenylcarbene complexes can be explained by the scheme shown below. The key step is the formation of a metallacyclohexadiene (**37**), which can either undergo reductive elimination to form indene complexes or CO insertion to give naphthol complexes.

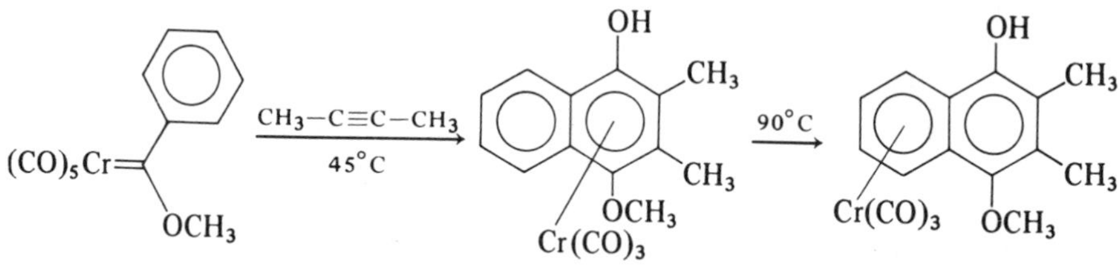

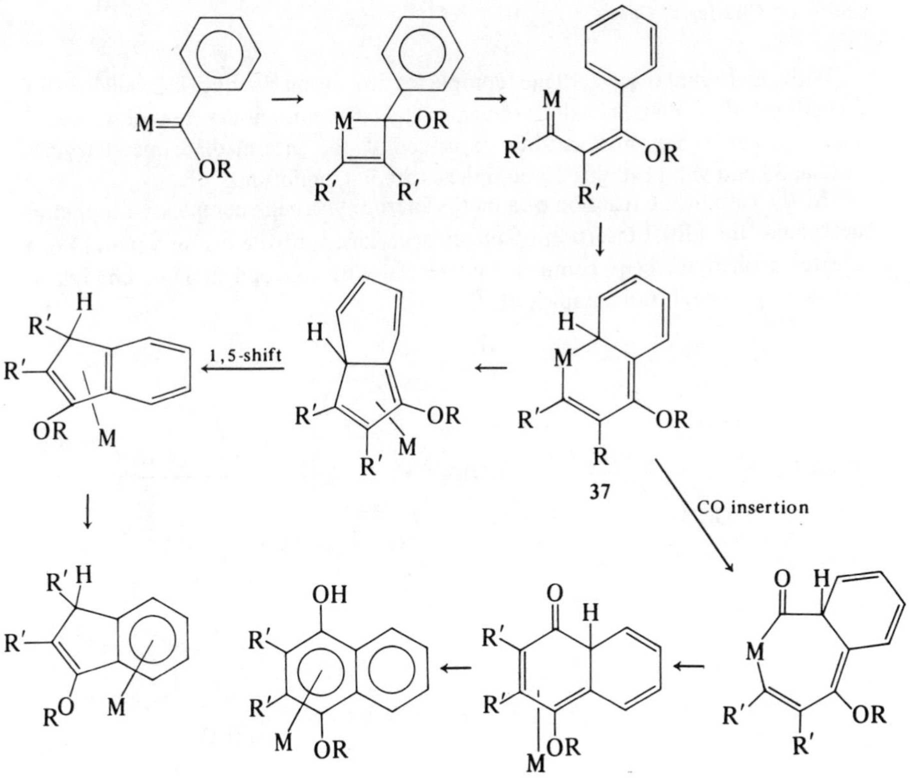

1,5-shift

37

CO insertion

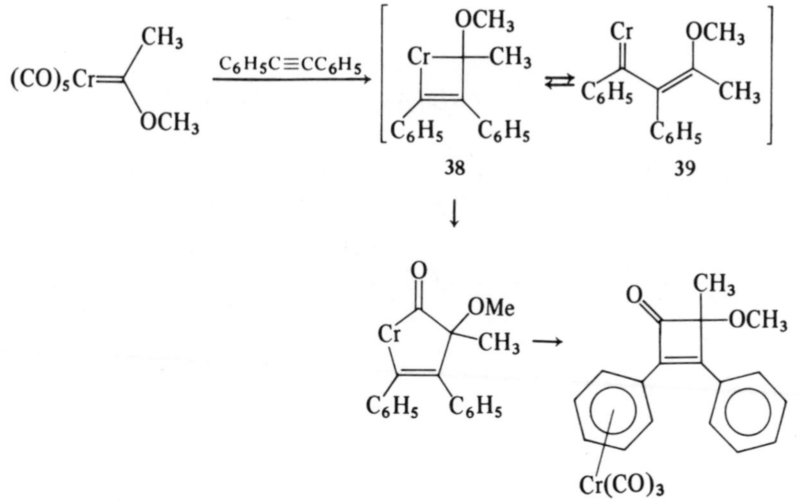

$(CO)_5Cr=\overset{CH_3}{\underset{OCH_3}{}}$ $\xrightarrow{C_6H_5C\equiv CC_6H_5}$

38 39

$Cr(CO)_3$

155

With methylmethoxycarbene complexes no aromatic ring is available for formation of a metallacyclohexadiene, and cyclobutanones are observed.[65] These products are most readily explained if the intermediate metallacyclobutene **38** and vinyl carbene **39** complexes are in equilibrium.

In the case of the reaction of a methylmethoxycarbene complex with phenylacetylene, the initial insertion of phenylacetylene into the carbene–metal bond creates a phenylcarbene complex that reacts with a second mole of phenylacetylene to give naphthol complex **40**.[66]

40

C. Catalysis of Acetylene Polymerization

Katz has found that $(CO)_5WC(OCH_3)C_6H_5$ and $(CO)_5WC(C_6H_5)_2$ are efficient catalysts for the polymerization of a variety of acetylenes.[67] For example, $(CO)_5WC(C_6H_5)_2$ catalyzes the reaction of 2-butyne at $40°C$ to give a 75% yield of a polymer with a molecular weight of 78,000 that was characterized by ^{13}C NMR to be poly-2-butyne. Similarly, t-butylacetylene is converted to a polymer of molecular weight 570,000. The reaction is proposed to proceed by sequential formation and ring opening of metallacyclobutenes.

VII. CARBENE–CARBYNE INTERCONVERSIONS

The first carbyne complex was prepared by Fischer in 1973 by reaction of BBr_3 with $(CO)_5CrC(OCH_3)C_6H_5$, which gave $trans$-$Br(CO)_4Cr≡C−C_6H_5$.[68] The reaction most likely proceeds via Lewis acid abstraction of the methoxy group to give a cationic carbyne complex. The cationic carbyne complex can react with halide to give either a halocarbene complex or a neutral carbyne complex with a $trans$-halogen ligand.

$$(CO)_5Cr{=}C \begin{matrix} OCH_3 \\ C_6H_5 \end{matrix} \xrightarrow{BBr_3} \left[(CO)_5Cr{\equiv}C{-}C_6H_5 \right]^+ \rightleftharpoons \left[(CO)_5Cr{=}C \begin{matrix} Br \\ C_6H_5 \end{matrix} \right]$$

$$\downarrow$$

$$Br{-}\underset{CO\ CO}{\overset{CO\ CO}{Cr}}{\equiv}C{-}C_6H_5$$

Halocarbene complexes have been isolated in a number of cases and have been shown to be converted to carbyne complexes.[69]

$$(CO)_5Cr{=}C \begin{matrix} Cl \\ N(Et)(Pr) \end{matrix} \rightarrow Cl{-}\underset{CO\ CO}{\overset{CO\ CO}{Cr}}{\equiv}C{-}N$$

The addition of nucleophiles to the carbyne carbon atom produces carbene complexes. Addition of ethoxide to a manganese–phenylcarbyne complex gave phenylmethoxycarbene complex **41** (70). In another interesting example, reaction of $Sn(C_6H_5)_3^-$ with $(CO)_5Cr≡CNEt_2$ gave the tin-substituted carbene complex **42**.[71] At room temperature, **42** rearranges to carbyne complex **43**. The rate-determining step for the reaction does not involve dissociation of the $Sn(C_6H_5)_3^-$ unit since the reaction proceeds at approximately the same rate in octane or nitromethane. The slow step in the reaction might be loss of CO followed by rapid ionization of $Sn(C_6H_5)_3^-$ and formation of Sn—Cr bond. Alternatively, CO loss and tin migration could be concerted, followed by rapid rearrangement of the initial *cis* carbyne complex to the observed *trans* product.

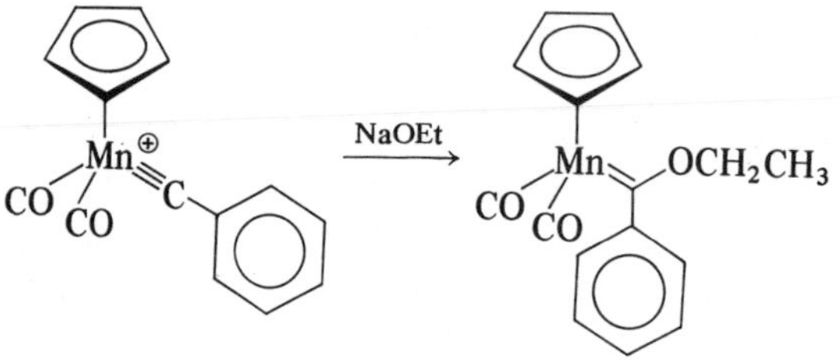

41

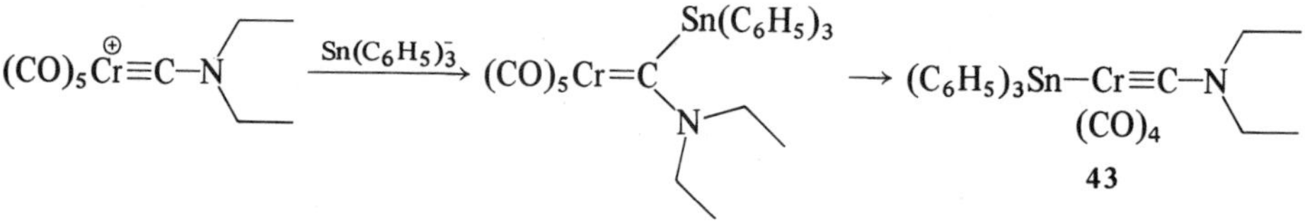

42   **43**

The bridging carbene complex **44** was obtained from reaction of $P(CH_3)_3$ with tungsten–carbyne complex **45**.[72]

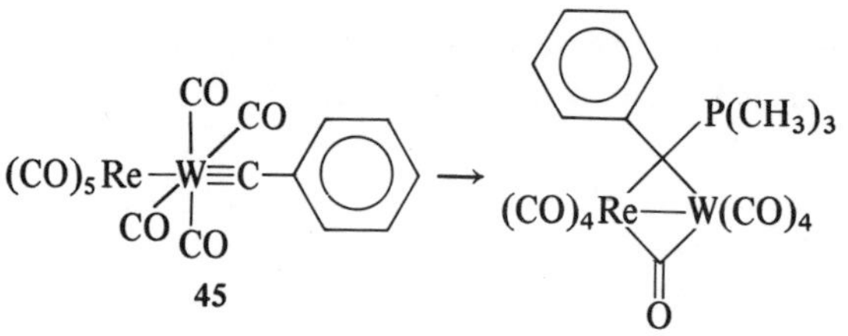

45   **44**

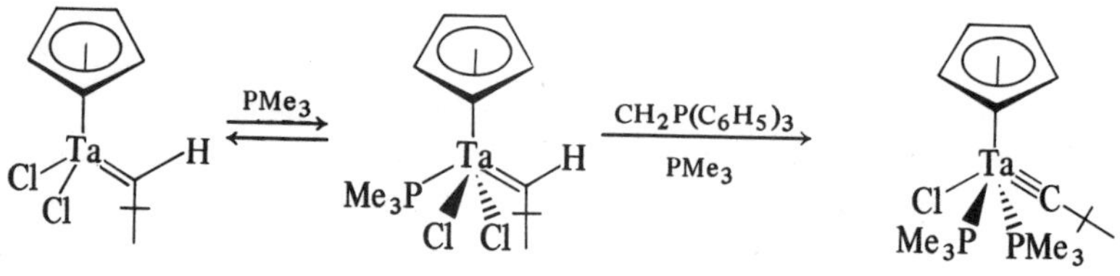

Alkylidyne complexes of tantalum have been prepared by deprotonation of alkylidene complexes. The reactions are greatly accelerated by $P(CH_3)_3$, which presumably generates a five-coordinate intermediate from which proton abstraction or α-elimination is more facile.[73]

The conversion of an alkyl–alkylidyne complex to a bisalkylidene complex also is a feasible reaction. For example, attempted generation of alkyl–alkylidyne complex **46** led to isolation of bisalkylidene complex **47**.[74] On the other hand,

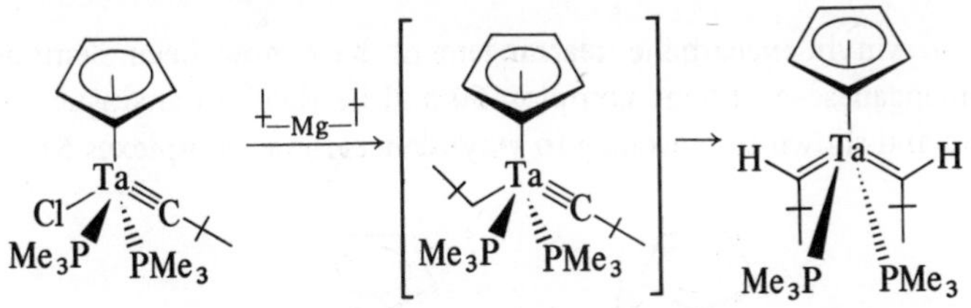

46   **47**

Schrock has prepared the remarkable complex **48**, which contains an alkyl, an alkylidene, and an alkylidyne ligand[75]; at 80°C there is no interconversion of the ligands on **48** observable by NMR spectroscopy. The related complex **49**, in which

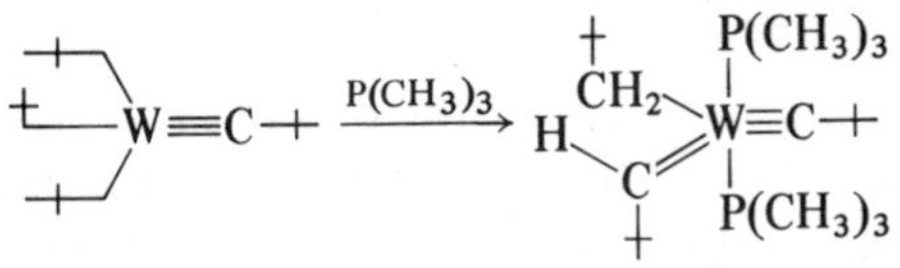

48

the phosphine ligands are constrained to be *cis*, is much less stable than **48** and decomposes to *trans*-di-*tert*-butylethylene possibly via a trisalkylidene complex.[75]

VIII. ACETYLENE–VINYLIDENECARBENE REARRANGEMENTS

Chisholm and Clark proposed that the conversion of an acetylene–platinum complex to a vinylidenecarbene complex was an important step in the prepara-

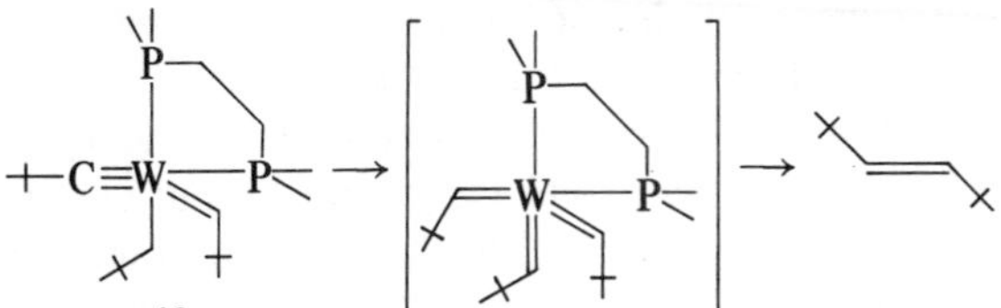

49

tion of platinum-carbene complexes from acetylenes.[76] Several examples of the

$$\underset{\underset{L}{\overset{L}{|}}}{Cl-Pt-CH_3} + RC{\equiv}CH \xrightarrow[CH_3OH]{Ag^+} \underset{\underset{L}{\overset{L}{|}}}{CH_3-Pt}{=}\overset{OCH_3}{\underset{CH_2-R}{\big<}} {}^+$$

acetylene to vinylidenecarbene rearrangement have now been demonstrated. Both the manganese–acetylene complex **50** and its rhenium analogue have been synthesized and shown to rearrange to vinylidenecarbene complexes **51**.[77]

The acetylene to vinylidenecarbene rearrangement could proceed via oxidative addition of the acetylene C—H bond to the metal. The resulting metal acetylide hydride could rearrange to the vinylidenecarbene complex either by a direct 1,3-shift (the reverse of a β-hydride elimination) or via a deprotonation-reprotonation process. Several examples of protonation of metal acetylides at the β-carbon atom to give vinylidenecarbene complexes have now been reported. Bruce has observed the reversible protonation of ruthenium acetylide (**52**).[78] In the presence of methanol, the resulting cationic vinylidenecarbene complex is converted to cationic carbene complex **53**; in water, the intermediate hydroxycarbene complex is not observed but apparently decomposes to give benzyl ruthenium carbonyl complex **54**.

Davison and Solar have examined the protonation of a series of iron acetylides, $C_5H_5Fe(CO)_n(PR_3)_{2-n}C{\equiv}CR$. Protonation of dicarbonyl compound **55** led to the formation of dimeric species **56**, probably by cycloaddition of an intermediate vinylidenecarbene complex to a second molecule of acetylide.[79] Reaction of $(C_5H_5)Fe(CO)_2(THF)^+$ with phenylacetylene also gave **56**. Reaction of monophosphine complex **57** with acid led to isolation of unstable vinylidenecarbene complex **58**. Protonation of **57** in the presence of methanol led to the formation of methoxycarbene complex **59**.[79] Reaction of diphosphine complex

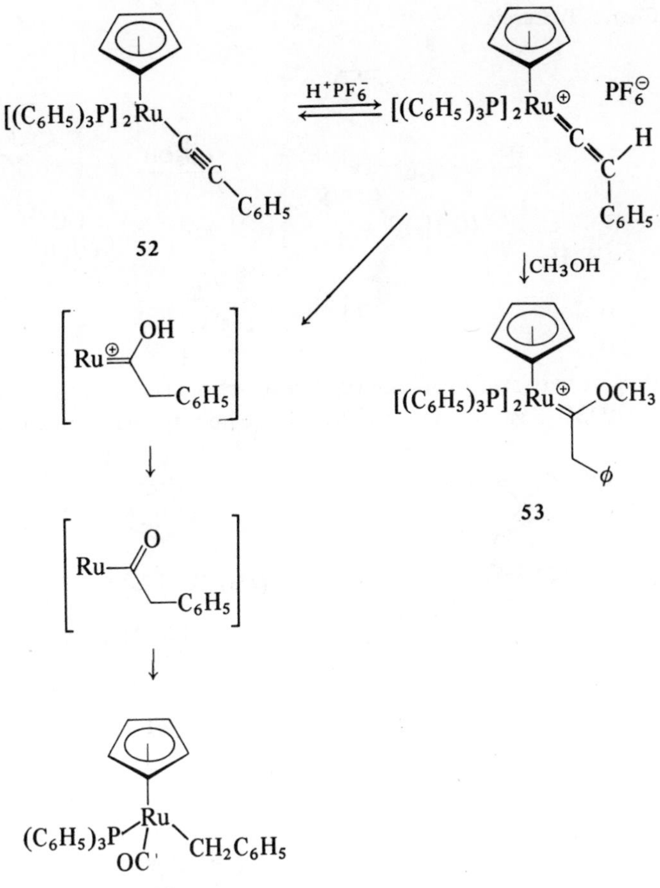

52 53 54

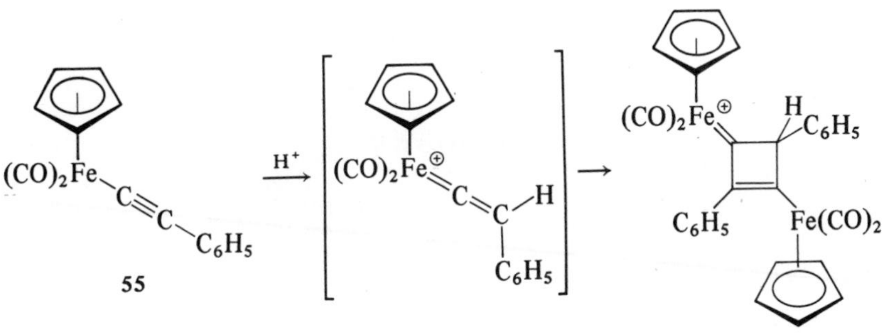

55 56

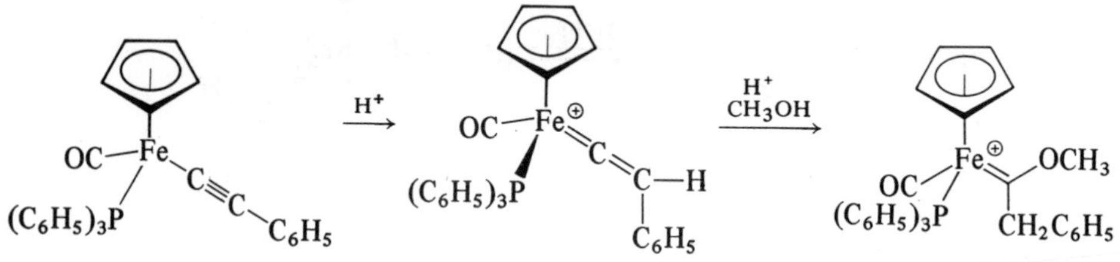

57   58   59

60 with acid gave very stable cationic vinylidenecarbene complex **61**.[80] Reaction of **60** with CH_3OSO_2F gave a mixture of **61** and the isobutylidene complex **62**.[81]

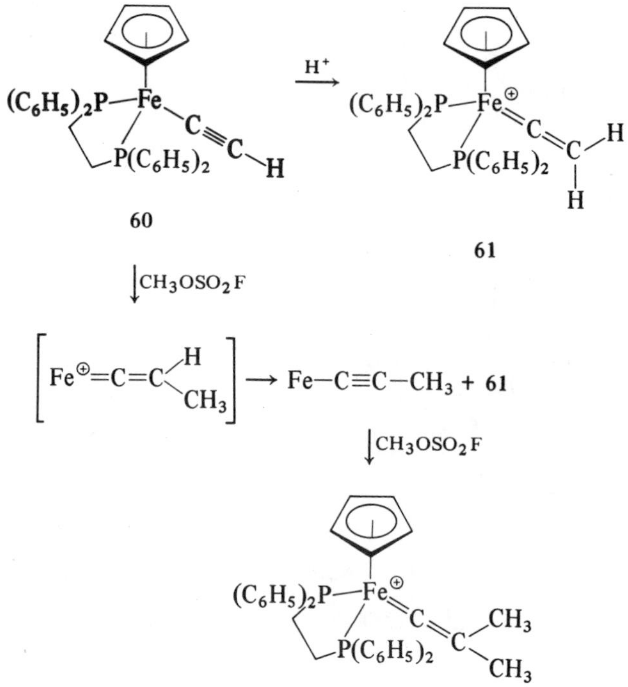

60

61

$$\Big| CH_3OSO_2F$$

$$\left[Fe^{\oplus}=C=C\diagdown_{CH_3}^{H} \right] \longrightarrow Fe-C\equiv C-CH_3 + 61$$

$$\Big| CH_3OSO_2F$$

62

Vollhardt has examined the reaction of excess $(CH_3)_3SiC\equiv CSi(CH_3)_3$ with $(C_5H_5)Co(CO)_2$ and found that a wide range of products, including butatriene (**63**), are obtained.[82] The formation of **63** was proposed to arise from decomposition of a bisvinylidenecarbene complex.

$(CH_3)_3Si-C\equiv C-Si(CH_3)_3$

$\xrightarrow{(C_5H_5)Co(CO)_2}$

C_5H_5Co (structure with multiple C and Si(CH$_3$)$_3$ groups)

$\longrightarrow$ (structure with $(CH_3)_3Si$ and $Si(CH_3)_3$ groups)

63

IX. INSERTION OF CO INTO CARBENE–METAL BONDS

In 1974 Fischer reported that the reaction of $(CO)_5Cr=C(OCH_3)C_6H_5$ with 1-(1-propenyl)-2-pyrrolidone under 150 atm of carbon monoxide gave a mixture of cyclobutanones and products derived from ring opening of the cyclobutanone.[83] The reaction was proposed to proceed via formation of methoxyphenyl-

$(CO)_5Cr=C(C_6H_5)(OCH_3)$ $\xrightarrow{CO}$ (reaction scheme with intermediates) $\rightarrow$ (cyclobutanone product) $\downarrow$ (ring-opened product)

ketene and subsequent 2 + 2 cycloaddition of the ketene and vinylpyrrolidone. Casey and Tuinstra have found that $(CO)_5WC(C_6H_5)_2$ reacts with CO in the presence of ethanol to give ethyl diphenylacetate, presumably derived from intermediate formation of diphenylketene.[84]

Herrmann has now reported that reaction of 650 atm of CO with manganese carbene complex **64** gave ketene complex **65**. The ketene complex reacts with H_2 under high pressure to cleave the ketene ligand.[85] Herrmann has reported the

photolysis of $(C_5H_5)Co(CO)_2$ and dimethyl diazomalonate at $-25°C$ to give adduct **66**.[86] The reaction was suggested to be a 2 + 3 cycloaddition of the dicarbomethoxycarbene to the cobalt-carbonyl bond. However, the reaction might also proceed via a ketene complex.

Diazo compounds normally react with $C_5H_5Mn(CO)_2(THF)$ to give carbene complexes.[87] Consequently, Herrmann was surprised to find that 9-diaza-anthrone reacted with $(C_5H_5)Mn(CO)_2(THF)$ to give ketene complex **67** in 4% yield.[88] The failure to observe anthronylcarbene complex **68** may be due to extreme steric interactions with the C_5H_5 ligand visible in molecular models. The reaction was examined in great detail and it was found that much higher yields (65–76%) of ketene complexes could be obtained if both $(C_5H_5)Mn(CO)_3$ and $(C_5H_5)Mn(CO)_2(THF)$ were present. Labeling experiments showed that

$(C_5H_5)Mn(CO)_3$ was converted to ketene complex **67**, but $(C_5H_5)Mn(CO)_2(THF)$ was not. In the absence of $(C_5H_5)Mn(CO)_2(THF)$, there was no reaction between $(C_5H_5)Mn(CO)_3$ and diazoanthrone. Thus the THF complex is a catalyst for ketene complex formation. Herrmann proposed that an unstable anthronyl-carbene complex **68** is formed from reaction of the THF complex and that anthronylidene is then transferred to a carbonyl unit of $(C_5H_5)Mn(CO)_3$. It is also possible that a nitrogen atom of diazoanthrone displaces THF from $(C_5H_4CH_3)Mn(CO)_2(THF)$ to give a reactive intermediate **69**, which then transfers a carbene unit to $(C_5H_5)Mn(CO)_3$.

A number of examples of the related coupling of carbyne ligands with co-ordinated CO have been observed and lead to formation of metal-substituted ketenes and hydroxyacetylene complexes.[89]

X. NOVEL METAL-CARBENE COMPLEXES

In 1968 Öfele prepared a stable cyclopropenylidene complex (70) that apparently owed its stability to the aromaticity of the cyclopropenium cation unit.[90] Jones has now reported the synthesis and remarkable stability of cyclo-heptatrienylidene complexes such as 71.[91] The high dipole moment of 71 (7.7 D) is undoubtedly due to resonance forms involving a tropyl cation unit.

70   71

In contrast to the synthesis of these stable aromatic systems, Herrmann's attempt to prepare cyclopentadienylidene complex 72 led to the isolation of dimer 73.[92] The antiaromaticity of the cyclopentadienyl cation may be responsible for the rapid dimerization of 72.

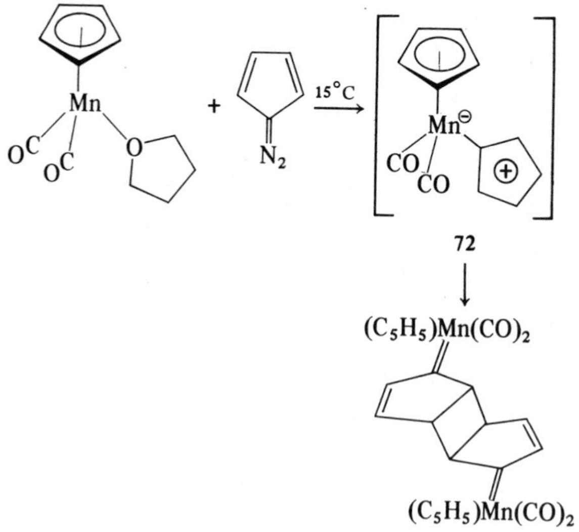

72

$(C_5H_5)Mn(CO)_2$

$(C_5H_5)Mn(CO)_2$

73

Several dihalocarbene complexes have now been synthesized. Reger reported that SbF$_5$ abstracts fluoride from (C$_5$H$_5$)Mo(CO)$_3$CF$_3$ to give difluorocarbene complex 74, which was characterized in solution.[93] Roper has reported the synthesis of the osmium dichlorocarbene complex 75 from the reaction of OsHCl(CO)[P(C$_6$H$_5$)$_3$]$_3$ and Hg(CCl$_3$)$_2$.[94]

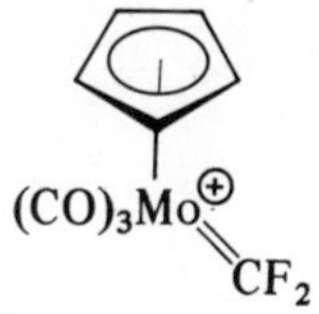

74

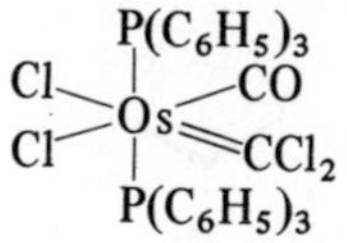

75

A number of carbene complexes possessing unusual combinations of ligands have been reported. Rosenblum has prepared an iron–carbene complex (76) that also possesses an alkyl ligand.[95] Roustan has prepared a molybdenum-

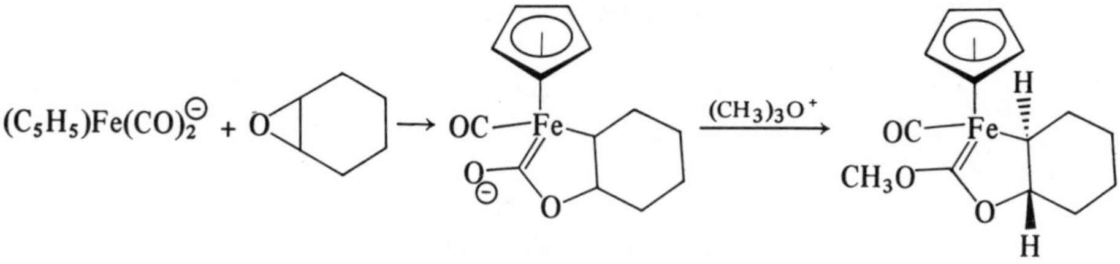

76

carbene complex (77) that also possesses a chelating enolate ligand as a result of the acid-catalyzed addition of methanol to a propargyl molybdenum compound.[96] Mitsuda has prepared iron–carbene complexes such as 78 in which a vinyl group attached to the carbene carbon atom is also π-bonded to iron.[97]

Other interesting new carbene complexes are 79, which possesses a cumulene unit[98], and 80, which was prepared by reduction of coordinated carbon monoxide.[99]

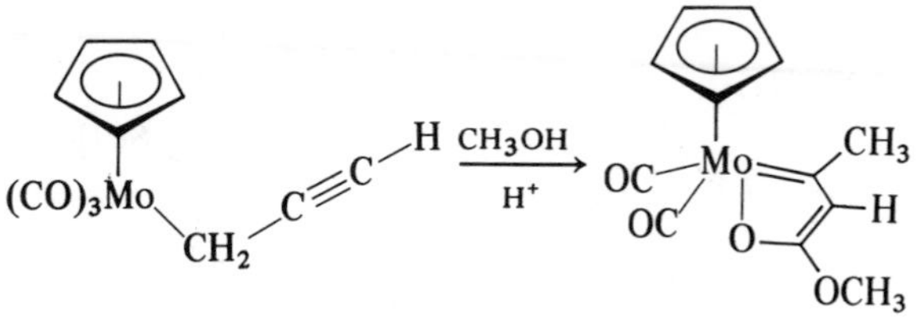

77

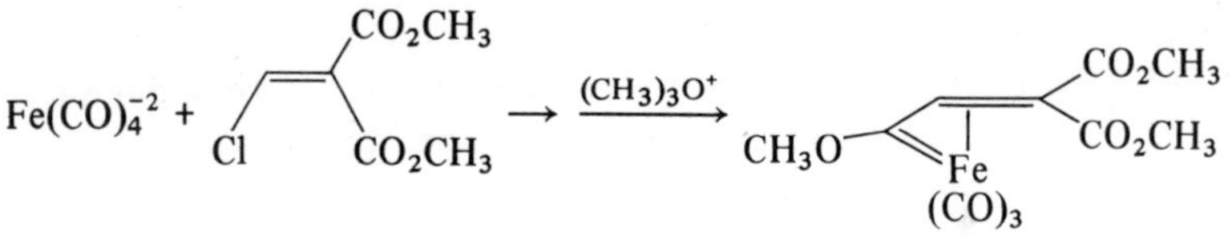

78

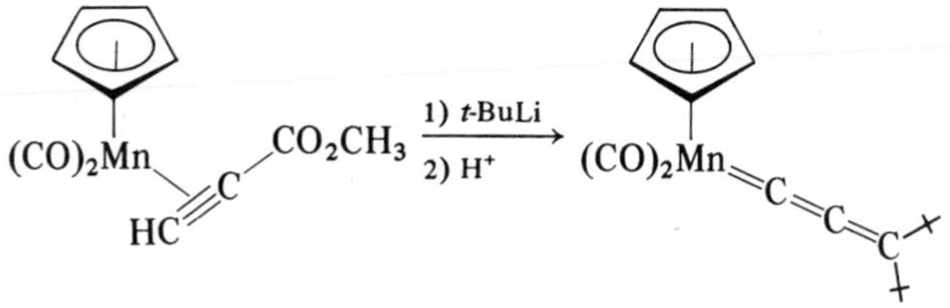

79

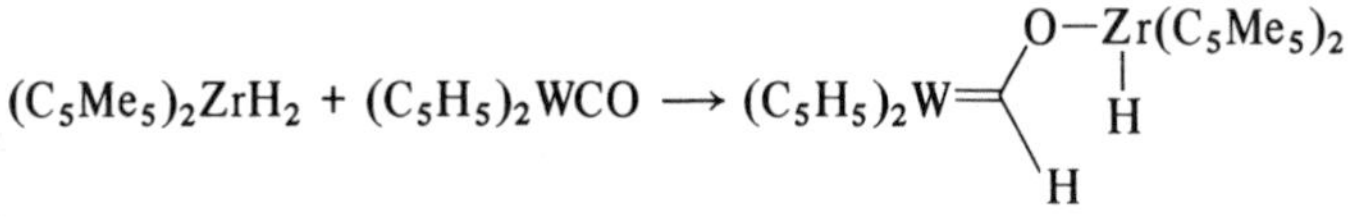

$$(C_5Me_5)_2ZrH_2 + (C_5H_5)_2WCO \longrightarrow (C_5H_5)_2W$$

80

XI. NEW ROUTES TO METAL-CARBENE COMPLEXES

Several interesting syntheses of carbene complexes have used CS_2 metal-carbene complexes as starting materials. The reaction of CH_3I with a platinum-carbon disulfide complex led to formation of carbene complex **81**.[100] Dixneuf has reported a very interesting cycloaddition of an acetylene to a complexed CS_2 ligand that leads directly to carbene complex **82**.[101]

Stone has investigated the possibility of complexing metal-carbon multiple bonds to Pt(0) in the same way that Pt(0) forms stable alkene and alkyne com-

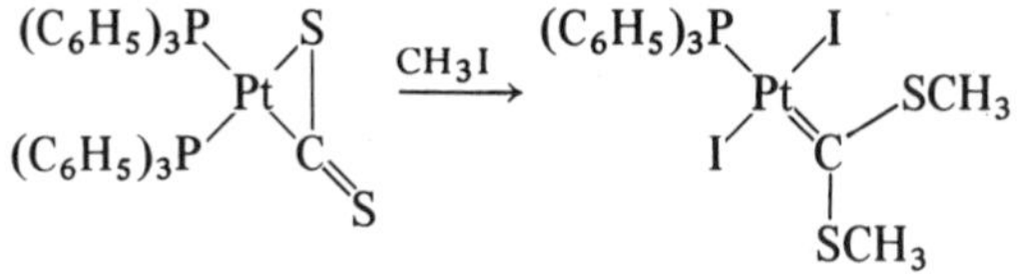

81

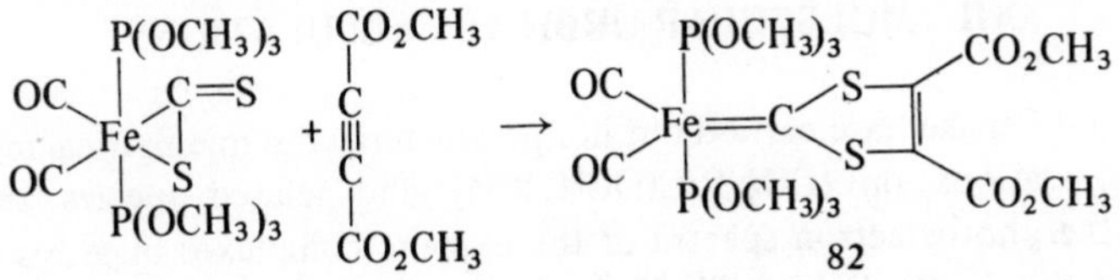

plexes. This new synthetic approach has proven remarkably successful for the preparation of heterobimetallic carbene complexes. Reaction of $(CO)_5WC-(OCH_3)CH_3$ with a platinum–ethylene complex led to the displacement of ethylene and the formation of bridging carbene complex **83**.[102] Reaction of $(CO)_5WC-$

83

$(OCH_3)C_6H_5$ with a related platinum–ethylene complex led to transfer of the carbene ligand to platinum and the formation of complex **84**, in which the carbene ligand bridges between two platinum atoms.[103] Carbyne complexes

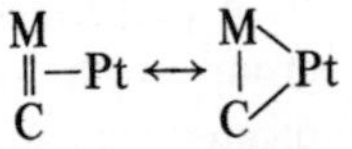

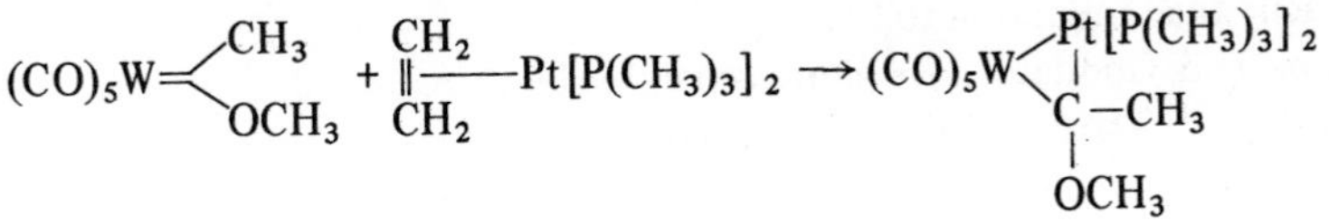

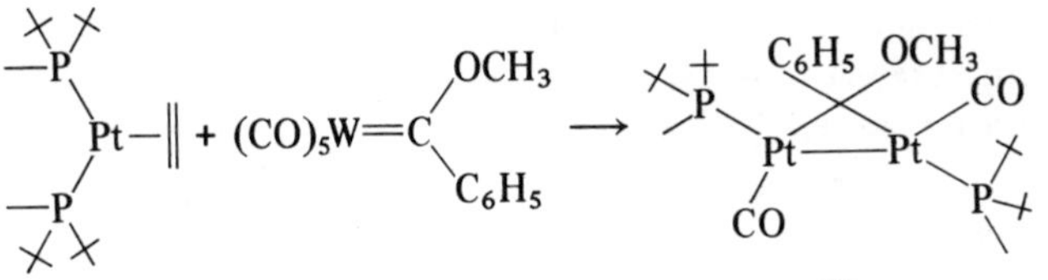

84

react with Pt(0) compounds in a similar way to give bridging carbyne complexes such as **85**.[104]

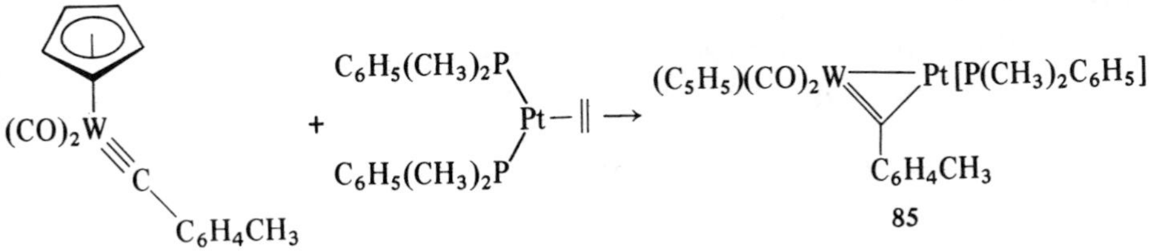

85

XII. MOLECULAR ORBITAL CALCULATIONS

Block and Fenske have carried out nonparameterized semiempirical molecular orbital calculations on $(CO)_5CrC(OCH_3)CH_3$ and related species and have discussed the photoelectron spectra of the carbene complexes in terms of their molecular orbital calculations.[105] The calculations showed that the π-acceptor ability of ligands decreased in the order $CO > :C(OR)R' > :C(NR_2)R'$.

Brooks and Schaefer have done single configuration SCF calculations with an expanded basis set on $MnCH_2$.[106] The ground electronic state was calculated to be of $^8B_1$ symmetry with a $Mn-CH_2$ bond length of 2.16 Å and a $H-C-H$ bond angle of 109°. Dissociation to $^3B_1$ CH_2 and ^6S Mn required 33 kcal/mol.

Berke and Hoffmann have performed extended Hückel calculations on the migration of a methyl group from manganese to a coordinated methylene group.[107] Their results indicated that migration to the carbene ligand should be more facile than migration to CO.

Rappe and Goddard have done general valence bond calculations of $NiCH_3$ and $NiCH_2$.[108] The bond dissociation energy for $Ni=CH_2$ was calculated to be 65 kcal/mol, only slightly larger than the bond dissociation energy for $Ni-CH_3$, which was calculated to be 60 kcal/mol.

REFERENCES

1.   E. O. Fischer, *Pure Appl. Chem.*, 24, 407 (1970).

2.   E. O. Fischer, *Pure Appl. Chem.*, 30, 353 (1972).

3.   E. O. Fischer, *Angew. Chem.*, 86, 651 (1974).

4.* E. O. Fischer, U. Shubert, and H. Fischer, *Pure Appl. Chem.*, 50, 857 (1978).

5.   D. J. Cardin, B. Cetinkaya, and M. F. Lappert, *Chem. Rev.*, 72, 545 (1972).

6.   D. J. Cardin, B. Cetinkaya, M. J. Doyle, and M. F. Lappert, *Chem. Soc. Rev.*, 2, 99 (1973).

7.   F. A. Cotton and C. M. Lukehart, *Prog. Inorg. Chem.*, 16, 487 (1972).

8.* C. P. Casey, in *Transition Metal Organometallics in Organic Synthesis*, Vol. 1, H. Alper, Ed., Academic, New York, 1976, Chap. 3.

9.   C. P. Casey, *J. Organomet. Chem. Libr.*, 1, 397 (1976).

10. C. P. Casey, *Chemtech*, 378 (1979).

11. K. H. Dötz, *Naturwissenshaften*, 62, 365 (1975).

12. C. P. Casey, W. R. Brunsvold, and D. M. Scheck, *Inorg. Chem.*, 16, 3059 (1977).

13. C. P. Casey and W. R. Brunsvold, *Inorg. Chem.*, 16, 391 (1977).

14. E. O. Fischer and H. J. Kalder, *J. Organomet. Chem.*, 131, 57 (1977).

15. C. P. Casey and T. J. Burkhardt, *J. Am. Chem. Soc.*, 96, 7808 (1974); C. P. Casey, H. E. Tuinstra, and M. C. Saeman, *J. Am. Chem. Soc.*, 98, 608 (1976).

16. J. McGinnis, T. J. Katz, and S. Hurwitz, *J. Am. Chem. Soc.*, **98**, 605 (1976); T. J. Katz, J. McGinnis, and C. Altus, *J. Am. Chem. Soc.*, **98**, 606 (1976).

17. S. P. Kolesnikov, N. I. Okhrimenko, and O. M. Nefedov, *Dokl. Akad. Nauk SSSR*, **243**, 1193 (1978).

18. J. Levisalles, H. Rudler, and D. Villemin, *J. Organomet. Chem.*, **146**, 259 (1978); J. Levisalles, H. Rudler, D. Villemin, J. Daran, Y. Jeannin, and L. Martin, *J. Organomet. Chem.*, **155**, C1 (1978).

19. C. P. Casey, S. W. Polichnowski, H. E. Tuinstra, L. D. Albin, and J. C. Calabrese, *Inorg. Chem.*, **17**, 3045 (1978).

20.* C. P. Casey, S. W. Polichnowski, A. J. Shusterman, and C. R. Jones, *J. Am. Chem. Soc.*, **101**, 7282 (1979).

21. C. P. Casey and S. W. Polichnowski, *J. Am. Chem. Soc.*, **99**, 6097 (1977).

22. M. Brookhart and G. O. Nelson, *J. Am. Chem. Soc.*, **99**, 6099 (1977); M. Brookhart, J. R. Tucker, T. C. Flood, and J. Jensen, *J. Am. Chem. Soc.*, **102**, 1203 (1980).

23. S. Brandt and P. Helquist, *J. Am. Chem. Soc.*, **101**, 6473 (1979).

24. A. E. Stevens and J. L. Beauchamp, *J. Am. Chem. Soc.*, **100**, 2584 (1978).

25. A. R. Cutler. *J. Am. Chem. Soc.* **101**, 604 (1979).

26. W.-K. Wong, W. Tam, and J. A. Gladysz, *J. Am. Chem. Soc.*, **101**, 5440 (1979); W. A. Kiel, G.-Y. Lin, and J. A. Gladysz, *J. Am. Chem. Soc.*, **102**, 3299 (1980).

27.* R. R. Schrock, *Acc. Chem. Res.*, **12**, 98 (1979).

28. R. R. Schrock, *J. Am. Chem. Soc.*, **96**, 6796 (1974).

29. R. R. Schrock and J. D. Fellmann, *J. Am. Chem. Soc.*, **100**, 3359 (1978).

30. S. J. McLain, C. D. Wood, and R. R. Schrock, *J. Am. Chem. Soc.*, **99**, 3519 (1977).

31. A. J. Schultz, J. M. Williams, R. R. Schrock, G. A. Rupprecht, and J. D. Fellmann, *J. Am. Chem. Soc.*, **101**, 1593 (1979).

32. P. R. Sharp and R. R. Schrock, *J. Organomet. Chem.*, **171**, 43 (1979).

33. J. Schwartz, *Pure Appl. Chem.*, **52**, 733 (1980).

34. L. J. Guggenberger and R. R. Schrock, *J. Am. Chem. Soc.*, **97**, 6578 (1975).

35. R. R. Schrock, L. W. Messerle, C. D. Wood, and L. J. Guggenberger, *J. Am. Chem. Soc.*, **100**, 3793 (1978).

36. M. R. Churchill, F. J. Hollander, and R. R. Schrock, *J. Am. Chem. Soc.*, **100**, 647 (1978).

37.* R. R. Schrock and P. R. Sharp, *J. Am. Chem. Soc.*, **100**, 2389 (1978).

38. R. R. Schrock, *J. Am. Chem. Soc.*, **98**, 5399 (1976).

39. S. H. Pine, R. Zahler, D. A. Evans, and R. H. Grubbs, *J. Am. Chem. Soc.*, **102**, 3270 (1980).

40. S. J. McLain, C. D. Wood, and R. R. Schrock, *J. Am. Chem. Soc.*, **101**, 4558 (1979).

41.* C. D. Wood, S. J. McLain, and R. R. Schrock, *J. Am. Chem. Soc.*, **101**, 3210 (1979).

42. R. H. Grubbs, *Prog. Inorg. Chem.*, **24**, 1 (1978).

43.* N. Calderon, J. P. Lawrence, and E. A. Ofstead, *Adv. Organomet. Chem.*, **17**, 449 (1979).

44. T. J. Katz, *Adv. Organomet. Chem.*, **16**, 283 (1977).

45. J. J. Rooney and A. Stewart, *Catalysis (Lond.)*, **1**, 277 (1977).

46. F. N. Tebbe, G. W. Parshall, and G. S. Reddy, *J. Am. Chem. Soc.*, **100**, 3611 (1978).

47.* F. N. Tebbe, G. W. Parshall, and D. W. Ovenall, *J. Am. Chem. Soc.*, **101**, 5074 (1979).

48. R. J. Puddephatt, M. A. Quyser, and C. F. H. Tipper, *J. Chem. Soc., Chem. Commun.*, 626 (1976).

49. C. P. Casey, D. M. Scheck, and A. J. Shusterman, *J. Am. Chem. Soc.*, **101**, 4233 (1979).

50. T. H. Johnson, *J. Org. Chem.*, **44**, 1356 (1979).

51. R. J. Al-Essa, R. J. Puddephatt, M. A. Quyser, and C. F. H. Tipper, *Inorg. Chem. Acta*, **34**, L187 (1979).

52. M. Ephritikhine, B. R. Francis, M. L. H. Green, R. E. Mackenzie, and M. J. Smith, *J. Chem. Soc., Dalton Trans.*, 1131 (1977).

53. M. L. H. Green, *Ann. N. Y. Acad. Sci.*, **333**, 229 (1980).

54. M. L. H. Green, *Pure Appl. Chem.*, **50**, 27, (1978).

55. C. P. Casey, D. M. Scheck, and A. J. Shusterman, in *Fundamental Research in Homogeneous Catalysis*, Vol. 3, M. Tsutsui, Ed., Plenum New York, 1979. p. 141.

56. S. J. McLain, J. Sancho, and R. R. Schrock, *J. Am. Chem. Soc.*, **101**, 5451 (1979).

57. C. P. Casey and H. E. Tuinstra, *J. Am. Chem. Soc.*, **100**, 2270 (1978).

58. K. H. Dötz, *Chem. Ber.*, **110**, 78 (1977); K. H. Dötz and C. G. Kreiter, *Chem. Ber.*, **109**, 2026(1976); K. H. Dötz, and C. G. Kreiter, *J. Organomet. Chem.*, **99**, 309 (1975).

59. H. Fischer and K. H. Dötz, *Chem. Ber.*, **113**, 193 (1980).

60. K. H. Dötz and I. Pruskil, *J. Organomet. Chem.*, **132**, 115 (1977).

61. K. H. Dötz, *J. Organomet. Chem.*, **140**, 177 (1977); K. H. Dötz and R. Dietz, *Chem. Ber.*, **111**, 2517 (1978).

62. K. H. Dötz and I. Pruskil, *Chem. Ber.*, **111**, 2059 (1978).

63. K. H. Dötz, R. Dietz, C. Kappenstein, D. Neugebauer, and U. Schubert, *Chem. Ber.*, **112**, 3682 (1979).

64. K. H. Dötz and R. Dietz, *Chem. Ber.*, **110**, 1555 (1977).

65. K. H. Dötz and R. Dietz, *J. Organomet. Chem.*, **157**, C55 (1978).

66. R. Dietz, K. H. Dötz, and D. Neugebauer, *Nouv. J. Chem.*, **2**, 59 (1978)

67.* T. J. Katz and S. J. Lee, *J. Am. Chem. Soc.*, **102**, 422 (1980).

68. E. O. Fischer, G. Kreis, C. G. Kreiter, J. Müller, G. Huttner, and H. Lorenz, *Angew. Chem., Int. Ed. Engl.*, **12**, 564 (1973).

69. E. O. Fischer, W. Kleine, and F. R. Kreissl, *J. Organomet. Chem.*, **107**, C23 (1976); H. Fischer, A. Motsch, and W. Kleine, *Angew. Chem., Int. Ed. Engl.*, **17**, 842 (1978).

70. E. O. Fischer, E. W. Meineke, and F. R. Kreissl, *Chem. Ber.*, **110**, 1140 (1977).

71. E. O. Fischer, H. Fischer, U. Schubert, and R. B. A. Pardy, *Angew. Chem., Int. Ed. Engl.*, **18**, 871 (1979).

72. F. R. Kreissl, P. Friedrich, T. L. Lindner, and G. Huttner, *Angew. Chem., Int. Ed. Engl.*, **16**, 314 (1977).

73. S. J. McLain, C. D. Wood, L. W. Messerle, R. R. Schrock, F. J. Hollander, W. J. Youngs, and M. R. Churchill, *J. Am. Chem. Soc.*, **100**, 5962 (1978).

74. J. D. Fellmann, G. A. Rupprecht, C. D. Wood, and R. R. Schrock, *J. Am. Chem. Soc.*, **100**, 5964 (1978).

75.   D. N. Clark and R. R. Schrock, *J. Am. Chem. Soc.*, 100, 6774 (1978).

76.   M. H. Chisholm and H. C. Clark, *Acc. Chem. Res.*, 6, 202 (1973).

77.   A. B. Antonova, N. E. Kolobova, P. V. Petrovsky, B. V. Lokshin, and N. S. Obezyuk, *J. Organomet. Chem.*, 137, 55 (1977); N. Y. Kolobova, A. B. Antonova, O. M. Khitrova, M. Y. Antipin, and Y. V. Struchkov, *J. Organomet. Chem.*, 137, 69 (1977).

78.   M. I. Bruce, A. G. Swincer, and R. C. Wallis, *J. Organomet. Chem.*, 171, C5 (1979).

79.   A. Davidson and J. P. Solar, *J. Organomet. Chem.*, 155, C8 (1978); see also N. Y. Kolobova, V. V. Skripkin, G. G. Alexandrov, and Y. T. Struchkov, *J. Organomet. Chem.*, 169, 293 (1979).

80.   A. Davison and J. P. Selegue, *J. Am. Chem. Soc.*, 100, 7763 (1978).

81.   A. Davison and J. P. Selegue, *J. Am. Chem. Soc.*, 102, 2455 (1980).

82.   J. R. Fritch, K. P. C. Vollhardt, M. R. Thompson, and V. W. Day, *J. Am. Chem. Soc.*, 101, 2768 (1979).

83.   B. Dorrer and E. O. Fischer, *Chem. Ber.*, 107, 2683 (1974).

84.   C. P. Casey and H. E. Tuinstra, unpublished observations; H. E. Tuinstra, Ph. D. Thesis, University of Wisconsin, 1978.

85.   W. A. Herrmann and J. Plank, *Angew. Chem.*, 90, 555 (1978).

86.   M. L. Ziegler, K. Weidenhammer, and W. A. Herrmann, *Angew. Chem., Int. Ed. Engl.*, 16, 555 (1977); W. A. Herrmann, *Chem. Ber.*, 111, 1077 (1978).

87.   W. A. Herrmann, *Chem. Ber.*, 108, 3412 (1975).

88.* W. A. Herrmann, J. Plank, M. L. Ziegler, and K. Weidenhammer, *J. Am. Chem. Soc.*, 101, 3133 (1979).

89.   F. R. Kreissl, P. Friedrich, and G. Huttner, *Angew. Chem., Int. Ed. Engl.*, 16, 102 (1977); E. O. Fischer and P. Friedrich, *Angew. Chem., Int. Ed. Engl.*, 18, 327 (1979).

90.   K. Öfele, *Angew. Chem., Int. Ed. Eng.*, 7, 950 (1968).

91.* N. T. Allison, Y. Kawada, and W. M. Jones, *J. Am. Chem. Soc.*, 100, 5224 (1978).

92.   W. A. Herrmann, J. Plank, M. L. Ziegler, and K. Weidenhammer, *Angew. Chem.*, 90, 817 (1978).

93.   D. L. Reger and M. D. Dukes, *J. Organomet. Chem.*, 153, 67 (1978).

94.   G. R. Clark, K. Marsden, W. R. Roper, and L. J. Wright, *J. Am. Chem. Soc.*, 102, 1206 (1980).

95.   P. Klemarczyk, T. Price, W. Priester, and M. Rosenblum, *J. Organomet. Chem.*, 139, C25 (1977).

96.   C. Charrier, J. Collin, J. Y. Merour, and J. L. Roustan, *J. Organomet. Chem.*, 162, 57 (1978).

97.   T. Mitsudo, H. Watanabe, Y. Watanabe, N. Nitani, and Y. Takegami, *J. Chem. Soc., Dalton Trans.*, 395 (1979).

98.   H. Berke, *Angew. Chem.*, 88, 684 (1976).

99.   P. T. Wolczanski, R. S. Threlkel, and J. E. Bercaw, *J. Am. Chem. Soc.*, 101, 218 (1979).

100. D. H. Farrar, R. O. Harris, and A. Walker, *J. Organomet. Chem.*, 124, 125 (1977).

101. H. Le Bozec, A. Gorgues, and P. H. Dixneuf, *J. Am. Chem. Soc.*, 100, 3946 (1978).

102. T. V. Ashworth, M. Berry, J. A. K. Howard, M. Laguna, and F. G. A. Stone, *J. Chem. Soc., Chem. Commun.*, 43 (1979).

103. T. V. Ashworth, M. Berry, J. A. K. Howard, M. Laguna, and F. G. A. Stone, *J. Chem. Soc., Chem. Commun.*, 45 (1979).

104. T. V. Ashworth, J. A. K. Howard, and F. G. A. Stone, *J. Chem. Soc., Chem. Commun.*, 42 (1979).

105. T. F. Block and R. F. Fenske, *J. Am. Chem. Soc.*, 99, 4321 (1977); T. F. Block and R. F. Fenske, *J. Organomet. Chem.*, 139, 235 (1977).

106. B. R. Brooks and H. F. Schaefer III, *Mol. Phys.*, 34, 193 (1977).

107. H. Berke and R. Hoffmann, *J. Am. Chem. Soc.*, 100, 7224 (1978).

108. A. K. Rappe and W. A. Goddard III, *J. Am. Chem. Soc.*, 99, 3966 (1977); W. A. Goddard III, S. P. Walch, A. K. Rappe, T. H. Upton, and C. F. Melius, *J. Vac. Sci. Technol.*, 14, 416 (1977).

109. H. D. Empsall, E. M. Hyde, M. Markham, W. S. McDonald, M. C. Norton, B. L. Shaw, and B. Weeks, *J. Chem. Soc., Chem. Commun.*, 589 (1977).

110. C. Crocker, R. J. Errington, W. S. McDonald, K. J. Odell, and B. L. Shaw, *J. Chem. Soc., Chem. Commum.*, 498 (1979); C. Crocker, R. J. Errington, R. Markham, C. J. Moulton, K. J. Odell, and B. L. Shaw, *J. Am. Chem. Soc.*, 102, 4373 (1980).

5

DIRADICALS

WESTON THATCHER BORDEN

University of Washington, Seattle, Washington 98195

175

I. INTRODUCTION

A diradical is literally a molecule with two unpaired electrons. Nevertheless, organic chemists use the word diradical to describe molecules in singlet, as well as in triplet, spin states. A somewhat broader definition is therefore required. One suggested by Salem and Rowland[1] is used here. A diradical is defined as a species in which two electrons occupy degenerate or nearly degenerate molecular orbitals (MOs). This definition is formulated narrowly enough to exclude electronically excited molecules that would not be classified as diradicals in the ground state, yet it is sufficiently broad to include molecules in which the partially filled MOs are truly degenerate only at geometries that may not correspond to a minimum for any state.

Carbenes, nitrenes, and silylenes may be termed one-centered diradicals, since in the simplest cases the nearly degenerate MOs are mostly confined to a single atom. These one-center diradicals are treated in other chapters of this volume, so that they are not discussed further here.

Two-centered diradicals are those in which, for the simplest examples, the approximately degenerate MOs principally span two atoms. Trimethylene and tetramethylene are two examples of this class. Developments in the study of these two diradicals from 1977 to 1979 are discussed in Sections II and III of this chapter.

One- and two-centered diradicals to which unsaturated groups have been appended might, for obvious reasons, be termed multicentered diradicals. Here, however, this appellation is reserved for those classes of molecules whose *simplest* members contain degenerate MOs that are delocalized over more than just two atoms. There are two types of multicentered diradicals. One of these is the antiaromatic annulenes. Progress in determining the structure of cyclobutadiene, the most thoroughly studied diradical belonging to this class, is discussed in Section IV. The other class of multicentered diradicals is comprised of fully conjugated molecules for which no classical Kekulé structures can be written. Significant developments from 1977 to 1979 in the chemistry of the simplest member of this class, trimethylenemethane, are recorded in Section V. Section VI treats two other molecules belonging to this class, 1,8-naphthoquinodimethane and *m*-quinomethane.

Space limitations in this chapter have precluded the inclusion of many topics that deserve a place in a comprehensive review of diradicals. The selection reflects the author's own interests in this area of chemistry. His occupation as a part-time theoretical chemist accounts for the prejudices that may be evident in the sections that follow.

II. TRIMETHYLENE

A. Trimethylene in Cyclopropane Stereomutations

Much of the experimental work on diradicals has been inspired by theoretical predictions. Trimethylene ($\cdot CH_2 - CH_2 - CH_2 \cdot$) provides an excellent illustration of this generalization. In 1968 Hoffman[2] published the results of a theoretical study of trimethylene using the extended Hückel method. The calculations revealed a strong intereaction between the symmetric combination of *p*-orbitals on the terminal carbons and a pseudo-π C—H combination on the central carbon in what Hoffmann called the 0,0 geometry of trimethylene (see Figure 1). In this early example of through-bond coupling, the antisymmetric combination of terminal *p*-orbitals was found to lie below the symmetric combination for C—C—C bond angles greater than 100°. With the highest occupied MO (HOMO) of 0,0 trimethylene being the antisymmetric combination of terminal *p*-orbitals, a conrotatory closure to cyclopropane was predicted. This prediction was consistent with experimental results obtained by Crawford and Mishra[3] and McGreer et al.[4] on pyrazoline decomposition, where one of the terminal carbons had been shown to undergo preferential inversion of configuration on cyclopropane formation.

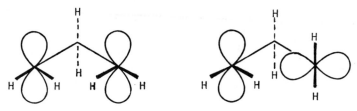

FIGURE 1.   Trimethylene geometries: 0,0 (left) and 0,90 (right).

Hoffmann further found that 0,0 trimethylene exhibited an energy minimum at a $C-C-C$ bond angle of $125°$, and that a $90°$ rotation of one terminal methylene, affording the 0,90 trimethylene geometry, was disfavored by 10 kcal/mol. Only a very small barrier, about 1 kcal/mol, was found to prevent 0,0 trimethylene from closing to cyclopropane. Therefore, Hoffmann proposed 0,0 trimethylene as the intermediate or transition state in the stereomutation of cyclopropane. The implication of Hoffmann's proposal was that cyclopropane stereomutation should occur preferentially by simultaneous conrotation of two methylene groups.

Subsequent, more sophisticated calculations[5-7] did not support Hoffmann's confidence in the existence of *sizable* rotation barriers in trimethylene. The energy difference between the 0,0 and 0,90 forms was reduced to less than 1 kcal/mol in the calculations carried out by Salem et al.[6] and by Goddard et al.[7] Nevertheless, these theoretical studies of the singlet trimethylene energy surface, and subsequent results regarding the dynamics on the surface,[8] did provide qualitative support for Hoffmann's proposal that coupled conrotation of two methylene groups should furnish the lowest energy pathway for cyclopropane stereomutation.

Experimental tests of Hoffmann's prediction were soon forthcoming. Several of these[9-13] used optically active 1,2-di-, tri-, and tetrasubstituted cyclopropane derivatives as a probe for coupled methylene group rotation. Assuming that only the bond between the two substituted ring carbons cleaves, racemization of 1 ($\rightarrow$1') should be much faster than *cis–trans*-isomerization to 2 or 2' according to the Hoffman mechanism. Experimentally, however, the rates were found to be competitive, with the latter even faster in some cases than the former.

A slightly different experiment[14] utilized *cis*-2,3-dideuterio-*trans*-1-vinylcyclopropane (3). Assuming the correctness of the Hoffmann mechanism and bond cleavage such that only an allylically stabilized radical was created, the *trans*-2,3-dideuterio-isomer (4) should have been formed at a much greater rate than the all-*cis*-isomer (5). In fact, 4 and 5 were obtained in the 2 : 1 ratio expected on purely statistical grounds.

Despite these discouraging results and the fact that calculations more sophisticated than Hoffmann's predicted only a very small preference for coupled methylene rotation, Berson and co-workers[15] undertook elegant studies of the stereomutation of optically active *trans*-1,2-dideuteriocyclopropane (6) and 1-phenyl-2-deuteriocyclopropane (7). For the former no assumption can be made about which bond cleaves. However, a full mechanistic analysis is possible

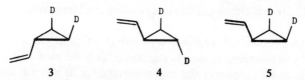

from the observed rate constants for *trans* ⇌ *cis* isomerization and for loss of optical activity, provided some estimate is made as to the magnitude of the secondary isotope effect.

If the isotope effect is ignored, the Hoffmann coupled rotation mechanism predicts the ratio of the rate constant for *trans* ⇌ *cis* isomerization in **6** to that for racemization should be 1.0. Rotation of just one methylene group would lead to a ratio of rate constants of 2.0, and for a randomly rotating intermediate the ratio would be 1.5. Single methylene group rotation is not subject to an isotope effect, since the two methylene groups whose rotation leads to an observable change are identically labeled with deuterium. If a secondary isotope effect of $1.1 = k_H/k_D$ is assumed, the ratio of isomerization to racemization rate constants becomes 1.05 in the Hoffmann mechanism and 1.53 for a freely rotating intermediate.

The ratio actually measured by Berson and his co-workers for **6** was 1.07 ± 0.04. Even in the case where k_H/k_D is taken as 1.0 and twice the standard deviation is added to 1.07, giving a ratio of isomerization to racemization rate constants of 1.15, coupled rotation, competing with a randomly rotating intermediate, still accounts for 70% of the reaction. If $k_H/k_D = 1.1$ and the ratio of rate constants for isomerization and racemization is really 1.07, the Hoffmann mechanism accounts for 98% of the reaction.

In 1-phenyl-2-deuteriocyclopropane (**7**) the data obtained by Berson and co-workers also indicate a distinct preference for the Hoffmann mechanism. If the isotope effect is assumed to be normal, their data show that no more than 4% of the rotation of the phenyl-bearing carbon can occur without an accompanying rotation of one of the two methylene groups. If the isotope effect is taken to be $k_H/k_D = 1.1$, the rate constant for the unaccompanied rotation of the phenyl-substituted carbon is zero. The data then reveal that 80% of the rotation at C_2 occurs synchronously with rotation at C_1. The cyclopropane ring in **7** is not sufficiently substituted to allow a determination of the separate contributions of unaccompanied C_2 rotation and synchronous rotation at both unsubstituted carbons to the remaining 20% of the stereomutation at C_2.

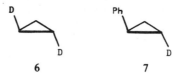

B. Recent Developments in Cyclopropane Stereomutations[16]

In all the studies of cyclopropane stereomutations carried out prior to Berson's, it was necessary to assume cleavage of only the most highly substituted ring bond in order to analyze the experimental data. Berson was able to analyze the data for both 6 and 7 without making this assumption. The results for 7 allow as much as 20% of the "wrong" bond to have cleaved upon synchronous rotation of C_2 and C_3, *if* one-center rotation at C_2 contributes negligibly to the mechanism. Although the much lower temperature required to effect the stereo-mutation of 7 versus 6 would then require an unprecedented equal weakening of the bonds remote from and adjacent to the phenyl group in 7, the "wrong" bond hypothesis was certainly deserving of an experimental test.

An unsuccessful attempt was made by Berson and co-workers[17] to use optically active *trans*-1-phenyl-2,3-dideuteriocyclopropane for this purpose. Unfortunately, their synthesis resulted in production of appreciable amounts of d_3-labeled molecules, whose presence precluded an accurate IR analysis of the rate of *trans* $\longrightarrow$ *cis* isomerization. However, the rate of racemization did provide another indication that unaccompanied phenyl rotation is negligible.

Berson and co-workers also found that the major products of structural isomerization of phenylcyclopropane are *cis*- and *trans-β*-methylstyrene and allylbenzene. No α-methylstyrene, the product expected from "wrong" bond cleavage, was found. Thus, at least in the structural isomerization, ring cleavage appears to occur only at the benzylic carbon.

Successful tests of the "wrong" bond hypothesis in cyclopropane stereo-mutations were carried out in two laboratories. Baldwin and Carter[18] prepared (+)-(1S,2S,3S)-*r*-1-cyano-*t*-2-phenyl-1,*c*-3-dideuteriocyclopropane (8). Through the use of an optically active shift reagent they established that the configuration at C_3 is maintained when 8 is equilibrated with the isomers that differ in configuration at C_1 and C_2. Doering and Barsa[19] used the four diastereomers of optically active 1-cyano-3-methyl-2-(*cis*-propenyl)cyclopropane (9) to again demonstrate that only the $C_1 - C_2$ bond is cleaved when 9 undergoes stereo-mutation. Neither group found evidence for the operation of the Hoffmann coupled rotation mechanism.

Similarly, Kirmse and Zeppenfeld[20] found no evidence for synchronous rotation of the methoxyl-bearing carbons in 1,2-dimethoxy-3-methylcyclopropane (10). The hundredfold slower rate of stereomutation of all-*cis*-1-methoxy-2,3-dimethylcyclopropane (11) was taken as evidence for the exclusive formation of the diradical stabilized by both methoxyl groups in the cleavage of 10.

H$_3$CO ... CH$_3$ ... OCH$_3$ H$_3$CO ... CH$_3$ ... CH$_3$

        10   **11**

The finding that it is only the "right" bond that cleaves in **8** and **9** has been capitalized upon in further experiments by the groups at both Oregon and Harvard. Baldwin and Carter[21] measured deuterium isotope effects on the *trans* $\longrightarrow$ *cis* isomerization rate of *trans*-1-cyano-2-phenylcyclopropane (**12**). It is, indeed, the isotope effects on the one-center epimerization of C$_1$ and of C$_2$ that are measured in this study, since coupled two-center rotations involving C$_3$ are excluded by the results obtained with **8**.

Comparison of the rate of rearrangement of **12a** and **12b** gave k_H/k_D = 1.07 at 241.2°C. If the entire change in rate of isomerization is due to slowing the rotation at C$_1$, the site of deuterium substitution, and if rotation at C$_1$ accounts for about 70% of the isomerization reaction, as it does in **8**, then k_H/k_D = 1.09 for epimerization of C$_1$. This is about the size of the isotope effect on coupled methylene rotation that is favored by Berson for analyzing the data on **6** and **7**.

However, comparing the rates of *trans* $\longrightarrow$ *cis* isomerization of **12a** and **12c**, Baldwin and Carter found a β-isotope effect of k_H/k_D = 1.13, even larger than the α effect. An independent determination of the isotope effect associated with deuterium substituted at C$_3$ was obtained from the rate of rearrangement of **12d** versus **12a** and the isotope effect at C$_1$, measured in the **12a**/**12b** comparison. This less direct measurement also gave k_H/k_D = 1.13 for the β-isotope effect.

Baldwin and Carter noted that the substantial isotope effect associated with deuterium substitution at C$_3$ indicated a large interaction in the transition state for isomerization between the C—H bonds at this carbon and the orbitals of the C$_1$—C$_2$ bond that was being cleaved. Such an interaction would, of course, be present in 0,0 trimethylene, although considerable hyperconjugative interaction might also be expected in 0,90 trimethylene. The authors further observed that their finding of a substantial β-isotope effect in a system bearing radical-stabilizing substituents suggested that less substituted cyclopropanes might exhibit even larger β-isotope effects.

In 1,2-dideuteriocyclopropane (**6**) a large β-isotope effect could lead to coupled rotation at C$_1$ and C$_2$ being faster than that at C$_1$ and C$_3$ and at C$_2$

Ph ... X
Y ... CN

12a, X = Y = H
  b, X = D, Y = H
  c, X = H, Y = D
  d, X = Y = D

and C_3. In analyzing the experimental data on **6**, Berson considered only α-isotope effects. Thus he expected rotation of a pair of deuterated carbons to be slower than a pair in which only one of the carbons was deuterated. Baldwin and Carter's results imply that the isotope effect on coupled methylene rotation might actually be inverse, $k_H/k_D < 1$, with rotation of the deuterated pair of methylene groups being fastest.

If the isotope effect were actually inverse in **6**, the Hoffmann mechanism would predict a ratio of isomerization to racemization that is less than 1.0. For instance, with $k_H/k_D = 0.95$, the coupled methylene rotation mechanism leads to an expected ratio of isomerization to racemization of 0.98, compared to the experimental ratio of 1.07. Although coupled methylene rotation would still be the dominant process in the stereomutation of **6**, a random intermediate would account for a larger percentage (~20%) of the reaction than if $k_H/k_D > 1$. An inverse isotope effect on coupled rotation would also reduce the dominance of the Hoffmann mechanism in **7**, but to a lesser extent than in **6**.

Doering and Barsa[22] have undertaken a study of whether one-center rotational preferences are transferrable from one disubstituted cyclopropane to another. Optically active *trans*- and *cis*-1-cyano-2-phenylcyclopropane (**13**) and 1-isopropenyl-2-phenylcyclopropane (**14**) were prepared, and the rate constants for one-center epimerization in these compounds were measured and compared with those obtained previously[13] for 1-cyano-2-isopropenylcyclopropane (**15**). Doering and Barsa found in **13** that rotation of cyano is favored over that of phenyl by a factor of 2.47 at 217.8°C. In **15** the ratio favoring rotation of cyano over isopropenyl had previously been found to be 2.20 at the same temperature. If rotational preferences were transferable, the predicted ratio favoring isopropenyl over phenyl rotation would be 1.12 in **14**. The observed ratio was 1.26, measured at 169.5°C. Doering and Barsa noted that, based on a comparison of their data for **13** at 217.8°C with those of Baldwin and Carter[18] at 242.1°C, rotational preferences appear to decrease as the temperature is raised. Therefore, the rotational preference in **14** that would be found at 217.8°C might be in even better agreement with the predicted value of 1.12 than the value of 1.26 measured at 169.5°C.

Doering and Barsa also measured the one-center rotational preference in 1-cyano-2-(phenylethynyl)cyclopropane (**16**). The preference for cyano over phenylethynyl rotation was found to be 1.76, smaller than that for cyano

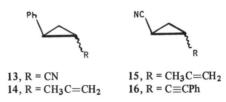

13, R = CN   15, R = $CH_3C{=}CH_2$
14, R = $CH_3C{=}CH_2$ 16, R = $C{\equiv}CPh$

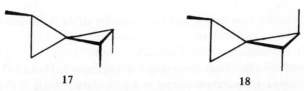

17   18

rotating in competition with either phenyl or isopropenyl. If moments of inertia had a large effect on rotational preferences, then, because phenylethynyl is a longer and heavier substituent than either phenyl or isopropenyl, the preference for cyano rotation observed in **16** should have been greater than that in **13** or **15**. Since just the reverse was found, Doering and Barsa suggested that steric "width" of substituents might be a more important factor than moment of inertia in determining rotational preferences.

The rate constants tabulated by Doering and Barsa for isomerization and enatiomerization in **13–16** again fail to reveal any strong preference for the latter process, involving coupled methylene rotations. Nevertheless, it is true that in all four *trans*-isomers, enantiomerization is slightly faster than one-center epimerization at the carbon that rotates more slowly. In **15**, the molecule previously studied by Doering and Sachdev,[13] coupled rotation in the *trans*-isomer is actually competitive with both one-center rotations that produce the *cis*-isomer. This fact is masked in the Doering-Barsa paper by two typographical errors in the tabulated rate constants. The correct values for the enantiomerization rate constants in **15** are $(k_e)_t$ = 70.3 and $(k_e)_c$ = 92.0, not 22.3 and 29.2.

Gajewski and co-workers[23] have studied the interconversion of the *cis*- and *trans*-1,2,4-trimethylspiropentanes (**17** and **18**) and their C_4 epimers. Assuming cleavage of only the C_1-C_2 peripheral bond, a kinetic analysis reveals a dominance of double over single rotation in ratios ranging from about 1 to 3 for the various interconversions. Correcting the ratios of double to single inversion for generation and destruction of 1,4 proximal methyl interactions leads the authors to conclude that, if there is no preference for *trans*-over *cis*-1,2-dimethyl groups, double inversion is favored by a factor of 3.0 in both the *cis*- and *trans*-compounds.

The stereomutation of 1,2,4-trimethylspiropentane represents the first case where there appears to be a clear preference for coupled methylene rotation in a cyclopropane in which the bond undergoing cleavage is disubstituted. However, the fact that the *cis*- and *trans*-isomers show an equal predilection for double over single rotation does not support the Hoffmann mechanism for methylene motions coupled in the conrotatory sense. If conrotation were preferred, the *trans*-isomers should manifest a greater preference than the *cis* for double over single rotation, since conrotation in the *cis*-isomer entails greater steric interactions than in the *trans*. The fact that the ratio of double to single inversion is apparently similar in both isomers led Gajewski to suggest that conrotation is preferred only in the *trans*-isomer and that, contrary to the Hoffmann prediction, methylene rotation is disrotatory in the *cis*-isomer. This hypothesis requires the same 0,0 trimethylene to be formed from both the *cis*- and *trans*-isomers, but to "remember" the isomer from which it was formed. Thus a 0,0 trimethylene that is formed from a *trans*-isomer in a conrotatory mode is hypothesized to continue to rotate preferentially in this mode until it recloses. As Gajewski points out, such a memory effect cannot be a feature of the potential surface, but must be embodied in the dynamics of the motions on the surface.

C. Trimethylene in Pyrazoline Decompositions

As is noted above, Hoffmann drew support for his proposal of a preferred conrotatory mode of closure in 0,0 trimethylene from the results of Crawford[3] and McGreer[4] on the stereochemistry of cyclopropane formation in pyrazoline decompositions. Both groups found that inversion at one center predominated over retention. If 0,0 trimethylene were formed by pyrazoline decomposition, this would be precisely the stereochemical result predicted by Hoffmann's extended Hückel calculations.

More recently, Bergman and co-workers[24] have carried out a complete analysis of the stereochemistry of pyrazoline decomposition using optically active *cis*- and *trans*-3-ethyl-5-methyl-1-pyrazoline. The absolute stereochemistries and optical purities of reactants and products were established, so that the stereochemical outcome of the reaction could be analyzed quantitatively. The results are summarized in Figure 2.

As found previously by Crawford and McGreer, single inversion predominates. With both the *cis*- and *trans*-pyrazoline this represents roughly 70% of the stereochemical outcome. Since the *cis*-cyclopropane formed from the *trans*-pyrazoline is almost racemic, this product could have been formed from an achiral species like 0,0 trimethylene. However, one enantiomer of the *trans*-

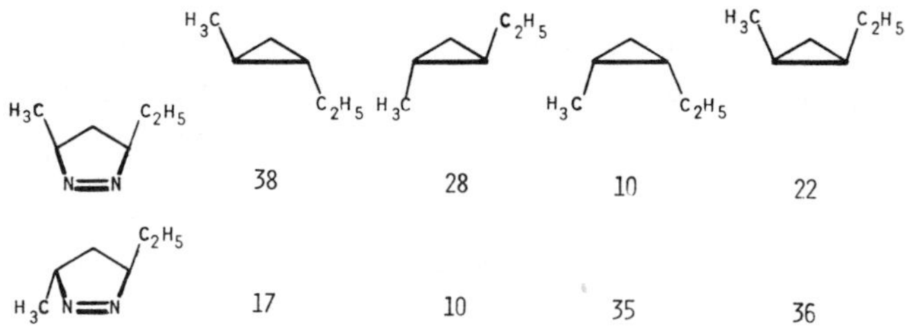

FIGURE 2. Percentage of cyclopropane products formed on pyrolysis of optically active *cis*- and *trans*-3-ethyl-5-methyl-1-pyrazoline.

cyclopropane is formed in excess from the *cis*-pyrazoline. As pointed out by Bergman, the *cis*-pyrazoline should give rise to a sterically less congested 0,0 trimethylene than the *trans*-pyrazoline. Therefore, one would expect more racemic product from the *cis*- than from the *trans*-pyrazoline. Just the reverse is observed. The putative participation of 0,0 trimethylene in these reactions also fails to provide a rationale for the predominant double retention observed in the conversion of *cis*-pyrazoline to *cis*-cyclopropane and for the contrasting double inversion in the transformation of the *trans*-pyrazoline to the *trans*-cyclopropane.

Bergman states, "It is very difficult to devise a unified diradical mechanism of the conventional type (i.e., one involving relatively long-lived, easily described intermediates that undergo a series of product forming steps in competition with various conformational changes) to account for the observations described in this paper." Thus, instead of a mechanism that emphasizes the topological features of the surface on which such intermediates lie, Bergman formulates a possible explanation of his results in more dynamical terms. Bergman seems to imply that the flatness of the energy surface for trimethylene is responsible for the failure of a more traditional mechanistic interpretation of pyrazoline thermolysis.

D. Trimethylene—Intermediate or Transition State?

A thermochemical calculation places the heat of formation of trimethylene about 10 kcal/mol below the transition state for stereomutation of cyclopropane.[25] The calculation equates the heat of formation of trimethylene to that of propane, plus twice the known C—H bond dissociation energy in propane, minus the bond dissociation energy of H_2. The calculation assumes no interaction between the terminal atoms in the trimethylene diradical.

In contrast, Hoffmann's extended Hückel calculations on trimethylene[2] showed only a 1 kcal/mol barrier to closure to cyclopropane. *Ab initio* calculations by Salem et al.[6] and Goddard et al.[7] detected no barrier at all to conrotatory closure of 0,0 trimethylene. Thus quantum mechanics suggests that trimethylene is a transition state in cyclopropane stereomutations, whereas thermochemistry indicates the biradical to be an intermediate with a 10 kcal/mol barrier to closure.

No information on whether trimethylene is an intermediate or transition state in cyclopropane stereomutations is provided by the stereochemical studies discussed above. Indeed, there really are no experimental data in the literature that are relevant to deciding this question unequivocally for the parent system. However, studies on a diradical that may be viewed as a derivative of trimethylene suggest the absence of a large barrier to closure in the lowest singlet state of the parent diradical.

E. 1,3-Cyclopentanediyl

A full account of the study of 1,3-cyclopentanediyl (**20**) by Buchwalter and Closs has recently appeared.[26]   When 2,3-diazabicyclo[2.2.1]heptane (**19**) was irradiated at 5.5 K in a hydrocarbon matrix, or as a powder, an EPR spectrum assigned to the triplet state of **20** was observed. When the sample was warmed, the triplet EPR signal was found to disappear. The kinetics of the signal decay were nonexponential and temperature independent in the range 5.3–20 K. The nonexponential decay was attributed to the distribution of the triplet among different sites in the matrices and the temperature independence to a tunneling process.

The triplet appears to lie below the singlet in **20**, as expected from *ab initio* calculations on the planar diradical.[27] The irreversible decay of the EPR signal prevented a Curie-Weiss study of the signal intensity as a function of temperature, but the triplet spectrum was observed at temperatures as low as 1.3 K. CIDNP experiments provided additional evidence for a triplet ground state. Enhanced absorptions were observed for all the protons in the NMR spectrum of the bicyclopentane (**21**) produced when **19** was photolyzed at room temperature in the presence of a triplet sensitizer. This result was interpreted as indicating a triplet biradical precursor to **21** in which intersystem crossing to the lowest singlet state occurs from the *upper* triplet sublevel.

Wilson and Geiser[28] have found that when the sensitized photolysis of **19** is carried out in the presence of oxygen, the formation of **21** is suppressed, and 2,3-dioxabicyclo[2.2.1]heptane is obtained. Since oxygen apparently has little effect on the production of **21** in the direct photolysis, the endoperoxide is indicated to be formed from reaction of triplet **20** with O_2. These experiments provide independent evidence for the formation of the triplet diradical on sensitized photolysis of **19** and show that the lifetime of **20** is sufficiently long that the triplet diradical can be trapped chemically before closing to **21**.

Buchwalter and Closs fitted the temperature dependence of the rate of triplet EPR signal decay with an Arrhenius equation containing a tunneling correction. The authors noted that although sizable uncertainties existed in the A factor and barrier width required to fit their data, an activation energy of 2.3 kcal/mol was necessary to obtain the onset of temperature dependence for EPR signal decay around 20 K. They interpreted this value as the energy required for the metastable triplet to cross to the singlet surface leading to **21**. As noted by Buchwalter and Closs, the existence of a 9 kcal/mol barrier to ring closure in **20**, predicted by a thermochemical analysis, is difficult to reconcile with the observed facility of ring closure at very low temperatures.

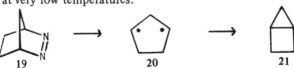

19   20   21

F. Summary

The most sophisticated quantum mechanical calculations on trimethylene predict the 0,0 species to be rather floppy stereochemically, with coupled conrotatory methylene motions only marginally favored over disrotation or rotation of a single methylene. In 1,2-dideuteriocyclopropane, a molecule unperturbed by the presence of substituents, coupled methylene rotation does, indeed, appear to be the dominant process. The extent to which other processes compete cannot be determined without knowing the isotope effect on the coupled rotation mechanism. Based on the isotope effects found in 12 for single rotations, it seems quite possible that $k_H/k_D < 1.1$ in 6 and is perhaps even inverse. If CHD does rotate in preference to CH_2, then double rotation is not the sole process for stereomutation in 6. Given the small amount by which conrotation is predicted to be favored over dis- and monorotation, competition from these other processes is expected at the temperatures required to effect stereomutation. Provided that the coupled methylene rotation observed is conrotation and not disrotation, experiment and theory appear to be in excellent agreement.

In di-, tri- and tetrasubstituted cyclopropanes there is evidence of coupled methylene rotation being dominant only in the 1,2,4-trimethylspiropentanes. However, it is hard to reconcile the preference for coupled methylene rotation observed in both the cis- and trans-isomers with conrotation. Similarly, while some of the stereochemical results observed in substituted pyrazoline decompositions can be rationalized by the postulate of conrotatory closure of a 0,0 trimethylene, all of the data cannot be adequately encompassed by such a mechanistic postulate.

Of course, in substituted trimethylenes there might be many effects present that could easily obscure the small preference for conrotation calculated and apparently found in the parent 0,0 trimethylene. Steric effects, not present in the parent diradical, might result in changes in both the potential surface and the dynamics on it. In addition, delocalization by substituents of the electrons in the bond being cleaved would be expected to decrease the interaction of these electrons with the central methylene group. Thus a diminished preference for conrotation might be anticipated. The results of extended Hückel calculations are consistent with this expectation.[29]

Perhaps the most surprising result in cyclopropane stereomutations is, therefore, the large preference for coupled rotations found in 1-phenyl-2-deuteriocyclopropane (7). If through-bond coupling is responsible for the observed preference for coupled methylene rotation in 6, this preference should be reduced by the presence of the phenyl in 7. However, rotation of the phenyl-bearing carbon is apparently almost invariably accompanied by rotation of one of the other two methylene groups. Even if, as seems nearly certain, "wrong" bond cleavage makes no contribution to the stereomutation of 7, four out of every five rotations at C_2 are accompanied by rotations at C_1. It would thus appear that substitution of phenyl for hydrogen does little to reduce the contribution of coupled rotations to the stereomutation.

Another surprising aspect of the distinct preference for coupled methylene rotations in **7** is that this finding stands in marked contrast to the apparently random rotations observed in 1-vinyl-2,3-dideuteriocyclopropane (**3**). Since phenyl and vinyl are usually considered to be rather similar substituents, the different behavior of **3** and **7** is, as noted by Berson et al.,[17] difficult to explain. Further studies on monosubstituted cyclopropanes would be most desirable.

The low activation energy found for ring closure from the triplet state of 1,3-cyclopentanediyl (**20**) can be taken as indicative of the correctness of another quantum mechanical prediction regarding trimethylene, the absence of a barrier to ring closure in the lowest singlet state. Nevertheless, it should be noted that the results on **20** really provide no direct information about the existence of a barrier on the singlet surface. Even if there were a barrier, crossing of the triplet to the singlet surface might occur at a geometry where the singlet could collapse without activation to **21**. However, the existence of a barrier on the singlet surface would then require a very steep initial increase in energy of the singlet as the geometry of **20** is distorted toward that of **21** and a very slow increase in the energy of the triplet. In the absence of any reason for expecting such a situation to obtain, it seems more likely that the thermochemical predictions of the existence of barriers to closure in both **20** and in the parent trimethylene are wrong. Several possible sources of error in the thermochemical predictions have been identified.[1, 6, 7, 13, 16]

III. TETRAMETHYLENE

A. Introduction

Hoffmann et al.[30] were again the first to calculate a potential surface for the tetramethylene diradical ($\cdot CH_2CH_2CH_2CH_2 \cdot$). Using the extended Hückel method, Hoffmann and his co-workers found two local minima, corresponding to conformations *gauche* and *trans* about the bond between the central carbons. Especially at the *trans* local minimum, a strong preference was found for the singly occupied AOs on the terminal carbons to eclipse the central C—C bond, presumably because this conformation allowed maximum through-bond interaction between these AOs. The two local minima were, however, found to be unstable with respect to fragmentation to two ethylenes, so that no true secondary minima appeared on the surface.

Ab initio CI calculations carried out by Segal[31] did find small barriers to bond breaking in the *gauche*- and *trans*-conformations of singlet tetramethylene,

22

so these conformations of the diradical do represent secondary minima on the *ab initio* surface. From the *gauche* minimum, closure to cyclobutane was found to require less energy (2 kcal/mol) than fragmentation to ethylene (3.6 kcal/mol). From the *trans* minimum, which was calculated to lie 1.1 kcal/mol below the *gauche*, fragmentation was found to require 2.3 kcal/mol, 2.4 kcal/mol less than rotation from the *trans* minimum to the gauche.

Rotational barriers for the terminal methylene groups were not reported. However, the optimal geometry for the *gauche* minimum has the AOs on these atoms rotated 65.4° away from eclipsing the central C—C bond to maximize the small amount of residual through-space interaction between these AOs. It seems probable, therefore, that in tetramethylene, as in trimethylene, extended Hückel calculations overestimate the energetic consequences of through-bond interactions, and *ab initio* CI calculations would find much smaller rotational barriers for the terminal methylene groups. The very large barrier found by an INDO calculation[32] almost certainly is an artifact, caused by the absence of CI. The similarly unphysical results obtained for the two C—H bond dissociation energies in forming tetramethylene from butane[33] have the same origin. CI is essential for removing the ionic terms from the wave functions for singlet diradicals.[1]

That the rotational barriers at the terminal carbons of tetramethylene are, in fact, small was suggested by the early studies of Gerberich and Walters[34] on the thermal fragmentation of *cis*- and *trans*-dimethylcyclobutane (22). As expected from preferential cleavage of the most substituted bond, the major product was propene; but ethylene and 2-butene were also obtained. The 2-butene formed was 64% *cis* in the decomposition of *cis*-22 and 85% *trans* in the fragmentation of the *trans*-22. Thus methylene rotation appeared to be competitive with fragmentation in the putative diradical intermediate.

The possibility that fragmentation of **22** occurs by an orbital symmetry-allowed [$_\sigma 2_s + _\sigma 2_a$] pathway[35] was unlikely because of the observation that *cis* $\rightleftarrows$ *trans* isomerization of **22** occurs along with olefin formation, albeit at a slower rate. Further evidence against the utilization of the orbital symmetry-allowed pathway was obtained by Frey et al.[36] with **23**, and by Baldwin and Ford[37]

23   **24**

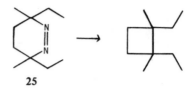

25

with **24**. The cycloalkene formed by fragmentation of either molecule is incapable of readily accommodating a *trans* double bond. Therefore, if the $[_\sigma 2_s + _\sigma 2_a]$ pathway were followed, the acyclic olefin fragmentation product would be formed with complete inversion of stereochemistry. Instead, Frey et al. actually observed a very slight preference for formation of *cis*-2-butene from *cis*-**23** and a 3:1 ratio favoring formation of *trans*-2-butene from *trans*-**23**. Interestingly, Baldwin and Ford did, in fact, find a preference for the formation of *trans*-dideuterioethylene from **24**, but the *trans/cis* ratio was only about 6:4.

Loss of stereochemical integrity does not always accompany formation of tetramethylene diradicals. Bartlett and Porter[38] studied the decomposition of the *cis*- and *trans*-isomers of the azo compound **25**. Formation of the corresponding *cis*- and *trans*-cyclobutane derivatives occurred stereospecifically in the thermal reaction and with >95% retention of configuration on direct photolysis. The marked drop in stereoselectivity observed on triplet-sensitized photolysis was attributed to a longer lifetime for the triplet diradical.

Berson and co-workers[39] found no geometrical isomerization or enantiomerization in *trans*-tetramethylcyclobutane-d_6 (**26**) during its thermal fragmentation to isobutylene-d_3. The contrast between this result and the geometrical isomerization observed by Gerberich and Walters during the fragmentation of **22** was attributed to the increase in substitution at the terminal atoms in **26**. Jones and Fantina[40] compared the behavior of **27** and **28**. As expected, geometrical isomerization was observed during fragmentation in **27**, but not in that of the more highly substituted **28**.

Similar effects of terminal substituents on stereochemistry have been observed in the 1,4-diradicals generated in Norrish type II reactions. Stephenson and co-workers[41] carried out the photolysis of the methyl esters of *erythro*- and *threo*-3,4-dimethyl-6-ketoheptanoic acid (**29**). In the presence of isoprene to

26, X = CH$_3$, Y = CD$_3$
27, X = H, Y = CO$_2$CH$_3$
28, X = CH$_3$, Y = CO$_2$CH$_3$

29, R = CH$_2$CO$_2$CH$_3$ 30, R = CH$_2$CO$_2$CH$_3$
31, R = D   32, R = D

quench the triplet photoreaction, the methyl esters of *cis*- and *trans*-3-methyl-3-pentenoic acid (30) were formed with 90:1 retention of stereochemistry. In the absence of quencher, the triplet photoreaction proceeded with nearly random stereochemistry, as might have been expected from the results of Bartlett and Porter on the sensitized decomposition of 25.

Casey and Boggs[42] also found that photolysis of *erythro*- and *threo*-4-methyl-2-hexanone-5-d_1 (31) gave a nearly stereorandom mixture of *cis*- and *trans*-2-butene-2-d_1 (32) in the absence of triplet quencher, but 5–10% of the olefin was still found to have lost its stereochemical integrity in the presence of piperylene. The latter result suggests that rotation is faster in the singlet 1,4-diradical formed from 31 than in the singlet diradical formed from 29. A rationale is again provided by the difference in substitution. Hydrogen abstraction by the excited carbonyl group in 29 generates a tertiary radical center, while in 31 the radical center produced is secondary.

B. Recent Developments in Tetramethylene Chemistry

Since tetramethylene diradicals can apparently be generated by cyclobutane fragmentations, azo decompositions, Norrish type II photoreactions, and alkene cycloadditions,[43] Stephenson and Brauman[44] raised the question of whether the behavior of the diradicals generated by different methods might, in fact, be different. The answer to this question, at least for 3-methyl-1,4-pentanediyl, is negative, based on comparison of Gerberich and Walter's results for fragmentation of 1,2-dimethylcyclobutane (22) with recent studies of the cycloaddition of ethylene to 2-butene, and with the decomposition of the appropriate azo compounds at the same temperature.

Scacchi, Richard, and Bach[45] studied the cycloaddition of ethylene to *cis*- and *trans*-2-butene at 12 atm pressure around 430°C. A simple kinetic scheme was proposed, involving the equilibration of cisoid and transoid diradical intermediates with the reactants and the *cis*- and *trans*-dimethylcyclobutane (22) products. Making the steady-state assumption for the concentrations of the diradicals, the authors derived an equality between the ratio of the 2-butene fragmentation products formed from *cis*- and *trans*-22 and the ratio of *cis*- to *trans*-22 formed by ethylene cycloaddition to each of the 2-butenes. Their experimental ratios of cycloaddition products gave a number that was the same as that obtained from the ratios of fragmentation products measured by Gerberich and Walters. This agreement was taken as evidence that the two reactions studied are, in fact, just the reverse of each other.

From the combined data, Scacchi et al. derived the ratios of rotation to cyclization, and cyclization to fragmentation for the cisoid and transoid diradicals. The ratios were all found to be close to unity, with the rates in the order fragmentation > cyclization > rotation for the transoid diradical. In the cisoid diradical fragmentation was found to occur at about the same rate as cyclization, with rotation to the transoid conformation being faster than both. From the temperature dependence of the equilibrium constant between the diradical conformations, Scacchi et al. deduced an enthalpy difference of 1.5 ± 1.0 kcal/mol, with the transoid conformation being lower in energy.

Dervan and co-workers[46] carried out the decomposition of *cis*- and *trans*-3,4-dimethyltetrahydropyridazine (33) in the gas phase and analyzed the data using the scheme shown in Figure 3. Since the ratio of *trans*-2-butene to *trans*-1,2-dimethylcyclobutane was higher when the *trans*-azo compound was decomposed than when the *cis* was pyrolyzed, they assumed that the excess *trans*-alkene was formed by a process that did not involve the diradical. A similar stereospecific component was assumed to account for the excess *cis*-alkene formed from the *cis*-azo compound. Correcting their data for the amount of alkene apparently formed without the intervention of a diradical intermediate, the authors obtained ratios of rate constants for diradical rotation, fragmentation, and closure that were in good agreement with those determined by Scacchi et al. at about the same temperature. Unless the agreement between the two sets of data is fortuitous, it seems reasonable to infer that a diazenyl radical is not a long-lived intermediate under Dervan's experimental conditions. The agreement is permissive evidence that tetramethylene diradicals behave similarly when they are generated under similar conditions from different precursors.

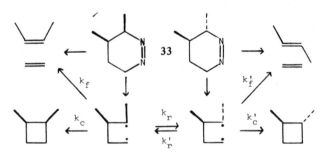

FIGURE 3. Mechanistic scheme for analyzing pyrolysis of *cis*- and *trans*-3,4-dimethyltetrahydropyridazine (33).

34   **35**   **36**

In the same paper,[40] Dervan and co-workers reported the decomposition of the *cis*- and *trans*-isomers of 3,6-dimethyltetrahydropyridazine (**34**). The ratio of retained to inverted stereochemistry in the 1,2-dimethylcyclobutane products was 1.7 for both isomers of **34** at 439°C in the gas phase. At 306°C the ratio increased to 1.9, starting from the *cis*-isomer of **34**, and to 2.2, starting from the *trans*-isomer. Since Bartlett and Porter decomposed **25** at 148°C in solution, some of the difference in stereochemical outcome between **34** and **25** might be attributable to the different experimental conditions under which these azo compounds were decomposed. Nevertheless, based on the rather modest temperature dependence of stereochemical outcome observed in **34**, it seems more probable that the difference in substitution is the major factor responsible for making cyclobutane formation so much more stereospecific in **25** than in **34**.

Dervan and Uyehara[47] also studied the decomposition of the *cis*- and *trans*-isomers of the pyrrolidyl nitrenes **35** and **36** in solution at 120°C. The ratios of retention to inversion of stereochemistry in the dimethylcyclobutane products were higher by factors of 2–4 than when the azo precursors, **33** and **34**, of the two biradicals were pyrolyzed in the gas phase at 439°C. However, when **35** and **36** were generated and decomposed in the gas phase at this temperature, the ratios of products with retained and inverted stereochemistry were very similar to those obtained with the azo compounds. These findings suggest that the tetramethylene diradicals, presumed to be generated from these two different types of precursors, behave similarly. Further, the results demonstrate the role played by temperature and phase on the stereochemistry observed in reactions involving tetramethylene diradicals.

Dervan's results bear on an earlier finding of Stephenson and Gibson,[48] that pyrolysis of the trimethylcyclobutanol stereoisomers **37** at 500°C in the gas phase leads to substantially less stereospecific formation of *cis*- and *trans*-2-butene than does Norrish type II cleavage of excited singlet **31** in solution at room temperature. Since the same 1,4-diradical should be generated in both reactions, Stephenson concluded that the behavior of this diradical depends on

37   **38**

39   40

its mode of generation. However, Dervan's finding that stereospecificity is decreased on going from low temperatures in solution to high temperatures in the gas phase tends to cast doubt on Stephenson's interpretation that thermally and photochemically generated diradicals behave differently. A difference in reaction temperatures may also account for the fact that the ratio of isomerization to cleavage is larger when 1,2-diphenylcyclobutane (38) is pyrolyzed at 200°C than when it is photolyzed at room temperature.[49]

Santilli and Dervan[50] have carried out a study of the stereochemical behavior of the parent tetramethylene diradical, using the azo compound 39, in which deuterium atoms have replaced the methyl groups of 33 as stereochemical markers. The data were analyzed using a scheme similar to that shown in Figure 3, except the rate constants for rotation, closure, and fragmentation for both conformations of the diradical were assumed to be the same. The amount of excess cis-1,2-dideuterioethylene, formed without intervention of the cisoid diradical, could then be established from the ratio of trans-olefin to trans-1,2-dideuteriocyclobutane. At 439°C the ratio of cleavage to closure in the diradical was found to be 2.2, slightly larger than the average value of 1.7 observed for the diradicals produced from 33 at the same temperature. However, the ratio of rotation to closure in the parent diradical was 12, an order of magnitude larger than the ratio observed[45,46] when one terminal atom is secondary and at least 2 orders of magnitude larger than when both terminal atoms are tertiary.[38]

Chickos[51] has also generated the parent tetramethylene diradical with two deuterium atoms as stereochemical markers. On pyrolysis of cis- and trans-1,2-dideuteriocyclobutane (40), he found the ratio of ethylene cleavage products to isomerized cyclobutane to be 3.0. This is slightly smaller than the value of about 5 expected from the ratios of rate constants obtained by Santilli and Dervan. Chickos' finding of complete isomerization in the dideuterioethylene formed also agrees qualitatively but not quantitatively with the results of Santilli and Dervan, after correction of their data for the fraction of ethylene formed without intervention of a diradical intermediate.

41   42   43

Doering and Guyton[52] have examined the stereochemical behavior of the presumed 1,4-diradical intermediates (42) in the dimerization of cis-1,2-dideuterioacrylonitrile (41) to the stereoisomers of the 1,2-dicyanocyclobutane 43. The stereoisomers of 43 with cis and with trans cyano groups were separated by GLC. The amount of material with each of the three possible proton stereochemistries was then determined by NMR for each of the two GLC-separated isomers. Since some cis → trans isomerization was detected in the 41 recovered from the reaction, the amounts of the six stereoisomers of 43 were corrected to 100% 41.

The results were analyzed in terms of rotations in the diradicals (42) formed with erythro and threo stereochemistry at the newly created C–C bond. It was found that the relative amounts of the three erythro products, as well as those of the three threo products, were predicted quite accurately by a purely statistical analysis. Thus, if x is the fractional probability of one cyano-bearing carbon in 42 rotating before closure, the ratios of products involving zero, one, and two such rotations are given by $(1 - x)^2 : x(1 - x) : x^2$. In the erythro series, the value of x that best fit the experimental data at 209.5°C was 0.38, and at 246.4°C the best value of x was found to be 0.41. In the threo series at these temperatures, the best values of x were found to be, respectively, 0.34 and 0.36. The larger values of x in the erythro series were attributed to a preference for products with trans cyano groups. The increase in x with temperature suggests that rotation requires a higher activation energy than closure.

Doering and Guyton also examined pyrolysis of the cyclobutane 44 to 2-deuterioacrylonitrile (45). They found cleavage to be twice as fast as cis → trans isomerization in 44. The cleavage product was a 60.2:39.8 mixture of cis- to trans-45 at short reaction times. This ratio is very close to that predicted for the reverse of the dimerization processes leading to products with cis cyano groups.

C. Summary

As in the case of trimethylene, substituents are not just stereochemical markers, but have a profound effect on the stereochemistry observed in reactions involving tetramethylene diradicals. In trimethylene, substituents mask the preference for coupled methylene rotations that is found in the parent diradical. In tetramethylene, terminal substituents actually increase stereoselectivity by slowing the rate of terminal methylene rotation relative to the rates of cleavage and closure.

It appears that tetramethylene diradicals are intermediates in alkene cyclo-additions and in the reverse of these reactions, cyclobutane fragmentations. The same diradicals can be generated by loss of nitrogen from azo compounds, although a stereospecific fragmentation pathway that circumvents the diradical is competitive. When similarly substituted tetramethylene diradicals are generated under similar conditions by any of the above three routes, their behavior is similar. At least at secondary centers, terminal methylene rotation is found to increase, relative to cleavage and closure, as the temperature increases. The higher stereoselection observed in singlet tetramethylene diradicals generated photo-chemically at room temperature compared to those formed by pyrolyses may be due entirely to this temperature effect. The low stereoselectivity observed when even highly substituted tetramethylene diradicals are produced in the triplet state is attributable to reduced rate constants for cleavage and closure, relative to rotation, since formation of ground state products requires intersystem crossing.

While much is now known about terminal methylene group rotation in tetra-methylene, and about how the relative rate is affected by substitution and by temperature, little experimental information is available regarding conformation and rotation about the central C—C bond. For instance, does cycloaddition of two alkenes lead initially to a *trans*- or *gauche*-geometry about this bond? That at least the latter is preferred to a *cis*-conformation may be inferred from the find-ing of Goldstein and Benzon[53] that cleavage of bicyclo[2.2.0]hexane takes place exclusively from the chair form of the presumed diradical intermediate. *trans*-Tetramethylene is calculated by Segal to be lower in energy than the *gauche* form, but to have a larger barrier to closure than to cleavage. Is formation of the *trans*-diradical in cycloadditions witnessed principally by isomerization of "unreacted" olefin? If, as suggested by Hoffmann's extended Hückel calculations, there were a sizable barrier to terminal methylene rotation in the *trans*-conformation, reversible formation of this tetramethylene geometry would, of course, go undetected. How does the ease of terminal methylene rotation actually depend on the conformation about the central C—C bond? Obtaining answers to these questions will further test the ingenuity of experimentalists.

IV. CYCLOBUTADIENE

A. Introduction

Unlike the case of tri- or tetramethylene, the fact that cyclobutadiene (CBD) is a diradical is not at all apparent from its structure (Figure 4). However, it should be recalled that [4n] annulenes at their most symmetrical geometries contain two nonbonding MOs that are occupied by two electrons.[54] Therefore, at a square geometry CBD is indeed a diradical.

FIGURE 4.   Square and rectangular cyclobutadiene (CBD).

Whether cyclobutadiene is, in fact, a square molecule has been a question of intense interest to theoreticians and experimentalists since the early seventies, when the IR spectrum of the matrix isolated molecule was first obtained independently by the groups of Krantz[55] and Chapman.[56] Only three bands were observed in the region below 2000 cm^{-1}, exactly the number expected for a square molecule. The fact that no EPR signal was observed for CBD could not be taken as evidence against a square molecule, for it was shown theoretically that, in violation of Hund's rule, CBD should have a singlet ground state, even at square geometries.[57]

B. Recent Results

Two reviews discuss recent developments in the fascinating story of CBD;[58, 59] therefore, only a brief recap is given here. In 1976, Maier et al.[60] and Masamune et al.[61] showed that the 653-cm^{-1} band is absent from the IR spectrum of matrix-isolated CBD when precursors are used that do not extrude CO_2. Therefore, this band does not belong to CBD but to CO_2 in proximity to CBD in the matrix. Krantz et al.[62] showed that both the 653-cm^{-1} band, and the 662-cm^{-1} band for uncomplexed CO_2 are shifted to lower frequency when a labeled precursor is used that generates $^{13}CO_2$.

With the assignment of the 653-cm^{-1} band to complexed CO_2, the IR spectrum of CBD contains too few bands for even a square molecule. In 1978 Masamune et al.[63] found two previously undetected weak bands, bringing the total number of bands below 2000 cm^{-1} to 4. This is one more than allowed for a square molecule. Masamune was able to fit the spectra of CBD and CBD-d_4, which he also prepared, with a set of force constants appropriate for a rectangular molecule with alternating single and double bonds. The good fit obtained is compatible with a rectangular equilibrium geometry for CBD, but the IR spectrum only serves rigorously to exclude a square one.

It should be noted that prior to the publication of Masamune's results, sophisticated *ab initio* calculations were carried out by three separate groups.[64-66]

        46a   R = C(CH$_3$)$_3$ 46b

All three groups used extended basis sets and included CI for both the σ- and π-electrons. The results of all three sets of calculations were generally in agreement with those obtained previously[67] with a minimal basis set and only π CI. A singlet was found to be the ground state at all geometries investigated, and the equilibrium geometry of this singlet was found to be rectangular. The barriers calculated for interconversion of the two rectangular minima via a square transition state were on the order of about 10 kcal/mol.

The major challenge currently facing experimentalists is the measurement of this barrier. Maier et al.[68] have found that the ^{13}C NMR spectrum of tri-*tert*-butyl-CBD (**46**) shows one line for C_1 and C_3 of the ring at temperatures from -60 to $-190°C$. It is possible that the absence of splitting of this resonance into two as the temperature is lowered is due to a much lower barrier in **46** than that computed for the parent molecule. Steric repulsion between the *tert*-butyl groups might be minimized at a square geometry, thus reducing the barrier in **46** from that in CBD itself. On the other hand, it may be that C_1 and C_3 just happen to have chemical shifts that differ negligibly, even when interconversion of **46a** and **46b** is frozen out.

V. TRIMETHYLENEMETHANE

A. Introduction

One of the reasons that trimethylenemethane (TMM) is such an interesting diradical is that, as a result of the Coulombic repulsion between the two open-shell electrons, the bonding in the triplet ground state ($^3A_2'$) differs from that in the lowest singlet ($^1E'$). In the triplet all three of the resonance structures shown in Figure 5 contribute equally. However, only the latter two contribute to one component.($^1E_x'$) of the degenerate singlet state, and the other component ($^1E_y'$) consists largely of the first structure with just a small admixture of the second and third.[69] Consequently, there is a strong first-order distortion away from D_{3h} symmetry in the lowest singlet state,[70] as predicted by the Jahn-Teller theorem. It is found that the $^1E_x'$ component can further lower its energy slightly by rotating the unique methylene out of conjugation with the rest of the π-system.

As is discussed in two recent reviews,[58, 71] the experimental data on TMM are in good qualitative agreement with the theoretical predictions regarding this

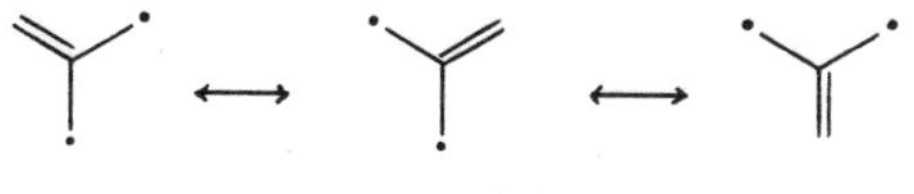

FIGURE 5.   Resonance structures for planar trimethylenemethane (TMM).

diradical. However, there is an apparent discrepancy between theory and experiment concerning the magnitude of the energy difference between the triplet ground state and the lowest singlet state. The best theoretical estimate of the energy difference between the planar triplet and the bisected singlet, where one methylene is twisted out of conjugation, is 14 kcal/mol.[72] Experimental measurements of singlet–triplet energy separations give much lower values.

B. Recent Results

Dowd and Chow[73] measured the temperature dependence of the rate of decay of the EPR signal from triplet TMM. Depending on the precursor used, they obtained different activation energies, but the highest value was only about half of the calculated difference between the planar triplet and bisected singlet.

It should be noted, however, that the bisected geometry is just a local minimum on the singlet surface; the global minimum occurs at the geometry of methylenecyclopropane. Dowd and Chow's experiment presumably measured the energy of activation required to reach the global singlet minimum from the triplet. This would correspond to the energy required to reach the lowest intersection of the two surfaces, assuming that triplet–singlet crossing has about the same efficiency at this point as at higher energy crossing points and that no further barrier to closure to methylenecyclopropane is encountered on the singlet surface. There is currently no compelling reason to believe that the surface intersection of lowest energy occurs at the bisected geometry. Therefore, Dowd's experiment places a lower limit on the separation between the planar triplet and bisected singlet; but the experiment does not necessarily measure the actual energy difference between them.

In a very different type of experiment Platz and Berson[74] obtained a maximum value of 3.5 kcal/mol for the singlet–triplet energy separation in the 6,6-dimethyl derivative of 2-methylenecyclopentane-1,3-diyl (48). The diradical was generated by photolysis of the azo compound 47 in a frozen solution containing fumarate or maleate as diylophile. On thawing of the solution, a mixture of fused (49) and bridged (50) adducts was obtained. From previous experiments it was known that the singlet diradical gives predominantly 49, while the triplet gives nearly an equal mixture of 49 and 50. From the composition of the mixture of 49 and 50 obtained, it was thus possible to deduce the amount of product formed from each state in what was assumed to be an equilibrated mixture of singlet and triplet diradicals (48). Since the rate constant for triplet trapping had been measured previously by EPR, it was possible to calculate a maximum value for the ratio of triplet to singlet diyl at equilibrium by assuming that the singlet was trapped at a diffusion-controlled rate. The equilibrium constant obtained from this elegant experiment yielded the maximum value of the free energy difference between the two states cited above.

47   48   X = CO₂CH₃   49   50

An important question about this experiment concerns the nature of the trappable singlet. If the trappable singlet were not **48**, but instead the ring-closed species 2-isopropylidenebicyclo[2.1.0]pentane (**51**), the experiment would only provide a lower limit to the singlet–triplet gap in **48**. Recent results from Berson's laboratory provide permissive evidence for the involvement of **51** in the Berson-Platz experiment.

Berson and co-workers[75] succeeded in generating **51** by two different routes. The first of these involved addition of dimethylvinylidene to cyclobutene at −78°C. Subsequent addition of methyl acrylate gave the trapping products previously thought to be characteristic of the singlet diyl. However, since it seemed very unlikely that the singlet diyl would persist for several minutes in solution, the intermediacy of **51** was indicated.

Confirmation of the persistence of **51** at −78°C was obtained by direct observation. Photolysis of the azo compound **47** at this temperature gave **51** in virtually quantitative yield. The NMR spectrum of **51** showed different chemical shifts for the two types of protons on the ethano bridge, indicating that **51** was not in rapid equilibrium with the planar diyl (**48**) at this temperature. However, on warming of the solution, **51** disappeared and the known dimers of **48** were formed. The rate of disappearance followed first-order kinetics, and from the temperature dependence, Arrhenius parameters of $E_a = 13.2$ kcal/mol and $\log A = 9.2$ were obtained. It was suggested that the low A factor signified a transition state for the rate-determining step in which opening of **51** to the triplet diyl occurred, with or without the intermediacy of singlet **48**. Some interesting substituent effects were observed on the rate of disappearance of three other derivatives of 2-methylenebicyclo[2.1.0]pentane, which were also prepared by photolysis of the corresponding azo compounds at −78°C.

51

Interestingly, when **47** was irradiated at much lower temperatures in rigid media, the dimers of **48** were the major products observed on thawing the solid solutions by warming to $-78°C$. Since **51** is stable at these temperatures, the very small amount of it detected suggests that the singlet diyl, presumed to be generated initially by photolysis of **47** under these conditions, undergoes intersystem crossing to the triplet faster than closure to **51**. A possible interpretation is that the activation energy for closure of singlet **48** to **51** is higher than that for intersystem crossing to triplet **48**, but the *A* factor for the former process is also larger than that for the latter. This would allow intersystem crossing to be the more rapid process at low temperatures, whereas closure would dominate at higher temperatures.

A methylated derivative (**55**) of **51** was found by Salinaro and Berson[76] to be generated by a third route. Intramolecular double bond addition of the carbenoid **52** gave **53** as a presumed intermediate. Under the reaction conditions, **53** dimerized and also opened to the diyl **54**, which itself gave dimers. The origin of each of the two different types of dimers was confirmed by labeling studies. At $-30°C$ where opening of **53** to **54** competed more effectively with dimerization of **53** than at $-78°C$, **52** was again generated, but the reaction mixture was then cooled to $-78°C$. After 8 min, methyl acrylate was added, and a 20% yield of trapping products of **55**, analogous to those formed from **51**, was obtained.

The finding that **51** is a stable molecule at $-78°C$ and gives the same trapping products as those previously attributed to the singlet diyl (**48**) has serveral ramifications. First, it suggests that **51** may have been the singlet species trapped in the Berson-Platz experiment. Since under their conditions any **51** formed in the photolysis might not have equilibrated with the triplet before trapping, their assumption of a spin-equilibrated mixture may have been invalid. Even if equilibration between triplet **48** and **51** had been achieved in the Berson-Platz experiment, the energy difference between them, although certainly an interesting number, could not be compared with the calculated energy difference between planar triplet and bisected singlet TMM.

Second, if **51** is, in fact, the trappable singlet in the cycloaddition reactions that have been studied, the interpretation[77] of these experiments in terms of **48** must be modified. The formation of fused adducts of **51** with olefins and bridged adducts with dienes is consistent with the expectations of orbital symmetry,[35] a fact that did not escape Berson et al.[77] The different pattern of regioselectivity that they observed when the methyl groups in **51** were replaced

52   **53**   **54**   **55**

by methoxyls could have several origins, one possibility being the increased intervention of the diyl in the cycloaddition reactions. The radical-stabilizing ability of methoxyls is well documented in thermal rearrangements.[20, 78] An increased stabilization of the diyl might also account for the rapid deazatization of the dimethoxyl analogue of azo compound 47.

The third and most exciting ramification of the isolation of 51 is that the chemistry of the highly strained 2-methylenebicyclo[2.1.0]pentane ring system can now be studied in detail. Berson et al.[75] have suggested that the "zero bridge" in 51 has a negative bond dissociation energy, since all the evidence points to the triplet diyl being more stable than the bicyclic compound. The facile opening of 51 and derivatives to singlet diyls provides a unique opportunity to examine thermal rearrangements involving TMM intermediates at low temperatures. Berson and co-workers[79] have already begun to exploit this opportunity.

They prepared a mixture of the labeled exo- and endo-bicyclo[2.1.0]-pentanes (57 and 58) by photolysis of the azo compound 56 at -78°C. Some scrambling of the label in the exocyclic methylene group of 56 was observed in 57 and 58. Both showed two resonances for the exocyclic methylene protons in a 2:1 ratio, instead of the single resonance expected if the bicyclopentanes had been formed stereospecifically. On warming the mixture to -60°C, the minor isomer, which was assigned structure 57 on the basis of its NMR spectrum, rearranged to the major isomer (58). No further scrambling of the label was detected during the rearrangement. At -40°C scrambling of the label in 58 was observed, and at temperatures above 5°C dimerization occurred.

Berson pointed out that the ability to detect scrambling of the label during the rearrangement of 57 and 58 is limited by the fact that 57 constitutes only 25% of the material and both 57 and 58 are already 50% scrambled. Nevertheless, the qualitative result that substantial stereointegrity about the exocyclic double bond is maintained during rearrangement and, perhaps to a lesser extent, during formation of 57 and 58 from 56, stands in contrast to previous results in the literature regarding similar reactions.

Roth and Wegener[80] and Gajewski and Chou[81] have carried out experiments on endo-exo-isomerization of substituted 2-ethylidenebicyclo[3.1.0]hexanes.

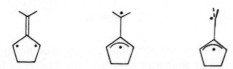

FIGURE 6. Depiction of the bonding in the $'A_1$ and $'B_2$ wave functions for planar 2-methylenecyclopentane-1,3-diyl and in the $'B_1$ wave function for the bisected diyl.

They both found that the stereointegrity about the exocyclic double bond was substantially lost during *endo–exo*-isomerization. Their results are consistent with the small preference calculated for the bisected geometry in singlet TMM.

There are a number of differences between Berson's experiment and those of Roth and Gajewski, any one of which could rationalize the apparent difference in results. If rotation of the exocyclic methylene group required an activation energy, the disparity in temperatures between the experiments would provide an explanation. Another difference is Berson's use of deuterium instead of methyl as the stereochemical probe at the exocyclic methylene group. The lesser substitution in Berson's diyl would be expected[70] to stabilize a $^1A_1$ wave function with respect to a $^1B_2$ (Figure 6), thus inhibiting exocyclic methylene rotation to the bisected geometry of the diyl. Finally, the difference in ring size might have some effect. The smaller distance expected between the nonbonded carbons in the five-membered ring of Berson's diyl should also serve to stabilize $^1A_1$ with respect to $^1B_2$.[70]

It must be noted that, even though the isolation of 2-isopropylidenebicyclo-[2.1.0]pentane (**51**) calls into question the previous interpretation of the Berson-Platz experiment, there exists independent evidence for a small singlet–triplet gap in diradical **48**. Turro, Berson, and co-workers[82] have carried out spectroscopic studies of the UV absorption and emission from **48** and the corresponding 6,6-diphenyl diyl. They found a short wavelength band in each whose position, allowing for the effect of the substituents, was in reasonable agreement with that calculated[83] for triplet TMM. However, they also observed a weak absorption at longer wavelengths from which emission was seen when the shorter wavelength absorption was excited. The fluorescent lifetime of the long wavelength band was sufficiently short that the weak intensity of this absorption was attributed to a low concentration of the absorbing species, not to an inherently small extinction coefficient. The position of the weak band was in reasonable agreement with that calculated[83] for the bisected geometry of TMM. Thus, unless **51** is found to have a UV absorption at unexpectedly long wave lengths, one reasonable interpretation of these experiments is that there is a detectable amount of bisected singlet diyl in equilibrium with the planar triplet at 77 K, the temperature at which these experiments were conducted.

C. Summary

The results discussed in the preceding section raise nearly as many questions as they answer. At what geometry does the lowest energy triplet–singlet surface crossing take place in TMM? Does efficient surface crossing occur only close to or at the bisected geometry? Is there an activation energy for closure of the singlet diyl, particularly in TMM derivatives such as **48**, where the closure product (**51**) contains appreciable strain? Is singlet **48** trappable, and how much of the chemistry previously attributed to the singlet diyl is actually that of **51**? Would replacement of deuterium by methyl in **56-58** lead to increased stereomutation at the exocyclic methylene group? Is the computed energy separation between the planar triplet and bisected singlet in **48** comparable to that found for the parent TMM? Preliminary results from Pitzer's group[84] suggest that the answer to this question will be in the affirmative. If this is the case, how can the predicted gap be reconciled with the results of the Turro-Berson experiment? Is the bisected singlet, in fact, responsible for the weak absorption at long wavelength that they observed? What process causes the reversible changes found by Dowd and Chow[85] in the EPR spectrum of the parent TMM diradical? Considerably more work, both experimental and theoretical, on TMM and derivatives is obviously warranted. Possible synthetic applications are just now beginning to be explored.[86]

VI. OTHER ALTERNANT HYDROCARBON DIRADICALS

Space limitations prevent a thorough review in this volume of other even alternant hydrocarbons for which no classical Kekulé structures can be written. Nevertheless, brief mention should be made of two diradicals of this type that have received attention recently from experimentalists and that will almost certainly be the subject of further studies.

A. 1,8-Naphthoquinodimethane

The parent diradical (**60**) has been prepared by Pagni and co-workers[87] from the azo compound **59** and, more recently, by Platz[88] from diazo compound **61**. The latter route involves hydrogen atom abstraction from the proximate methyl group. Derivatives in which the methylene groups in **60** are bridged by one and two carbon atoms have been synthesized, respectively, by Michl and co-workers[89] and by Pagni and co-workers.[90] The diradical in which one of the CH_2 groups of **60** has been replaced by NH has also been prepared by Platz.[91]

Perhaps the single most important development in the study of **60** and its derivatives is the finding by Platz[88] and by Wirz and co-workers[92] that the

59   60   61

parent diradical and its methano- and ethano-bridged derivatives are ground-state triplets. The curvature previously observed in the Curie-Weiss plots for these diradicals may have been due to saturation of the triplet EPR signals at very low temperatures. The discovery that **60** and its derivatives do have triplet ground states has an obvious bearing on the interpretation of the optical spectra that have been recorded for the methano-[89] and ethano-bridged[90] diradicals. The finding is in pleasing agreement with the naive prediction of a triplet ground state for **60**, based on the fact that the nonbonding Hückel orbitals of the diradical have atoms in common.[93]

B. *m*-Quinomethane

Berson and co-workers[94] have prepared **63** from the bicyclic isomer **62**. On thermolysis or photolysis of **62** in alcoholic solutions, the trapping product **64** was isolated. It would be interesting to establish whether the formal addition of ROH proceeds by an ionic mechanism, since, as depicted in resonance structure **63b**, the lowest singlet state of the diradical might be expected to have some ionic character.

Despite the possible stabilization of the lowest singlet state by accumulation of electron density on the electronegative oxygen atom, it would appear that the lowest triplet is either the ground state or very close to it. At 10 K a strong triplet EPR spectrum is observed that persists for at least an hour when **62** is irradiated either directly or in the presence of a triplet sensitizer. This finding suggests that *m*-quinodimethane, the diradical in which O in **63** is replaced by CH_2, almost certainly is a ground-state triplet. Such a result would again be in agreement with qualitative theoretical expectations,[93] based on the fact that the two Hückel nonbonding MOs in *m*-quinodimethane span common atoms.

62   63a   63b   64

VII. ACKNOWLEDGMENT

The author wishes to thank Professors J. A. Berson, P. B. Dervan, and M. S. Platz for reading this manuscript and for their comments. The work of these and the other experimentalists, whose names appear in the references, has stimulated much of the author's own efforts to understand the behavior of diradicals. Preparing this chapter has only served to increase the author's respect for those who accept the challenge of carrying out the difficult experiments required in this field of chemistry.

VIII. REFERENCES

1.* L. Salem and C. Rowland, *Angew. Chem., Int. Ed. Engl.*, 11, 92 (1972).

2. R. Hoffmann, *J. Am. Chem. Soc.*, 90, 1475 (1968).

3. R. J. Crawford and A. Mishra, *J. Am. Chem. Soc.*, 87, 3768 (1966); *J. Am. Chem. Soc.*, 88, 3963 (1966).

4. D. E. McGreer, N. W. K. Chiu, M. G. Vinje, and K. C. K. Wong, *Can. J. Chem.*, 43, 1407 (1965).

5. A. K. Q. Siu, W. M. St. John III, and E. F. Hayes, *J. Am. Chem. Soc.*, 92, 7249 (1970).

6. Y. Jean, L. Salem, J. S. Wright, J. A. Horsley, C. Moser, and R. M. Stevens, *Proc. 23rd Int. Cong. Pure Appl. Chem., Spec. Lect.*, 1, 197 (1971); *J. Am. Chem. Soc.*, 94, 279 (1972).

7. P. J. Hay, W. J. Hunt, and W. A. Goddard III, *J. Am. Chem. Soc.*, 94, 638 (1972).

8. Y. Jean and X. Chapuisat, *J. Am. Chem. Soc.*, 96, 6911 (1974); *J. Am. Chem. Soc.*, 97, 6325 (1975).

9. R. J. Crawford and T. R. Lynch, *Can. J. Chem.*, 46, 1457 (1968).

10. J. A. Berson and J. M. Balquist, *J. Am. Chem. Soc.*, 90, 7343 (1969).

11. W. Carter and R. G. Bergman, *J. Am. Chem. Soc.*, 90, 7344 (1968); *J. Am. Chem. Soc.*, 91, 7411 (1969).

12. A. Chmurny and D. J. Cram, *J. Am. Chem. Soc.*, 95, 4237 (1973).

13. W. von E. Doering and K. Sachdev, *J. Am. Chem. Soc.*, 96, 1168 (1974); *J. Am. Chem. Soc.*, 97, 5512 (1975).

14. M. R. Willcott III and V. H. Cargle, *J. Am. Chem. Soc.*, 91, 4310 (1969).

15.* J. A. Berson and L. D. Pedersen, *J. Am. Chem. Soc.*, 97, 238 (1975); J. A. Berson, L. D. Pedersen, and B. K. Carpenter, *J. Am. Chem. Soc.*, 97, 240 (1975); *J. Am. Chem. Soc.*, 98, 122 (1977).

16.* The literature up to 1977 has been reviewed by J. A. Berson, *Ann. Rev. Phys. Chem.*, 28, 111 (1977). For an earlier review see R. G. Bergman, in *Free Radicals*, Vol. 1, J. K. Kochi, Ed., Wiley-Interscience, New York, 1973, p. 191.

17. J. T. Wood, J. S. Arney, D. Cortes, and J. A. Berson, *J. Am. Chem. Soc.*, 100, 3855 (1978).

18.* J. E. Baldwin and C. G. Carter, *J. Am. Chem. Soc.*, 100, 3942 (1978).

19. E. A. Barsa, Ph. D. thesis, Harvard University, 1976; *Diss. Abstr. Int.*, 37, 5077-B (1977); W. von E. Doering and E. A. Barsa, *Proc. Natl. Acad. Sci.*, 77, 2355 (1980) .

20.   W. Kirmse and M. Zeppenfeld, *J. Chem. Soc., Chem. Commun.*, 124 (1977).

21.* J. E. Baldwin and C. G. Carter, *J. Am. Chem. Soc.*, 101, 1325 (1979).

22.* W. von E. Doering and E. A. Barsa, *Tetrahedron Lett.*, 2495 (1978).

23.* J. J. Gajewski, R. J. Weber, and M. J. Chang, *J. Am. Chem. Soc.*, 101, 2100 (1979).

24.* T. C. Clarke, L. A. Wendling, and R. G. Bergman, *J. Am. Chem. Soc.*, 99, 2740 (1977).

25.   S. W. Benson, *J. Chem. Phys.*, 34, 521 (1961).

26.* S. L. Buchwalter and G. L. Closs, *J. Am. Chem. Soc.*, 101, 4688 (1979).

27.   M. P. Conrad, R. M. Pitzer, and H. F. Schaefer III, *J. Am. Chem. Soc.*, 101, 2245 (1979).

28.   R. M. Wilson and F. Geiser, *J. Am. Chem. Soc.*, 100, 2225 (1978).

29.   A. Gavezzotti and M. Simonetta, *Tetrahedron Lett.*, 4155 (1975).

30.   R. Hoffmann, S. Swaminathan, B. D. Odell, and R. Gleiter, *J. Am. Chem. Soc.*, 92, 7091 (1970).

31.   G. A. Segal, *J. Am. Chem. Soc.*, 96, 7892 (1974).

32.   L. M. Stephenson and T. A. Gibson, *J. Am. Chem. Soc.*, 94, 4599 (1972).

33.   L. M. Stephenson, T. A. Gibson, and J. I. Brauman, *J. Am. Chem. Soc.*, 95, 2849 (1973).

34.   H. R. Gerberich and W. D. Walters, *J. Am. Chem. Soc.*, 83, 3935 (1961); *J. Am. Chem. Soc.*, 83, 4884 (1961).

35.   R. B. Woodward and R. Hoffmann, *Angew. Chem. Int. Ed. Engl.*, 8, 781 (1969).

36.   A. T. Cocks, H. M. Frey, and I. D. R. Stevens, *J. Chem. Soc., Chem. Commun.*, 458 (1969).

37.   J. E. Baldwin and P. W. Ford, *J. Am. Chem. Soc.*, 91, 7192 (1969).

38.   P. D. Bartlett and N. A. Porter, *J. Am. Chem. Soc.*, 90, 5317 (1968).

39.   J. A. Berson, D. C. Tompkins, and G. Jones II, *J. Am. Chem. Soc.*, 92, 5799 (1970).

40.   G. Jones II and M. E. Fantina, *J. Chem. Soc., Chem. Commun.*, 1213 (1972).

41.   L. M. Stephenson, P. R. Cavigli, J. L. Parlett, *J. Am. Chem. Soc.*, 93, 1984 (1971).

42.   C. P. Casey and R. A. Boggs, *J. Am. Chem. Soc.*, 94, 6457 (1972).

43.   Fluorinated olefins have been extensively used by Bartlett and co-workers to investigate the 1,4-diradicals formed in cycloaddition reactions. For leading references see P. D. Bartlett and J. J.-B. Mallet, *J. Am. Chem. Soc.*, 98, 143 (1976).

44.   L. M. Stephenson and J. I. Brauman, *J. Am. Chem. Soc.*, 93, 1988 (1971).

45.* G. Scacchi, C. Richard, and M. H. Bach, *J. Chem. Kinet.*, 9, 513 (1977); *J. Chem. Kinet.*, 9, 525 (1977).

46.* P. B. Dervan and T. Uyehara, *J. Am. Chem. Soc.*, 98, 1262 (1976); P. B. Dervan, T. Uyehara, and D. S. Santilli, *J. Am. Chem. Soc.*, 101, 2069 (1979).

47.* P. B. Dervan and T. Uyehara, *J. Am. Chem. Soc.*, 98, 2003 (1979); *J. Am. Chem. Soc.*, 101, 2076 (1979).

48.   L. M. Stephenson and T. A. Gibson, *J. Am. Chem. Soc.*, 96, 5624 (1974).

49.   G. Jones II and V. L. Chow, *J. Org. Chem.* 39, 1447 (1974).

50.* D. S. Santilli and P. B. Dervan, *J. Am. Chem. Soc.*, 101, 3663 (1979); *J. Am. Chem. Soc.*, 102, 3863 (1980).

51.* J. S. Chickos, *J. Org. Chem.*, 44, 780 (1979).

52.* W. von E. Doering and C. A. Guyton, *J. Am. Chem. Soc.*, 100, 3229 (1978).

53.* M. J. Goldstein and M. S. Benzon, *J. Am. Chem. Soc.*, 94, 5119 (1972).

54. See, for instance, W. T. Borden, *Modern Molecular Orbital Theory for Organic Chemists*, Prentice-Hall, Englewood Cliffs, N.J., 1975, pp. 95–98.

55. C. Y. Lin and A. Krantz, *J. Chem. Soc.*, *Chem. Commun.*, 1111 (1972).

56. O. L. Chapman, C. L. McIntosh, and J. Pacansky, *J. Am. Chem. Soc.*, 95, 614 (1973).

57. W. T. Borden, *J. Am. Chem. Soc.*, 97, 5968 (1975).

58.* W. T. Borden and E. R. Davidson, *Ann. Rev. Phys. Chem.*, 30, 125 (1979).

59.* T. Balley and S. Masamune, *Tetrahedron*, 36, 343 (1980).

60. G. Maier, H.-G. Hartan, and T. Sayrac, *Angew. Chem.*, *Int. Ed. Engl.* 15, 226 (1976).

61. S. Masamune, Y. Sugihara, K. Morio, and J. E. Bertie, *Can. J. Chem.*, 54, 2679 (1976).

62. R. G. S. Pong, B.-S. Huang, J. Laureni, and A. Krantz, *J. Am. Chem. Soc.*, 99, 4153 (1977).

63.* S. Masamune, F. A. Souto-Bachiller, T. Machiguchi, and J. E. Bertie, *J. Am. Chem. Soc.*, 100, 4889 (1978).

64. H. Kollmar and V. Staemmler, *J. Am. Chem. Soc.*, 99, 3583 (1977).

65. W. T. Borden, E. R. Davidson, and P. Hart, *J. Am. Chem. Soc.*, 100, 388 (1978).

66. J. A. Jafri and M. D. Newton, *J. Am. Chem. Soc.*, 100, 5012 (1978).

67. R. D. Buenker and S. D. Peyerimhoff, *J. Chem. Phys.*, 48, 354 (1968).

68. G. Maier, U. Schafer, W. Sauer, H. Hartan, and J. F. M. Oth, *Tetrahedron Lett.*, 1837 (1978).

69. W. T. Borden, *J. Am. Chem. Soc.*, 98, 2695 (1976); E. R. Davidson and W. T. Borden, *J. Chem. Phys.*, 64, 663 (1976).

70.* E. R. Davidson and W. T. Borden, *J. Am. Chem. Soc.*, 99, 2053 (1977).

71.* J. A. Berson, *Acc. Chem. Res.*, 11, 446 (1978).

72. D. M. Hood, H. F. Schaefer, and R. M. Pitzer, *J. Am. Chem. Soc.*, 100, 8009 (1979).

73.* P. Dowd and M. Chow, *J. Am. Chem. Soc.*, 99, 6438 (1977).

74.* M. S. Platz and J. A. Berson, *J. Am. Chem. Soc.*, 99, 5178 (1977).

75.* M. Rule, M. G. Lazzara, and J. A. Berson, *J. Am. Chem. Soc.*, 101, 7091 (1979).

76.* R. F. Salinaro and J. A. Berson, *J. Am. Chem. Soc.*, 101, 7094 (1979).

77. R. Siemionko, A. Shaw, G. O'Connell, R. D. Little, B. Carpenter, and J. A. Berson, *Tetrahedron Lett.*, 3529 (1978).

78. W. Kirmse and H. R. Murawski, *J. Chem. Soc.*, *Chem. Commun.*, 122 (1977).

79.* M. G. Lazzara, J. J. Harrison, M. Rule, and J. A. Berson, *J. Am. Chem. Soc.*, 100, 7092 (1979).

80.* W. R. Roth and G. Wegener, *Angew. Chem.*, *Int. Ed. Engl.*, 14, 758 (1975).

81.* J. J. Gajewski and S. K. Chou, *J. Am. Chem. Soc.*, 99, 5696 (1977).

82.* N. J. Turro, M. Mirbach, N. Harrit, J. A. Berson, and M. S. Platz, *J. Am. Chem. Soc.*, 100, 7653 (1978).

83. J. H. Davis and W. A. Goddard, *J. Am. Chem. Soc.*, 99, 4242 (1977).

84. R. Pitzer, private communication.

85.* P. Dowd and M. Chow, *J. Am. Chem. Soc.*, 99, 2825 (1977).

86.* R. D. Little and G. W. Muller, *J. Am. Chem. Soc.*, 101, 7129 (1979).

87. R. M. Pagni, M. N. Burnett, and J. R. Dodd, *J. Am. Chem. Soc.*, 99, 1972 (1977).

88.* M. S. Platz, *J. Am. Chem. Soc.*, 101, 3308 (1979).

89. J.-F. Muller, D. Muller, H. J. Dewey, and J. Michl, *J. Am. Chem. Soc.*, 100, 1629 (1978).

90. C. R. Watson, Jr., R. M. Pagni, J. R. Dodd, and J. E. Bloor, *J. Am. Chem. Soc.*, 98, 2551 (1976).

91.* M. Platz, *J. Am. Chem. Soc.*, 101, 4425 (1979).

92.* M. Gisin, E. Rommel, J. Wirz, M. N. Bernett, and R. M. Pagni, *J. Am. Chem. Soc.*, 101, 2216 (1979).

93. W. T. Borden and E. R. Davidson, *J. Am. Chem. Soc.*, 99, 4587 (1977).

94.* M. Rule, A. R. Matlin, E. F. Hilsinki, D. A. Dougherty, and J. A. Berson, *J. Am. Chem. Soc.*, 101, 5098 (1979).

6

CARBOCATIONS

D. BETHELL and D. WHITTAKER

Robert Robinson Laboratories, University of Liverpool,
Liverpool, England

I. INTRODUCTION

This chapter is devoted in large measure to an account of recent studies of carbocations in superacidic media in which they are often quite stable, at least at low temperatures. The inclusion of such material in a book on *reactive* intermediates perhaps needs a little justification.

The suggestion in the 1920s and 1930s that carbenium ions were involved transiently in a number of organic reactions was one of the major developments in mechanistic organic chemistry. It was based on earlier quantitative study of highly stabilized carbenium ions in solution in, for example, concentrated sulfuric acid, coupled with painstaking experimentation and interpretative flair. Carbocation chemistry developed rapidly, stretching established physical principles to their limit in its attempts to interpret observations of product structures and proportions and rates of reaction in terms of the structure and properties of the transient intermediates. The more recent development of chemistry in superacidic media has presented carbocation chemists with the opportunity to generate some of these intermediates, hitherto having only a fleeting existence, in stable solutions. NMR spectroscopy continues to be the most important technique for structural work, but the application of other quantitative physical techniques now permits the reexamination of some of the less securely based postulates to which rapid development of any subject inevitably gives rise. Since much organic chemistry takes place in the liquid phase, the development of gas-phase techniques for the study of carbocation chemistry now also affords a clearer appreciation of which aspects of the behavior of carbocations require a structural and which require an environmental interpretation.

II. THE ENERGETICS OF GENERATION AND REARRANGEMENT OF CARBENIUM IONS

A. Heats of Ionization in Solution and in the Gas Phase

In an important investigation with implications in many areas of carbenium-ion chemistry, Arnett and Petro[1] have measured directly the heat of solution (ΔH_s) of a series of organic chlorides, RCl, in solutions of SbF_5 in the solvents SO_2F_2, SO_2ClF, CH_2Cl_2, and SO_2 using a low-temperature calorimetric method. Subtraction of the observed value of ΔH_s in the absence of SbF_5 yields the heat of ionization (ΔH_i) for the process $RCl + SbF_5 \longrightarrow R^+ SbF_5Cl^-$. Their values, including the uncertainties (which may prove crucial in future discussions) are in Table 6.1 A feature of the work was the careful establishment, using NMR spectroscopy and quenching, that the carbocation was indeed the reaction product.[2]

Table 6.1   Heats of Ionization for Organic Chlorides, RCl, in Solutions of SbF$_5$ at $-55°C$ and Some Related Quantities

RCl	R'	$\Delta H_i{}^a$ SO$_2$ClF	$\Delta H_i{}^a$ CH$_2$Cl$_2$	$\Delta H_i{}^a$ Gas phase[b]	$\Delta H_{cap}{}^a$ CH$_2$Cl$_2$	$\Delta G^{\ddagger}_{solv}$ (25°C)^{a,c} EtOH
(structure: t-butyl-type, Cl on C with R')	H	-15.3 ± 0.9	-7.5 ± 1.5	251.7	–	31.7 (37.6)^d
	Me	-25.4 ± 0.8	-15.5 ± 0.3	235.9	-18.1	27.1
	Ph	-30.3 ± 0.3	-19.0 ± 0.6	–	–	22.1
(structure: sec-chloride)		{ -15.7 ± 0.8^e	–	247.9^f	–	31.4 (36.3)^d
		{ -30.0 ± 0.8^g	–	–	–	–
(structure: cyclopentyl with R')	H	-17.3 ± 0.9	-9.0 ± 1.0^e	245.8		30.5 (34.9)^d
	Me	-27.1 ± 0.6	-17.8 ± 0.5	229.5	-15.4	24.6
(structure: adamantyl-Cl)		-21.6 ± 0.8	-11.1 ± 0.8	225.1^f	-21.0	31.2
(structure: cyclohexyl-Cl)		-22.5 ± 0.6^h	–	–	–	32.9 (36.2)^d
(structure: norbornyl with R')	H	-23.6 ± 0.8	-11.1 ± 0.4	234.4	-23.9	30.1
	Me	-31.0 ± 1.5	-19.6 ± 0.8^i	228.4	-16.3	23.6
	Ph	-37.0 ± 1.2	-25.9 ± 1.2^i	–	-10.3	18.5
PhCCl$_3$		–	-16.4 ± 0.7^i	–	-18.5	–
Ph$_2$CCl$_2$		–	-23.7 ± 0.5^i	–	-12.5	–
Ph$_3$CCl		–	-27.1 ± 0.3	–	-6.0	–

^{a}Kcal/mol.
^{b}Ionization of RH.[3]
^{c}Reference 4.
^{d}Corrected for nucleophilic solvent assistance.
^{e}Determined at -75°C.
^{f}J. L. Beauchamp, quoted by Arnett and Petro.[1]
^{g}Determined at -25°C.
^{h}Formation of rearranged ion.
^{i}Determined at 0°C.

Arnett and Petro also measured the heat of capture of their carbocations in CH$_2$Cl$_2$ using Me$_4$N$^+$Cl$^-$ as the scavenger (see Table 6.1) Not suprisingly ΔH_{cap} is related linearly to ΔH_i by the expression $\Delta H_i + \Delta H_{cap} = 34.9 \pm 1.0$, the constant representing the heat of the reaction Me$_4$N$^+$Cl$^-$ + SbF$_5$ $\longrightarrow$ Me$_4$N$^+$-SbF$_5$Cl$^-$. Taken in conjunction with the observation that values of ΔH_i were not affected by the use of excess SbF$_5$, this relationship sustains the belief that the values of ΔH_i obtained are indeed measures of the heat change associated with the ionization process alone.

As can be seen in Table 6.1 ΔH_i varies with the solvent. For t-butyl chloride the variation is equal and opposite to the heat of solution of SbF$_5$ in the solvent, suggesting that the solvent effect on ΔH_i is related to the heat required to displace the Lewis acid from the solvent. That electrostatic solvation of the organic cation, though clearly important, is not very sensitive to structure is shown by the good linear correlation of ΔH_i (RCl in SO$_2$ClF) with Solomon and Field's values of ΔH_i (RH in the gas phase).[3] This is in line with findings in other solvents reviewed in Volume 1 of this series.

B. Correlation of Heats of Ionization and Solvolysis Rates

Free energies of activation for the ethanolysis at 25°C of RCl are included in Table 6.1. With the exception of the results for norbornyl and cyclohexyl chlorides (see below), these give excellent linear correlation with ΔH_i in both CH$_2$Cl$_2$ and SO$_2$ClF.[4] The linearity is improved in the case of SO$_2$ClF (but not CH$_2$Cl$_2$) if $\Delta G^{\ddagger}_{solv}$ is corrected by removing the contribution due to nucleophilic solvent assistance, which is important in the solvolysis of the secondary chlorides, and the slope of the line is then 0.89.

From this finding, the authors reach two conclusions of great significance to carbocation chemistry. The first of these is that the correlation of $\Delta G^{\ddagger}_{solv}$ with ΔH_i provides retrospective justification for the time-honored use of rates of solvolysis as a guide to the stability of carbenium ions. Since ΔH_i values demonstrably refer to the production of carbenium ions, the transition state in the solvolysis must be close in structure to those same carbocations. The authors almost resist the temptation to equate the slope of their linear correlation with the degree of charge development in the transition state (cf., the analogous use of Brønsted exponents in the proton transfer), even though the value appears to be close to estimates made in other ways for t-butyl chloride.[5]

The second conclusion is based on the improvement in the $\Delta H_i/\Delta G^{\ddagger}_{solv}$ correlation when the solvolysis rates of secondary chlorides are corrected for nucleophilic assistance. This is taken to validate the idea of nucleophilic solvent assistance in solvolysis and the method developed for its estimation.[6]

Inferences that accord with prejudices should be the most closely scrutinized. It should therefore be asked whether the correlation of $\Delta G^{\ddagger}_{solv}$ with ΔH_i implies that the transition state for ethanolysis of RCl is structurally close to the carbocation observed in SO$_2$ClF + SbF$_5$. Since the correction of $\Delta G^{\ddagger}_{solv}$ for nucleophilic solvent assistance means that the values refer to the limiting mechanism of solvolysis, gross differences in the entropy of activation for solvolysis along the series of RCl have been eliminated; the correlation is thus $\delta_R \, \Delta H_{solv}$ versus $\delta_R \, \Delta H_i$. Therefore, structural effects on the two measured quantities are linearly related. What is not known is how sensitive such enthalpy correlations are to the structures involved. For example, would values of $\Delta G^{\ddagger}$ (or $\Delta H^{\ddagger}$) for the hydration in aqueous acid of olefins related to RCl correlate with ΔH_i?

This example is of some significance in view of the picture of the solvated
t-butyl cation emerging from Kebarle's gas-phase results.[7] According to this, the
energy released on interaction of $C_4H_9^+$ with one molecule of water (11.2 kcal/
mol) is substantially less than that released by addition of further water molecules
to the cluster [e.g., 17.7 kcal/mol for addition of the third water molecule, giving
$C_4H_9^+(H_2O)_3$, which is very close to the value for generation of $H^+ (H_2O)_3$].
The carbenium ion thus appears essentially as an olefin molecule surrounded by
a hydrogen-bonded network of water molecules accommodating the additional
proton. This structure seems to accord not only with simple models of hyperconjugation in alkyl cations, but also fairly well with formulations of the transition state for isobutene hydration. It should be remembered, however, that it
must bear a similar relationship to the transition state for formation of the t-
butyl cation from the corresponding protonated alcohol. Such a picture of
carbocations is rather different from that suggested for carbenium ions in less
basic solvents than water or ethanol, such as SO_2ClF. It should be added that
the role of ion association, both in the transition-state formulation and in the
calorimetric studies, has not been explicitly discussed.

Notwithstanding the caution expressed above, the work must rank as one of
the most important contributions to carbocation chemistry in recent years.
Further exploration of the relationship between thermodynamic parameters for
ionization in superacidic media and for activation in related processes is desirable,
however, before the "fundamental soundness of the carbocation theory" can
truly be claimed to be established.

C. Rearrangements of Alkyl and Cycloalkyl Cations

The values of ΔH_i in SO_2CIF (Table 6.1) for 2-butyl chloride at $-75°C$, where
the 2-butyl cation is formed, and at $-25°C$, where the product is the t-butyl
cation, differ by -14.3 kcal/mol, which represents the enthalpy of the isomerization 2-Bu$^+ \longrightarrow t$-Bu$^+$.[1] Such an analysis of the heat changes is not possible for
ionization of cyclohexyl chloride and 1-methylcyclopentyl chloride because,
even at the lowest attainable temperatures, the unrearranged cyclohexyl cation
could not be observed; both chlorides yielded the 1-methylcyclopentyl cation.
A problem of some significance is how to divide the observed ΔH_i for cyclohexyl chloride $\longrightarrow$ 1-methylcyclopentyl cation (-22.5 kcal/mol[1]) between the
heat of ionization to the cyclohexyl cation and the heat of carbenium-ion
isomerization. A typical ΔH_i for a secondary chloride is about -17 kcal/mol,
which would leave -5.5 kcal/mol for isomerization; this is much less than is
found for 2-Bu$^+ \longrightarrow t$-Bu$^+$. On the other hand, taking -14.3 kcal/mol for the
heat of the isomerization $c-C_6H_{11}^+ \longrightarrow c-C_5H_9Me^+$ would leave only about
-8 kcal/mol for the heat of ionization. The fit of $\Delta G^{\ddagger}_{solv}$, corrected for nucleophilic solvent assistance, and ΔH_i suggests that for ionization of cyclohexyl

chloride ΔH_i is -16.1 kcal/mol, leaving -6.4 kcal/mol for isomerization. This sort of division, of course, presupposes that the generation of the rearranged ion is stepwise and involves the cyclohexyl cation as an intermediate (see below).

In an important paper on the rates and thermodynamics of rearrangement of 1-alkylcycloalkyl cations generated from the corresponding tertiary alcohol using 1: FSO_3H/SbF_5 usually in SO_2ClF, Kirchen, Sorensen, and Wagstaff[8] have shown that the direction of rearrangement (whether ring expansion or contraction) is dependent on ring size as shown in Figure 1. As indicated in the figure (but not explicitly demonstrated in the paper), the rearrangements are usually, but not always, stepwise. The overall rates of isomerization, all measurable by NMR in the temperature range -105 to $-60°C$, lead to free energies of activation that show remarkably little variation with structure (12.1–15.4 kcal/mol).

If we assume that $\Delta S^{\ddagger}$ for rearrangement is small, the value of $\Delta H^{\ddagger}$ for rearrangement is then very similar to the value of the heat of isomerization of a tertiary to a secondary butyl cation, as estimated by Arnett. It is tempting to suggest that the rate-limiting step in the 1-alkylcycloalkyl cation rearrangement is an isomerization such as that indicated in Scheme 1, with the transition-state structure close to that of the secondary ion.

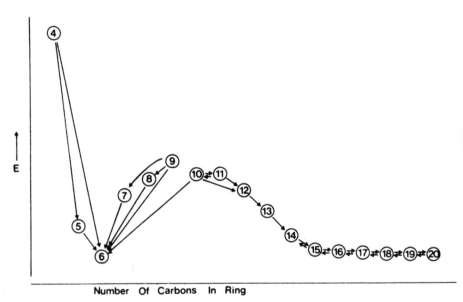

Number Of Carbons In Ring

FIGURE 1. Dependence of the direction of rearrangement of tertiary cyclo-alkyl cations on ring size.[8] (Revised with permission. Copyright by the American Chemical Society.) Source of figure is Ref. 8 as indicated, i.e., R. P. Kirchen, T. S. Sorensen and K. M. Wagstaff, *J. Am. Chem. Soc.* **100**, 5134 (1978).

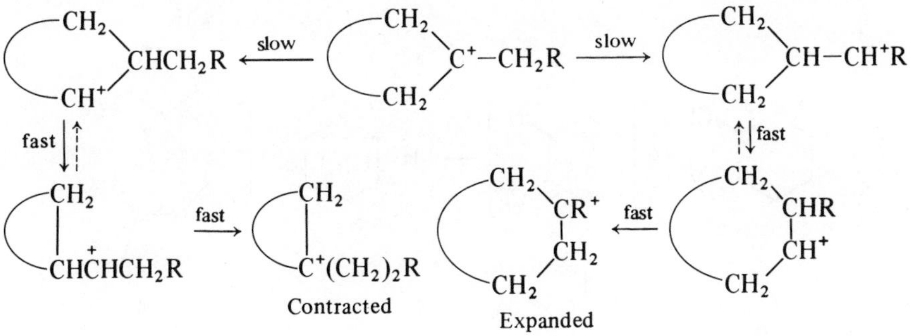

SCHEME 1. Stepwise ring contraction and expansion in 1-alkylcycloalkyl cations.

The transformation of the secondary cyclohexyl cation into 1-methylcyclopentyl cation is apparently very much faster than the processes examined by Kirchen et al., and it has been claimed to be fast also in the gas phase (W.J. Hehre quoted by Arnett and Petro[1]). Of necessity a different pathway must connect the cyclohexyl and 1-methylcyclopentyl cations, since one analogous to that for ring contraction in Scheme 1 would require conversion of a secondary into a primary carbenium ion, a process for which the enthalpy change would be likely to be much larger than 15 kcal/mol. Two possible pathways are shown in Scheme 2. One of these involves intermediate formation of protonated

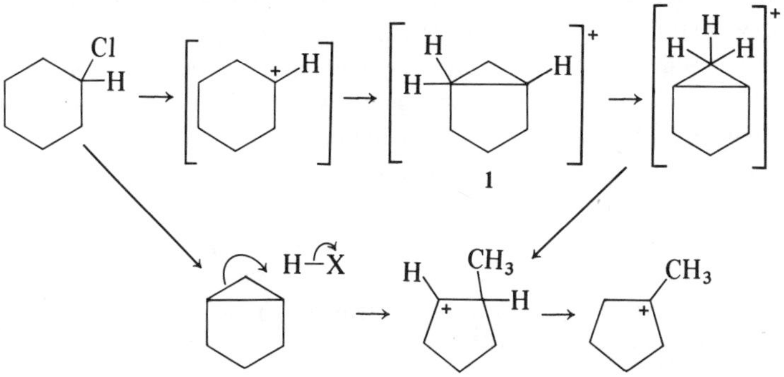

SCHEME 2. Possible pathways for the isomerization of cyclohexyl into 1-methylcyclopentyl cation.

cyclopropanes,[9] although 1-methylcyclopentyl cation could be formed in a concerted fashion without prior generation of the cyclohexyl cation. This pathway readily accommodates the observation of Fărcasiu[10] that trans-1,2-dichloro cyclohexane (2) in superacidic solution at $-70°C$ yields 3-chloro-1-methylcyclopentyl cation (3) and none of the 2-chloro-1-methyl isomer (4) or 1-chloromethyl isomer (5) (See Scheme 3).

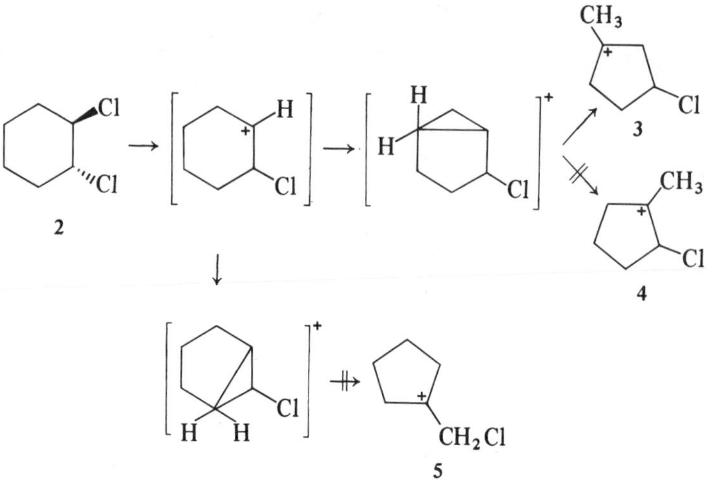

SCHEME 3

In contrast to the cyclohexyl cation itself, 1-alkylcyclohexyl cations are the most stable of the 1-alkylcycloalkyl cations.[8] 1-Chlorocyclohexyl cation has also been observed at $-120°C$, but is rapidly converted into the 1-methylcyclopentenyl cation when the temperature is raised to $-90°C$.[11] The problem of the mechanism of rearrangement of cyclohexyl cation deserves more detailed study. Sorensen's demonstration[12] that the twist-boat conformation of tertiary cyclohexyl cations is preferred over the chair form by 0.5 kcal/mol helps to make the rapidity of skeletal rearrangement understandable, since the stereoelectronic requirements for rearrangement, whether by 1,2-shift or by way of protonated cyclopropanes, are more readily met.

In conclusion it is worth pointing out that if a methylene group were to span C_3 and C_6 of structure 1 (Scheme 2), the carbocation would represent the σ-bridged ("nonclassical") formulation of the 2-norbornyl cation, the structure of which is considered in the next section. 1,3-Bridging in six-membered carbocyclic rings is also germane to the problems of the generation and bonding in trishomocyclopropenyl cations.[13]

III. SOME CARBOCATION STRUCTURAL PROBLEMS

A. The 2-Norbornyl Cation

Despite the recent publication of a book[14] largely devoted to the problem, the question of whether the 2-norbornyl cation is best represented as a σ-bridged (carbonium) structure (7) or whether 7 represents merely the transition state for interconversion of conventional carbenium ions **6** and **8** continues to challenge the intellect, the ingenuity, and the patience of chemists. If the latter situation obtains, then **7** can be no more than 5 kcal/mol higher in energy than the classical structures. It is because the energy differences are so small that the problem has proved so difficult to resolve. New developments both in gas-phase and solution studies have illuminated the problem and point the way to a possible consensus.

The thermodynamics of hydride-transfer equilibria in the gas phase between *t*-butyl cation and norbornane and related hydrocarbons[3] and of proton-transfer equilibria involving norbornene[15,16] yield three independent estimates that agree[17] that the heat of formation of the 2-norbornyl cation is 182 kcal/mol. Comparison with values for related systems suggests that the 2-norbornyl cation has 6–10 kcal/mol of extra stabilization because of special structural factors. The heterolytic bond dissociation energies in Table 6.1 provide an example of the sort of argument that is used. Values of ΔH_i for 1-methylcyclopentyl and 2-methylnorbornyl cations (both generally agreed to be classical) differ by only 1.1 kcal/mol, but the dissociation of norbornane needs some 11.4 kcal/mol less energy than dissociation of cyclopentane. Of course, the gas-phase experiments do not provide direct evidence of structure, but a feature of these equilibrium studies is that entropy changes can be monitored, and these are consistent with reactions involving no gross structural change on converting norbornane or norbornene to the product cation. These gas-phase studies are of considerable importance because they provide data against which the plethora of theoretical studies (see e.g., Ref. 18) can be evaluated. Moreover, since the point for *exo*-2-norbornyl chloride falls on the linear correlation of ΔH_i (SO_2ClF) versus ΔH_i (RH, gas phase), the findings are relevant to the 2-norbornyl cation in solution, suggesting that in SO_2ClF the 2-norbornyl cation has the same structure as in the gas phase. (It is fair to say, however, that a deviation from the correlation line of a few kcal/mol could easily have been overlooked.) In more nucleophilic solvents than SO_2ClF the more localized charges of **6** and **8** could be more strongly solvated than the charge in **7** and the situation could be reversed.[18c] We pointed out earlier that solvation of carbenium ions does not show any strong structural dependence, but in this instance the differential effect need only be quite small.

Whether in superacidic media the 2-norbornyl cation is best represented as **7** or **6**⇌**8** could perhaps be decided by using a NMR method devised by Saunders and co-workers.[19] In cases where two degenerate carbenium ions interconvert rapidly on the NMR time scale giving an averaged spectrum, the degeneracy can be lifted by introduction of deuterium close to one of the carbenium centers (e.g., as in 9). In the ^{13}C spectrum, C_1 and C_2 give signals showing a separation δ (14.9 ppm at 67.9 MHz) that is determined by the equilibrium isotope effect and Δ, the estimated chemical shift difference of C_1 and C_2 in a static ion.[20] For 1,2-dimethylcyclopentyl cation δ/Δ is 0.18. Where the carbenium ion has a symmetrical delocalized structure as in 10, however, the introduction of deuterium attached at one of the equivalent carbenium centers leads to a downfield shift of the undeuterated carbenium carbon relative to the deuterated carbon.[21] This splitting is referred to as isotopic perturbation of resonance and for 10 implies that the positive charge density is greater at the undeuterated carbenium center. This effect arises not from any change in the potential energy surface nor from differences in electronegativity (which is the same for isotopes), but because NMR chemical shifts are averages over zero-point motion and this is isotopically sensitive. Values of δ/Δ per deuterium atom turn out to be much smaller in cases of perturbation of resonance than of perturbation of degeneracy. For 10, δ/Δ is 0.0035, and for 11 the value of 0.0058 is taken to indicate significant σ-bridging. A weakness of the method is the need to estimate Δ in cases of very fast rearrangement, and this makes it difficult to define with certainty the borderline of δ/Δ values between equilibration and delocalization. For example, in the case of 9, an equilibrating system, δ/Δ is 0.032. More examples are needed if this method of structural diagnosis is to be applied in difficult cases such as 2-norbornyl cation.

9

10

11

The main thrust of Brown's work has been to reexamine the solvolysis data upon which the original suggestion that 7 represents the structure of the 2-norbornyl cation was based. In recent years he has applied a method developed by Gassman and Fentiman[22] (now referred to as the tool of increasing electron demand) to the problem. The method rests on the premise that the rate of ionization of tertiary benzylic systems, m- or p-$XC_6H_4C(Y)RR'$, to the corresponding benzylic cation will show reduced sensitivity to the substituent X when R and R' themselves provide strong stabilization for the positive charge. The sensitivity is judged from the magnitude of the (negative) slope (ρ^+) of plots of log k_{solv} versus σ_X^+. Alternatively, in cases where a change in the manner of internal charge dissipation occurs within the series of substituents X, changes in ρ^+ or deviations from linearity may signal the onset of participation as X becomes less able to contribute to the stabilization of the positive charge. The second approach has been found to be appropriate in interpreting NMR chemical shift data[23] in superacidic media, and indeed Farnum et al.[24] and Olah et al.[25] have presented results suggesting that, in the 2-arylnorbornyl series δ_{C_1} for cations having aromatic substituents that are powerfully electron withdrawing deviates from the linear plot against σ_X^+, indicating the incursion of bridging.

A selection of Brown's values for ρ^+ obtained from kinetics of solvolysis of bicyclo[2.2.1]heptyl and heptenyl systems is in Table 6.2. The remarkable finding is that the variation in ρ^+ is remarkably small, although, in the bicyclo-heptenyl series, the change in ρ^+ with variation in the remote substituent R″ is in the direction expected for homoallylic 2,6-bridging. On the other hand, norbornyl compounds yield ρ^+ values that are slightly smaller in the *endo* than in the *exo* series, which, if the small differences are significant, is the reverse of expectation if σ participation selectively assists *exo* ionization. The only clearcut examples in Table 6.2 of large changes in ρ^+ are in systems containing cyclopropane rings. In particular, Coates' system (14) gives a dramatic decrease in ρ^+ from the value observed in the 7-norbornyl (12) and even the 7-tricyclyl (13) series. This is a clear indication of participation in ionization by the σ-bond of the cyclopropane ring *anti* to the leaving group in 14.

How significant, then, are the ρ^+ values from solvolysis of the 2-arylnorbornyl derivatives and how do they relate to solvolysis of norbornyl compounds themselves? Schleyer[14] has questioned the sensitivity of ρ^+ to bridging, citing the absence of a reduction in ρ^+ in the *exo*-2-arylnorbornenyl *p*-nitrobenzoate series. Jorgensen[34] has attempted to answer this criticism by a perturbational MO approach in which the interaction energy of the LUMO of RR'CH$^+$ (L) and the HOMO of an aryl substituent (H) were calculated in terms of the positive charge on the carbenium center (Q_L) and the HOMO-LUMO energy separation ($\epsilon_L - \epsilon_H$). The charge on the *ipso*-carbon of the aryl substituent (Q_H), ϵ_H, and β, the resonance integral, were taken as parameters, and ρ^+ was assumed to be a linear function of $Q_L/(\epsilon_L - \epsilon_H)$, giving two further adjustable parameters. Values of the key calculable quantities, Q_L and $-\epsilon_L$, were obtained for 13 secondary carbenium ions by the MINDO/3 method; low values yield low values of ρ^+. The adjustable parameters were optimized using both classical carbenium ions and ions for which controversy concerning σ-bridging remains. Consequently, only very large effects are likely to be revealed by this highly empirical approach, and, as we saw earlier, the problem of the 2-norbornyl

TABLE 6.2 Values of ρ^+ for Solvolysis[a] of Bicyclo[2.2.1]heptyl and heptenyl p-Nitrobenzoates

Compound	R''	ρ^+ exo	ρ^+ endo	Ref.
	H	-4.21	-4.17	26
	Me	-3.27	-4.19	27
	H	-4.50	-4.51	28
	OMe	-3.71	-4.10	29
	H	-3.82	-3.72	30
	Me	-3.65	-3.47	31
(12)		-5.27		32
(13)		-3.27		32
(14)		-2.05		33

[a] In 80% aqueous acetone at 25°C.

cation revolves around quite small energy differences. Nevertheless Jorgensen has estimated that the ρ^+ value for a σ-delocalized 2-norbornyl cation should be -2.09.

The question of the relevance of the conclusions from the application of the tool of increasing electron demand in tertiary benzylic systems to the corresponding secondary alkyl cation has been tackled by Peters.[35] He has attempted to place the substituent H (in place of Ar) on the σ^+ scale so that values of log k for solvolysis of secondary p-nitrobenzoates can be predicted by extrapolation of the linear plots obtained for the tertiary benzylic esters. By placing log k for solvolysis of 7-norbornyl p-nitrobenzoate on the extrapolated plot of log k for 7-aryl-7-norbornyl p-nitrobenzoate versus σ^+, he obtained a substituent constant for H (γ^+) of $+2.53$. Using this value, the predicted values of k for solvolysis of secondary p-nitrobenzoates were in good agreement with observation and, since the tertiary systems are now universally regarded as producing classical carbenium ions, this is taken to imply that the secondary ion is also classical. This procedure is not above criticism since it assumes that both secondary and tertiary systems are solvolyzed by the same mechanism and that no correction is necessary for differential solvent assistance, solvation, or steric influences on the orientation of the aryl substituent. Furthermore, γ^+ was obtained from the correlation having the highest ρ^+; 2.53 thus represents the smallest value γ^+ can have. In other steric situations, not constrained by a small $C-C-C$ bond angle, γ^+ may be larger, making the reported agreement of predicted and observed solvolysis rates for exo- and $endo$-2-norbornyl p-nitrobenzoates fortuitous.

The correlation of $\Delta G^{\ddagger}_{solv}$ and ΔH_i (SO_2ClF) includes points for exo- and $endo$-2-norbornyl chlorides. Interestingly, points for exo-norbornyl, exo-2-methylnorbornyl, and exo-2-phenylnorbornyl chlorides fit the correlation well. The corresponding $endo$-chlorides have $\Delta G^{\ddagger}_{solv}$ values that are substantially higher than expected on the basis of ΔH_i (SO_2ClF) for the exo-chlorides, and the deviations from the linear plot are equal for the three compounds. The authors have so far been reluctant to draw inferences from these observations, although advocates of both points of view in the controversy appear to think that the findings support their position. In the absence of a clearer understanding of the sensitivity of $\Delta G^{\ddagger}$ versus ΔH_i plots to transition state structure, it should simply be said that the transition state for ethanolysis of exo-2-norbornyl chloride appears to resemble the ion generated in SbF_5/SO_2ClF more closely than does the transition state for solvolysis of the $endo$-chloride.

We conclude that the balance of evidence available at present suggests that in the gas phase and perhaps also in superacidic media, the 2-norbornyl cation is probably best represented by the σ-bridged structure 7. In more nucleophilic solvents, including those used in solvolysis studies, bridging effects appear to be much less important. The debate will undoubtedly continue.

B. Pyramidal Cations.

One of the most interesting developments in carbocation chemistry during the last decade is the study of pyramidal cations and dications.[36] The early calculations of Stohrer and Hoffmann[37] provided a theoretical basis for the square pyramidal structure for the ion $(CH)_5^+$ which was observed experimentally as the dimethyl-substituted derivative by Masamune et al.[38] They found that the products of solvolysis of the benzoate ester of 15 were consistent with reaction through a structure such as 18, and that solution of any one of the three isomeric starting materials, 15, 16, or 17, in FSO_3H/SO_2ClF at $-78°C$ gave rise to a solution whose ^{13}C spectrum could be interpreted in terms of the ion 18.[39] The ^{13}C resonances observed are at higher field than is usual for

charged carbon atoms; the figures in parentheses around structure 18 are the ^{13}C chemical shifts in ppm, relative to Me_4Si. The products of quenching were identified, but were identical to the secondary products of solvolysis of the ester of 15; the primary products from this ester were acid labile.

Masamune et al.[40] extended the study by treating 19 with SbF_5 in SO_2ClF at $-110°C$ and obtaining ion 20; in this case, quenching gave 21, clearly by attack of the nucleophile at the basal position. Working on a similar system, except for extra methyl substitution, Hart and Kuzuya showed[41] that the product of methanolysis of 22 was 23. When the carbon atoms marked with an asterisk in 22 were labeled with deuterium, the label was found to be distributed over the asterisked carbons in 23. Again, this could be consistent with a pyramidal intermediate; reaction of the alcohol 24 with FSO_3H/SO_2ClF at $-78°$ gave a solution whose spectrum was consistent with 25. However, on

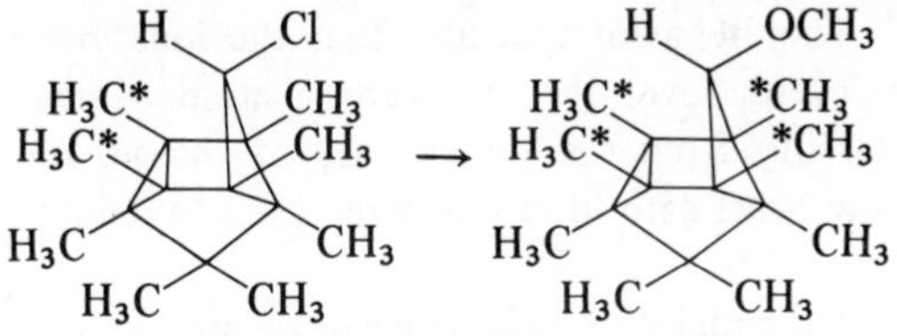

$$
\begin{array}{ccc}
\text{22} & \rightarrow & \text{23}
\end{array}
\tag{2}
$$

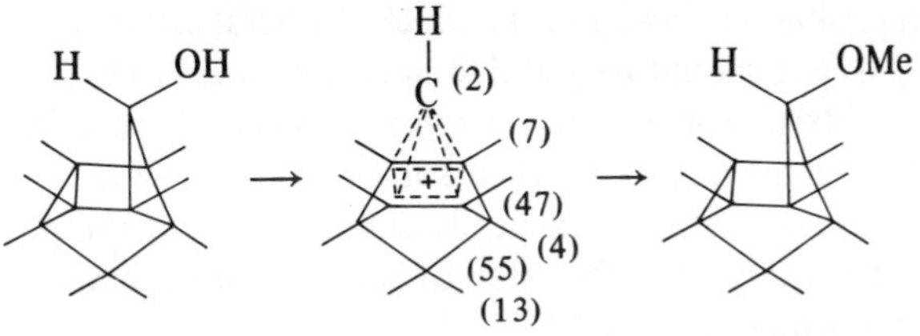

$$
\text{24} \quad \rightarrow \quad \text{25} \quad \rightarrow \quad \text{26}
\tag{3}
$$

quenching this ion, attack of the nucleophile took place at the apical carbon atom to give a nearly quantitative yield of 26. Since then other pyramidal cations have been prepared,[42,43] but the main trends of behavior are well illustrated by the examples quoted.

Consideration of the evidence presented above immediately provokes a number of questions, such as whether these ions really do have the pyramidal structure shown, or whether this is in fact a transition state for the interconversion of less symmetrical ions. If they do exist in the pyramidal form shown, then the position of the charge and the type of bonding involved are of great interest, particularly considering the high field positions of the ^{13}C resonances.

If we consider the ^{13}C spectrum of a simple cyclic secondary carbocation,[44] the carbon atom carrying the positive charge would be expected to resonate around 330 ppm downfield from Me_4Si; if the charge were to be spread over a number of carbon atoms, then it would be expected that the total downfield shift of the charged carbon atoms, compared to their uncharged state, would be around 300 ppm. There are a few exceptions to this rule, notably the trishomocyclopropenium ion,[45] in which the carbon resonances show an upfield shift on ionization, but it generally holds. Pyramidal carbocations appear to provide another exception to the rule, in that the downfield shifts are, in total, small.

Since the apical carbon resonance is shifted upfield, we might suggest, as a first assumption, that it carries less charge than the basal carbon atoms, but this assumption is not consistent with the chemistry of nucleophilic attack on **20** and **23**. These ions differ only in that **23** has methyl substituents on all the basal carbon atoms and on some surrounding carbon atoms. Nucleophilic attack is entirely basal in the unsubstituted ion,[40] and entirely apical when the basal carbon atoms carry methyl substituents, suggesting that it is controlled by steric rather than electronic factors.[41] This behavior indicates that the ion probably carries the charge spread fairly evenly over the five carbon atoms. Indeed, it may well draw some of its stability from this equivalence of charge, since phenyl substitution of two of the basal carbon atoms is reported to reduce the stability of the ion.[43]

If the charge is spread equally over the five carbon atoms, then it seems likely that the carbon atoms could become equivalent, which has not been reported. This would be impossible in those ions in which the basal carbon atoms are part of a bicyclic system, but should be possible in an ion such as **18**. If we consider that three isomeric starting materials, differing in the position of the methyl substituents, all give the same ion, then such an interconversion must be taking place, and **18** is the thermodynamically stable isomer of the possible dimethyl pyramidal carbocations. The first step of such a process may well be the formation of an intermediate such as **27**.

27

An intermediate corresponding to **27** has been proposed by Hart[41] to explain the rearrangement of ion **25** when the solution is warmed to $-50°C$, yielding **28**.

$$(4)$$

25   28

The suggested mechanism is consistent with the results of labeling the asterisked carbon atoms with deuterium.

If the apical carbon atom does carry a partial positive charge, then its chemical shift is unexpected, except perhaps in comparison with the trishomocyclopropenyl cation. Kwant[46] has discussed two possible explanations for this shift. The first is based on the fact that the pyramidal cation can be constructed from interaction of CH^+ and a cyclobutadiene molecule, as shown in Figure 2. The apical carbon atom can then be regarded as being influenced by the ring current generated in the cyclobutadiene. However, as Longuet-Higgins[47] has pointed out, it is reported that the internal protons of the $(4n + 2)$ annulenes resonate at high fields, whereas the internal protons of the $4n$ annulenes resonate at low field. These results are consistent with a ring current flowing "in the wrong direction." This may be because a paramagnetic current, which flows in the opposite direction to the diamagnetic current, is particularly large in antiaromatic systems. Thus any ring current effects would be expected to cause the apical carbon atom to resonate at an unusually low, rather than an unexpectedly high frequency. The position of the ^{13}C resonance of the bridging carbon in 25 at 55 ppm suggests that the effect is not large.

The alternative explanation considered by Kwant is steric strain due to crowding of the pentacoordinated apical carbon atom, but values of +3 and -17 for apical carbons carrying H, compared to -23 for methyl substituted apical carbon do not support this. However, in a system such as the pyramidal ion, the

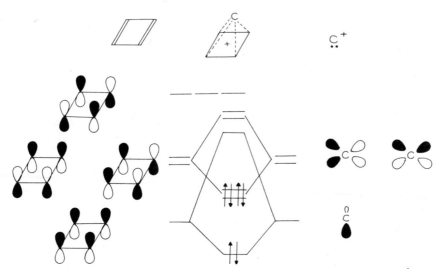

FIGURE 2. Interaction diagram showing how the orbitals of the $(CH)_5^+$ pyramidal cation are constructed from C^+-H and cyclobutadiene.

strain must come in the bonds to the corners of the cyclobutadiene, and it is reasonable to argue that the effect of substituents on this would be small.

The effect on the chemical shift of the apical carbon atom resonances of the change in substitution at the basal carbon atoms revealed by comparing 20 and 25 suggests that the bonding of the ions may have a significance that has not been properly considered. However, a recent experiment of Hart has shown that the coupling between the apical carbon atom and its hydrogen in 25 is 220 Hz, which falls close to the lower end of the range expected for sp hybridization,[48] possibly as a result of some effects of steric strain.

The other fundamental question is whether the pyramidal ion represents an energy minimum or is a transition state, and this is a difficult question to answer. It must be stated that the evidence to date falls short of convincing proof; the chemical shift of the apical carbon atom is not what would be expected on the basis of either of the theories discussed above.

Recently, Coates and Fretz[49] have investigated a system in which equilibrating ions, if formed, would themselves be trishomocyclopropenium ions. Solvolysis of 30 proceeds with rate enhancement of 2×10^7 over that of 29 and gives an apparently unrearranged product (31). However, labeling the atom marked with an asterisk in 30 gives a product that has the deuterium atom spread among the four carbon atoms marked with an asterisk in 31.

29   30   31   (5)

Dissolving the parent alcohol of 30 in SbF_5/SO_2ClF at $-100°C$, gave a solution whose spectrum was consistent with ion 32. The deuterium scrambling is

32

consistent with this ion, but both the scrambling experiment and the spectrum could also be consistent with a pair of interconverting trishomocyclopropenium ions. However, a feature of trishomocyclopropenium ions is that they show peaks due to carbon atoms carrying positive charge that are moved upfield compared to simple ions. If we seek to explain the position of the apical carbon atom by an upfield shift connected with the interconversion of a number of ions, then in this case it should show a further substantial upfield shift. Its position is only slightly upfield from those of previous pyramidal cations, indicative, though not conclusive, evidence in favor of the pyramidal species being an intermediate.

C. Pyramidal Dications

On treatment of the bishomocyclopropenyl cation (33) with FSO_3H/SbF_5 at $-60°C$, further reaction occurs to yield an ion that is believed to be the dication 34^{50} (Scheme 4). The same ion is obtained by treating the diol 35 with FSO_3H/SbF_5 in SO_2 at $-60°C$.

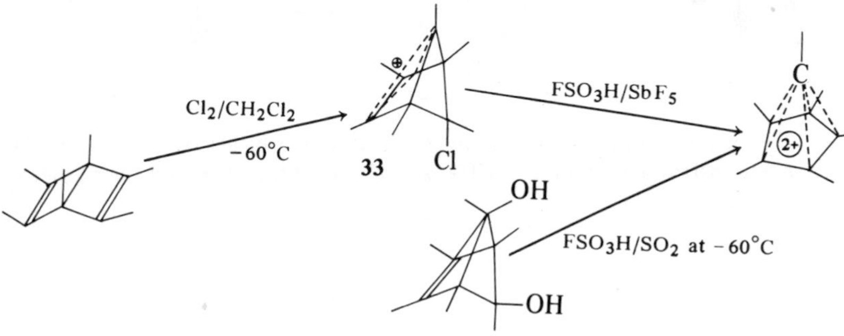

SCHEME 4

(6)

The ion shows ^{13}C resonances at 10.6 ppm (basal C) and -2.0 ppm (apical C), relative to TMS. Evidence that it is a dication comes from the greatly increased downfield shift of the basal carbon atoms, and the relative simplicity of the spectrum allows other possible formulations of the dication, plus any possible monocation structures, to be eliminated.

The dication can be regarded as a structure produced by the interaction of C^+-CH_3 with a pentamethylcyclopentadienyl cation, which, like cyclobutadiene, is antiaromatic (Figure 3). The dication, once formed, appears to be remarkably stable. There is very little deuterium incorporation from a deuterated acid,[36] and the spectrum of the ion shows no changes over the region -140 to $+40°C$. Above this temperature range, it decomposes to the hexamethylbenzenium ion. Since this is a monocation, the reaction probably occurs by reaction with suitable hydride donors. Such reactions can be induced at lower temperature by adding hydride donors, such as isopentane, when the monocations 36 and 37 can be observed.

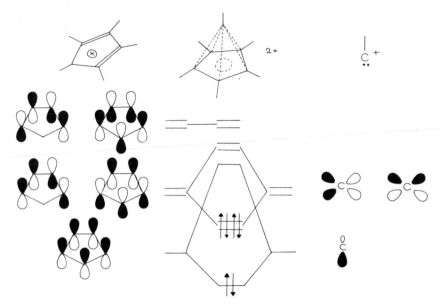

FIGURE 3. Interaction diagram showing how the orbitals of the pyramidal dication are constructed from C^+-CH_3 and the pentamethyl cyclopentadienyl cation.

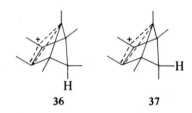

36   37

Quenching the dication 34 with sodium methoxide in methanol gives rise to 38, while reaction with potassium hydroxide gives rise to 39, reaction in both cases taking place at the basal carbon atoms. The different products obtained are again consistent with a two step mechanism for the reaction, the stereochemistry of the product depending on the lifetime of the monocation under the reaction conditions.

OCH_3 OCH_3 OH $-OH$
38   39

In a recent paper, Giodrano, Heldeweg, and Hogeveen[51] have been able to modify the precursors of the dications so that they can put different alkyl groups on the basal and apical positions. They found that, as with the mono-cations, the substituents did not move between basal and apical positions, but were able to show that putting an isopropyl group on the apical position decreased the stability of the ion, so that it completely decomposed during 48 hr at $-40°C$.

The fundamental questions of the dication work echo those of the mono-cation studies; they have not, to date, been studied to a comparable extent.

D. Cyclopropylcarbinyl/Cyclobutyl Cations

It has long been known that cyclopropylcarbinyl derivatives and cyclobutyl derivatives can be ionized to a common intermediate, but the nature of this intermediate remains uncertain, despite extensive studies. The species involved appears to be a nonclassical ion, the favored proposals being the bicyclobutonium structure (40) and the cyclopropylcarbinyl structure (41).

40   41

Recent studies[52] of the variation in ^{13}C NMR chemical shift with temperature in the cyclopropylcarbinyl cation showed a variation greater than that expected for a single cation; it was proposed that the observed spectrum was the average of at least two ions, and was tentatively suggested that the data are consistent with one of these being the bicyclobutonium ion (40). Staral and Roberts[53] have used a sample of cyclopropylcarbinol labeled with ^{13}C on the carbinyl carbon atom and made up the solution of the ion at $-125°C$. This gave the unexpected result of a spectrum showing the expected cation, labeled on all the methylene but not at the methine carbon atoms, together with another species recognized as protonated cyclobutanol, labeled on all carbon atoms except the carbinyl carbon atom. On warming to $-70°C$ for 20 min and then cooling to $-100°C$, the spectrum showed only the cyclopropylcarbinyl cation then labeled on all carbon atoms. These observations effectively establish the exis-tence of hydride shifts in the system, but leave open the question of the exact geometry of the actual intermediates.

The 1-methylcyclobutyl cation has also been a subject of some interest. Early work on this ion[54] showed considerable similarities between it and the unsubstituted cyclobutyl cation, in that it had a single ^{13}C line due to the methine carbon, a line due to the three CH_2 carbons, and a line due to the methyl carbon. It was suggested that the methyl-substituted ion was similar to the unsubstituted ion, except that it had a greater contribution from the classical cyclobutyl structure to the overall system.

(7)

42

This picture was consistent with the evidence available when first proposed, but was shattered by the experiments of Kirchen and Sorensen,[55] who lowered the temperature of the ion solution to $-158°C$. The single peak due to the methylene carbons broadened at $-136°C$, and by $-145°C$ had decoalesced into two separate peaks at 72 and -3 ppm. The other peaks in the spectrum were unchanged.

The problem with which any theory must cope is the presence of the methine carbon resonance at 163 ppm. The corresponding cyclopentyl ion has the appropriate carbon at 337 ppm, the cyclohexyl ion at 330 ppm. Sorensen suggests a simple classical ion, such as 42, and seeks to explain the chemical shift of the methine carbon in an ingenious way. He points out that the charge-carrying carbon atom in such an ion is usually shifted downfield on ionization by about 300 ppm, which is the result of a shift due to the change from sp^3 to sp^2 hybridization of about 120 ppm, and a shift due to the charge of about 180 ppm. He suggests, therefore, that the strain of the cyclobutane ring makes the ion more stable when this carbon atom retains its sp^3 hybridization, instead of going to sp^2 hybridization, and therefore, the chemical shift is reduced to the level observed.

Olah, et al.[56] on the other hand reject this theory. They point out that the only cyclobutyl cation that is believed to be a simple classical ion is the 1-phenylcyclobutyl cation, in which the methine carbon is at 272 ppm, a figure in good agreement with cyclic phenyl-substituted cations in general. Also, in other highly strained cations where the charged atom is nonplanar, the chemical shift of the charged atom is at high values, such as 299 (adamantyl)[57] and 356 ppm (manxyl).[58]

Olah suggests that the ion does not have structure 42, but is either a set of σ-delocalized nonplanar bicyclobutonium-like ions, or else is a single σ-delocalized puckered species such as 43. If this species is regarded as a modified pyramidal cation, then the presence of a β-methylene carbon atom resonating at -2.83 ppm is consistent with work on pyramidal cations. Such a species would be formulated as :CH_2 associated with a methyl-substituted allylic carbocation (44; Figure 4). The orbital interaction diagram, however, suggests that ion 44 has limited

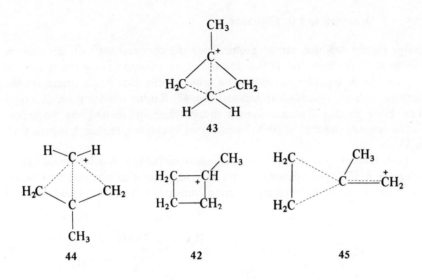

43

44   **42**   **45**

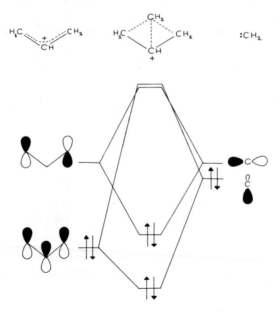

FIGURE 4. Interaction diagram showing how the orbitals of (**44**) are constructed from :CH_2 and an allylic cation.

charge on the methine carbon atom. Since the chemical shift of this atom is 163 ppm, this single ion is not a satisfactory explanation of the results. It is probably in equilibrium with some other species that has a charge on the methine carbine, possibly the classical ion **42**. Raising the temperature would then bring in contributions from a species that equilibrated the methylene carbon atoms, possibly a methyl-substituted cyclopropylcarbinyl system such as **45**.

Levi, Blurock and Hehre,[59] on the basis of *ab initio* molecular orbital theory using the 4-31G split-valence basis set have suggested that contributions to the structure of the methylcyclobutyl cation come from **46** and **47**. These may well

46   **47**

contribute, though neither provides any explanation of the observed chemical shifts in the ion.

It may be possible to obtain evidence for the existence of a modified pyramidal ion by using 1,2,4-trimethylcyclobutanol (**48**); the ion derived from this substrate should then be **49**, in which the extra methyl groups on the carbon atoms that carry a charge would stabilize this species, rather than any other that contributes to the overall picture.

48   **49**

In the unsubstituted cyclobutyl cation, an unsubstituted version of **44** could well contribute, though whether in place of the bicyclobutonium ion (**40**) or in addition to it is not clear.

E. The Trishomocyclopropenyl Cation

The intermediacy of a species of C_{3v} symmetry, believed to be the trishomocyclopropenyl cation **50** in the solvolysis of *cis*-bicyclo[3.1.0]hexan-3-yl tosylate (51), was proven many years ago by the deuterium scrambling experiment of Winstein *et al.*[60] The ion was observed[45] when *cis*-bicyclo[3.1.0]hexan-3-yl chloride was dissolved in SbF_5/SO_2ClF. Since then, other examples of these tris-homoaromatic species have been found,[61] but they have been looked for and not found in similar substrates. In general, trishomoaromatic ions are formed readily from substrates in which the bicyclo[3.1.0]hexyl system is locked rigidly into a chair, but they form much less readily in flexible systems, which prefer the boat form[62] in which the necessary interaction is absent (see 52).

The energy relationships among 50, 53 and 54 have now been studied by MINDO/3 calculations,[63] which indicate 50 to be more stable than 53 by 16.5 kcal/mol, with an energy barrier of 3.4 kcal/mol, while 54 has been found to be more stable than 53 by 12.6 kcal/mol, with an energy barrier of 1.5 kcal/mol. However, the presence of a methyl substituent on C_3 makes the classical ion more stable than the trishomoaromatic ion by 4.9 kcal/mol.

More recent observations[13] of the ions are not consistent with these values. While 50 has now been obtained from *cis*-bicyclo[3.1.0]hexan-3-ol in SbF_5/SO_2ClF, reaction of the *trans*-chloride (55) with SbF_5/SO_2ClF gives only the cyclohexenyl cation (56), presumably via 54. Substitution at the ring junction

also inhibits formation of the trishomoaromatic ion, both the 1-methyl and 1,3-diphenyl substrates giving rise to cyclohexenyl cations. Substituents at C_3 also blocked formation of the trishomocyclopropenyl cation, even though in the triphenyl-substituted species the classical ion was observed (eq. 8).

$$\text{(structure: Ph, OH cyclohexane with Ph, Ph)} \quad \xrightarrow[-140^\circ C]{FSO_3H/SO_2ClF} \quad \text{(cation structure, Ph, Ph, Ph)} \quad \xrightarrow{-70^\circ C} \quad \text{(cation structure, Ph, Ph, Ph)} \qquad (8)$$

These results require a very much higher energy barrier for rearrangement of **53** to **50** than the 3.4 kcal/mol predicted by MINDO/3 and support the view of Winstein et al.[60] that leakage from **53** to **50** is relatively inefficient. Formation of **50** is certainly dependent on the orientation of the leaving group. One difference between solvolytic and superacid results is that while the 1-methyl and 1,5-dimethylbicyclo[3.1.0]hexan-3-yl tosylates solvolyzed via trishomocyclopropenium ions, the 1-methyl substituent inhibited formation of a trishomoaromatic species in superacid conditions. This may be a steric requirement—the extra barrier to ring inversion to a chair form may tip the balance towards the classical ion, which is favored in superacid media by the extra solvation.

Like pyramidal cations, these homoaromatic species exhibit unusual upfield shifts for the carbon atoms carrying a positive charge, so that in the parent ion (**50**) they occur at 4.9 ppm. Since the species are homoaromatic, ring currents would be expected to cause an upfield shift, but doubts exist as to whether it is sufficient to give the large shift that is observed.

IV. ORGANIC CHEMISTRY IN SUPERACIDS

Although most studies of carbocations in superacid media concentrate on the structure and bonding of the ions, some of the reactions of these cations also have been studied. In many cases, the reactions parallel those in dilute acid media, but in others, different products are obtained. Reactions that show these differences can be divided into two types.

1. Reactions in which the same ion is generated in dilute acid and in superacid, but in the latter medium have an increased lifetime, so enabling slow reactions to compete successfully with ion capture.
2. Reactions of multifunctional substrates, in which reaction in superacid is initiated at a different functional group than reaction in dilute acid, and hence the reaction pathway followed is completely different.

In dilute acids, reactions are studied mainly by product analysis, kinetics, and labeling experiments; in superacids, kinetic studies are less commonly used, though energy barriers are readily measured, but spectroscopic studies of the intermediates are important. Product studies, which are readily carried out by pouring a solution of the ion in superacid into an aqueous alkaline medium, are subject to the serious criticism that it is impossible to be sure that the observed product is in fact the kinetically controlled product of reaction of the ion. One of the few cases in which there is evidence on this point is from the work of Masamune et al.,[38] referred to earlier, in which quenching a solution of 18 in methanol gives rise to 59 and 60 (Scheme 5).[39] However, solvolysis of 57, believed to proceed via 18, gives 59 and 58, but 58 rearranges readily in the presence of even a trace of acid to give 60.[38]

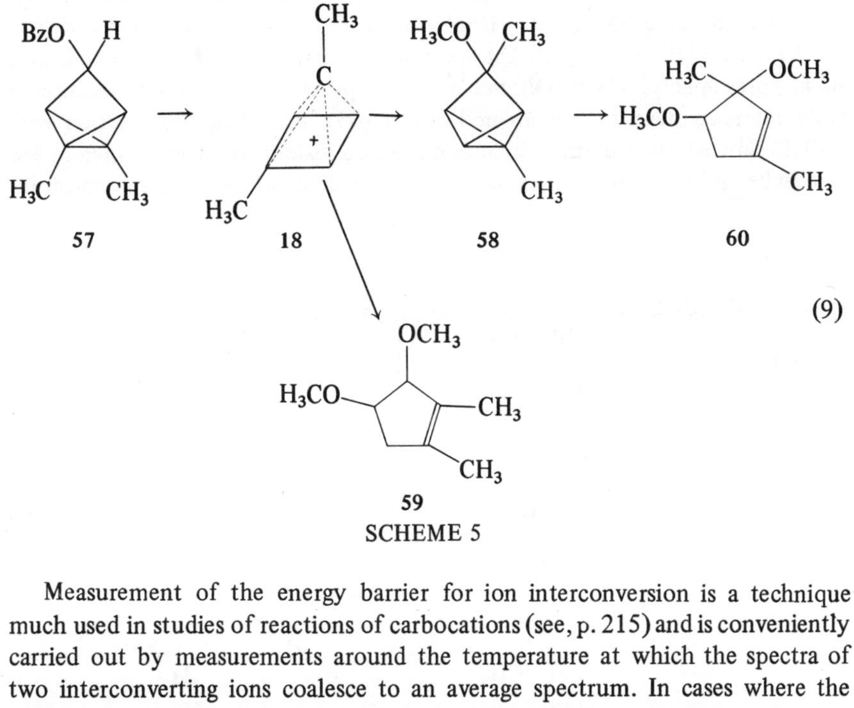

$$(9)$$

SCHEME 5

Measurement of the energy barrier for ion interconversion is a technique much used in studies of reactions of carbocations (see, p. 215) and is conveniently carried out by measurements around the temperature at which the spectra of two interconverting ions coalesce to an average spectrum. In cases where the coalescence temperature is too low to be achieved in the spectrometer, barriers can be measured by use of the technique of transfer of spin saturation. This method was originally developed by Forsen and Hoffman[64] for ^1H nuclei, but has been applied by Engdahl and Ahlberg[65] to ^{13}C spectroscopy. In this technique, certain atoms in the molecule being studied are irradiated so as to produce spin saturation, while the ^{13}C spectrum of the whole molecule is recorded. If the saturated carbons exchange with magnetically different carbons, then saturation is transferred, provided that the lifetime for chemical exchange is smaller than the lifetime of relaxation, and the ^{13}C peaks of the exchanged carbons are diminished or disappear. If exchange is slow compared to relaxation, the peaks are unaffected. When rates of exchange and relaxation are comparable, the rate of exchange can be estimated from the measured relaxation time.

The study of rearrangements of ions by use of ^{13}C and ^2H labels is particularly convenient in superacid solutions, since the positions of both labels can be monitored by the measurement of a single spectrum. The use of labeling techniques in this field was pioneered many years ago by Winstein,[66] who used ^2H labels in a study by ^1H NMR to demonstrate that the norbornenyl cation undergoes a degenerate rearrangement by the vinyl group moving round the five-membered ring. The technique has recently been used by Olah et al.[67] to investigate the mechanism of cyclization of the 1-phenylallyl cation to the indanyl cation (see Scheme 6). In $FSO_3H/SbF_5/SO_2ClF$ at $-120°C$, alcohol **61** yielded the 1-phenylallyl cation (**62**). On warming to $-80°C$, an intermediate formed via **63**, presumably, **64a** or **64b**, but it was not properly characterized because of rapid rearrangement to the indanyl cation (**65**). Repeating the reaction with FSO_3D showed no deuterium incorporation, consistent with species such as **64a** or **64b** being formed by a series of two 1,2-hydride shifts, rather than deprotona-

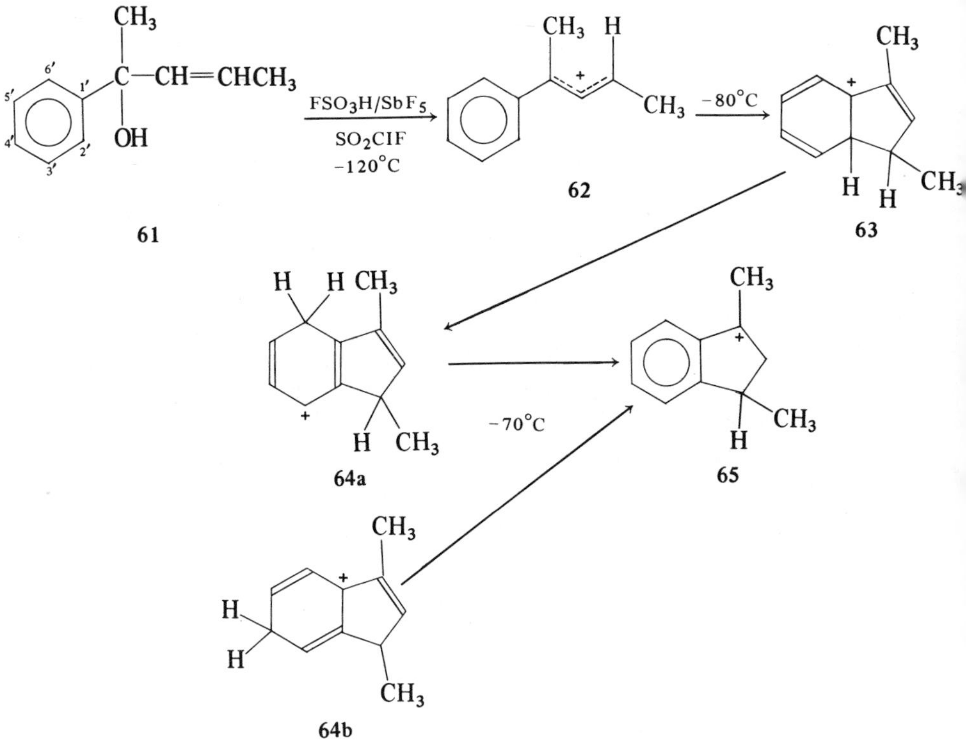

SCHEME 6

tion–reprotonation. To distinguish between **64a** and **64b**, the starting material was labeled with deuterium on $C_{5'}$. This gives rise to two possible label positions in **63**, and hence in either version of **64**. The proton on $C_{6'}$ in **64b** or on $C_{4'}$ in

D on $C_{5'}$ →   **63**   **64a**   or   **64b**   (10)

D on $C_{3'}$ →   or

64a is readily distinguishable from the others; in both versions of **64a** it should be a doublet, in one version of **64b** a doublet, and in the other a singlet. Observation of a doublet and a singlet confirmed that reaction does in fact proceed through **64b**.

Reactions that follow different pathways in dilute acid and superacid as a result of initiation of the reaction at different positions have been known for some time. A good example is from the work of Banthorpe et al.[68] on geraniol (**66**), which in dilute acids reacts at the hydroxyl group to give an allylic carbenium ion (**67**) that cyclizes to yield **68**, but in FSO_3H/SO_2 reaction by protonation of the lone double bond gives rise to the ion **69**, which cyclizes to **70** and

66   →   **67**   →   **68**   (11)

thence to the iridoid ether **71**.

Differences in the point of initiation of reactions such as these are the result of the enormous increase in the acidity of the medium. In dilute acid, an alco-

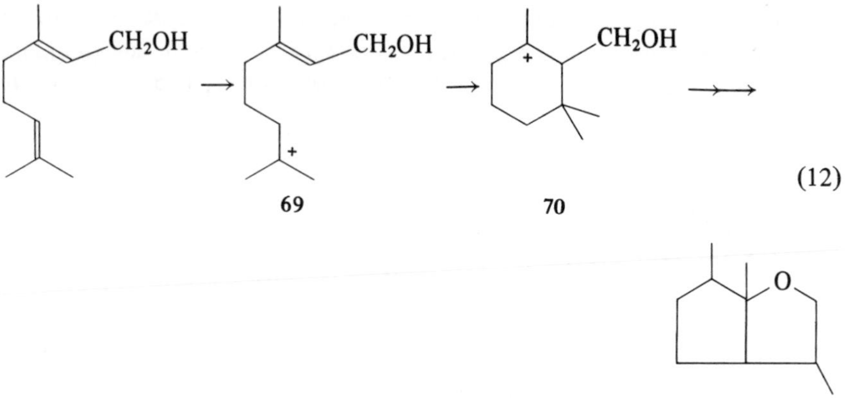

(12)

69   70

71

hol is protonated on the hydroxyl group to give the oxonium ion, which then undergoes rate-determining ionization to the carbocation and a water molecule. The rate of ionization depends on the concentration of protonated alcohol, which is proportional to acid protonating power. When the protonating power of the acid is increased by many powers of 10, protonation of the alcohol approaches completion, and the rate of ionization increases only slowly with increasing acidity. On the other hand, protonation of a double bond or a cyclopropane ring is believed to be rate determining so that the rate of formation of the ion should remain proportional to the protonating power of the acid over a much wider range of acid concentrations. The latter reaction is favored relative to the former in very strong acids.

This phenomenon proved to be the explanation for a rather puzzling set of results on the cyclohexenyl cations. The cyclohexenyl cation 72 was shown to rearrange to the methylcyclopentenyl cation 73 at around $-20°C$. The reaction was presumed to go via a hydride shift within 72 to yield the ion 74. However, experiments on the scrambling of deuterium in 72 showed that the hydride transfer reactions proceeded rapidly at $-65°C$. This led to Scheme 7. The next

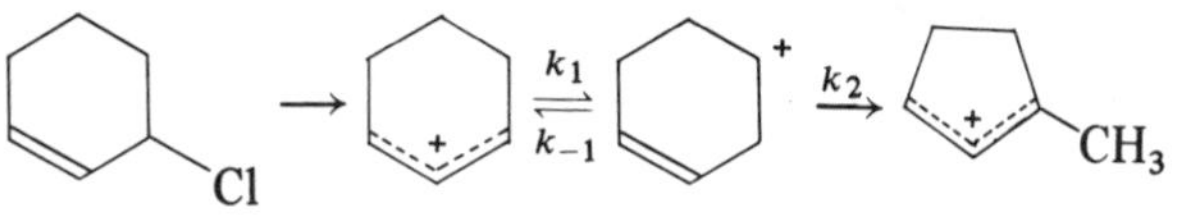

72   74   73

$k_1 \gg k_{-1}$ $k_2 \ll k_{-1}$

SCHEME 7

step of the study was to prepare ion 74 directly.[69] However, this ion was never obtained. All attempts to obtain it led directly to the methylcyclopentenyl ion

73, which is not consistent with $k_2 \ll k_{-1}$. The problem was solved[70] by comparing the reactions of the chlorides with FSO_3H/SbF_5, the solutions being made up under carefully controlled conditions at between -94 and $-70°C$. In this way, it was shown that reaction of the 3-chloride proceeded by ionization of the chloride to give the cyclohexenyl cation, but that reaction of the 4-chloride was by protonation of the double bond, which gave the 3-chloromethylcyclopentenyl cation (**75**) (Scheme 8).

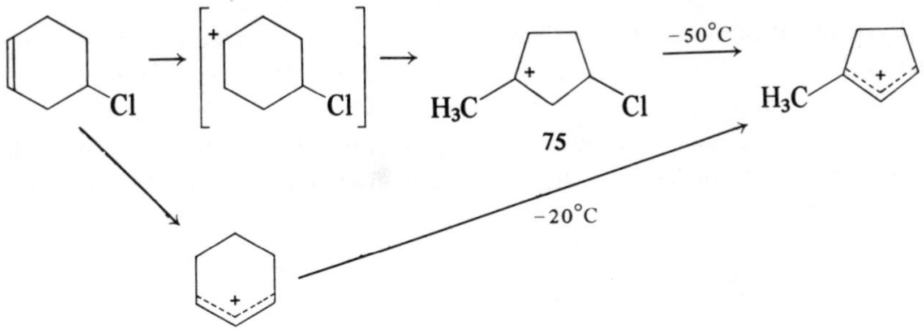

SCHEME 8

This work on the cyclohexenyl cations gives a clue as to how the direction of reaction can be controlled.[13] Reaction can be initiated at an alcohol function or a chloride by using antimony pentafluoride in the absence of fluorosulfuric acid. However, reaction at a double bond or at a cyclopropane ring can be initiated by using fluorosulfuric acid, either with or without antimony pentafluoride (see Scheme 9). An exception to this is a case where a tertiary alcohol has particularly good stabilizing groups, such as phenyl, when reaction occurs at the hydroxyl group rather than at the double bond, even with fluorosulfuric acid.

Although the details of these reactions are reasonably well established, the use of antimony pentafluoride as a reagent for generation of a carbenium ion at a hydroxyl function in the presence of double bonds yields the same ion as dilute

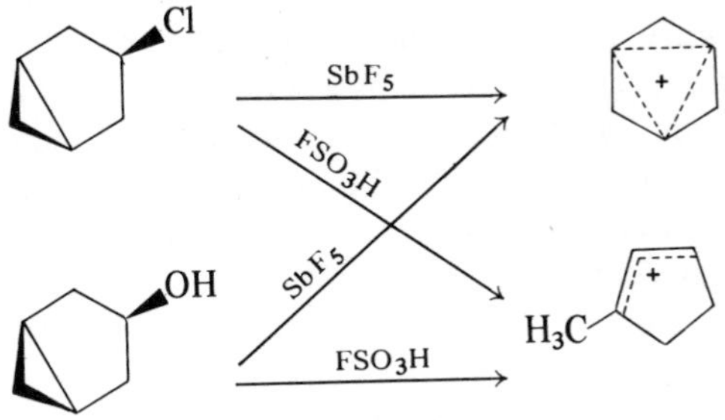

SCHEME 9

aqueous acid and has many disadvantages, particularly that it tends to cause polymerization. It does, however, have advantages when control over a system is required to produce a particular ion, and is indeed used extensively for this purpose.

V. ALKOXYCARBENIUM ("OXOCARBENIUM") IONS IN THE HYDROLYSIS OF ACETALS

Experimentation on the mechanism of hydrolysis of acetals continues un-abated, largely promoted by the continuing interest in the mechanism of action of the enzyme lysozyme. For many years it was believed that all acetals are hydrolyzed by an A-1 mechanism, the rate-limiting step being cleavage of the protonated substrate to yield an alkoxycarbenium ion (Scheme 10) since this

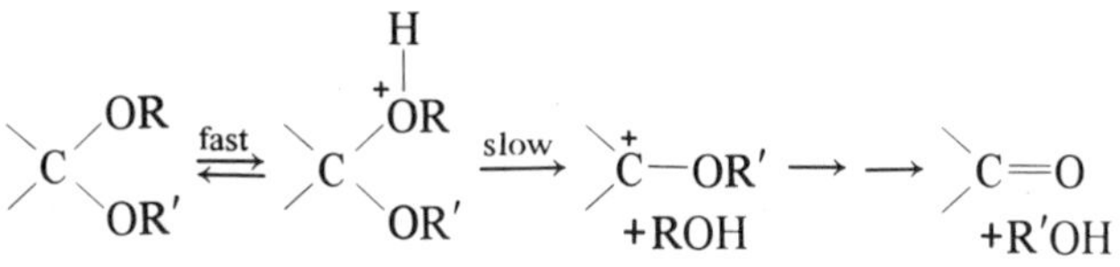

SCHEME 10

accorded with the observation of specific hydrogen ion catalysis. More recently, general acid catalysis has been observed, particularly in cases when the alkoxy-carbenium ion is particularly well stabilized, and this is now generally accepted to arise when protonation and cleavage of ROH are concerted to some degree. Lysozyme hydrolyzes specifically the glycosidic linkage between an N-acetyl-glucosamine (NAG) residue and an N-acetylmuramic acid (NAM) residue in a polysaccharide chain bound at the active site in a favorable conformation so that proton transfer from glutamic acid (Glu-35) to oxygen occurs in the rate-determining step.[71] To explain why this reaction falls in the group of acetal hydrolyses showing rate-limiting proton transfer, it has been suggested that formation of the glycosidic alkoxycarbenium ion (which is not intrinsically very stable) is facilitated by electrostatic stabilization of the developing positive charge by the ionized carboxyl group of a suitably located aspartic acid residue (Asp-52), as shown in 76. Molecular models suggest that the carboxylate group is located some 3 Å from the glycosidic carbon atom, a distance judged to be incompatible with covalent involvement either in cleavage of the C−O bond or in trapping the alkoxycarbenium intermediate. Evidence in support of this inter-pretation has come from (1) the pH dependence of the enzyme's reactivity, (2) the observation in the hydrolysis of 77 of an α-deuterium isotope effect, k_H/k_D = 1.11, a value comparable with those found in S_N1-type solvolyses, and (3) the failure to detect an acylal intermediate.

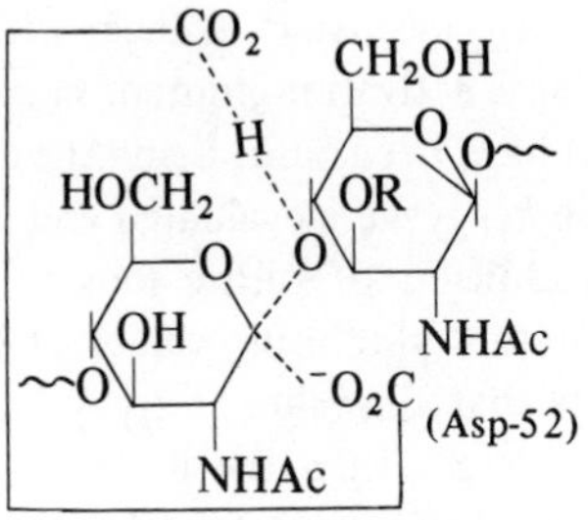

(Glu-35)

76 (R = —CHMeCO₂H)

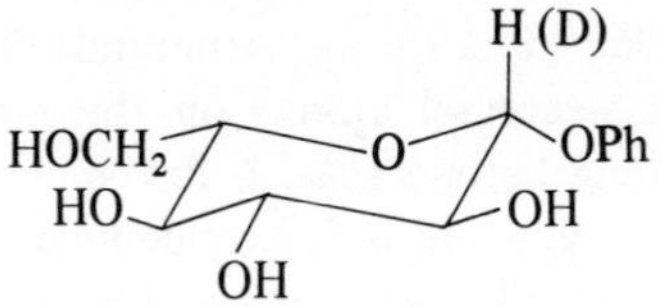

77

In the context of the present chapter, the important questions in lysozyme-mediated hydrolyses and in acetal hydrolysis generally are (1) whether it is possible in cases of general acid catalysis to distinguish formation of a discrete carbenium ion followed by reaction with an external nucleophile, from an intramolecular S_N2-type reaction followed by solvolysis of the intermediate, and (2) how, if a carbenium ion intermediate is formed in reactions involving lysozyme, to assess the degree to which its formation is assisted by the nearby carboxylate group.

The ground rules have been established by Young and Jencks[72] who have used sulfite ions to compete with water in trapping the intermediate $ArC^+(Me)OMe$ formed in the specific hydrogen ion catalyzed hydrolysis of acetophenone dimethyl ketals. Sulfite ion, which has no effect on the observed reaction velocity, yields the α-methoxysulfonic acid (78) and, significantly, the efficiency of its formation is the same if sulfite ion is allowed to intercept the reactive intermediate formed in the hydrolysis of the enol ether of acetophenone (Scheme 11). Clearly, the same intermediate, a free, solvent-equilibrated methoxycarbenium ion, is involved in both reactions. The product proportions yield partitioning ratios, k_S/k_{H_2O}. The quite small observed values, in the range

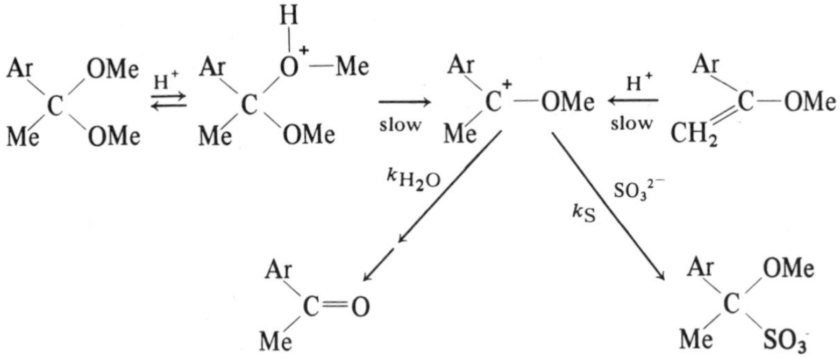

SCHEME 11   78

$10\text{-}10^3 \ M^{-1}$, are thought to indicate that the sulfite reaction is diffusion controlled. Moreover, a correlation of log (k_{H_2O}/k_S) against σ^+ is linear with a slope of 1.6, indicating that the reaction with water is activation limited, since, were it also a diffusion-controlled process, ρ^+ would have been zero. Using the diffusion-controlled value for k_S, structural effects on k_{H_2O} were evaluated and shown to parallel structural effects on the rate of addition of sulfite ions to the corresponding ketone (k_{SO_3}). Extrapolation of this plot using values of k_{SO_3} for formaldehyde and isobutyraldehyde suggests that the value of k_{H_2O} for the corresponding alkoxycarbenium ion would be larger than the limit of diffusion control. The same would be true in a glycoside hydrolysis, for which, therefore, two possibilities exist: first, that diffusion of the alcohol molecule and orientation of a water molecule become rate determining, and, second, that an alternative "preassociation" mechanism, involving weak nucleophilic participation by water, takes place (i.e., the alkoxycarbenium ion does not become free). Of these possibilities, the former requires specific hydrogen ion catalysis, so the latter seems to be necessary to interpret the catalysis of glycoside hydrolysis.

More recent evidence that lends some support to the second interpretation comes from Eliason and Kreevoy's interpretation of solvent isotope effects on acetal hydrolysis,[73] and more especially from some studies by Jensen et al.,[74] who have demonstrated that the hydrolysis of a series of acetals of substituted benzaldehydes show general acid catalysis with high, but substituent-dependent, values of the Brønsted exponent, α. This behavior indicates a late, alkoxy-carbenium-like transition state. The mechanism of hydrolysis, best described (as the authors do) using the Jencks-O'Ferrall diagram, is represented in Scheme 12.

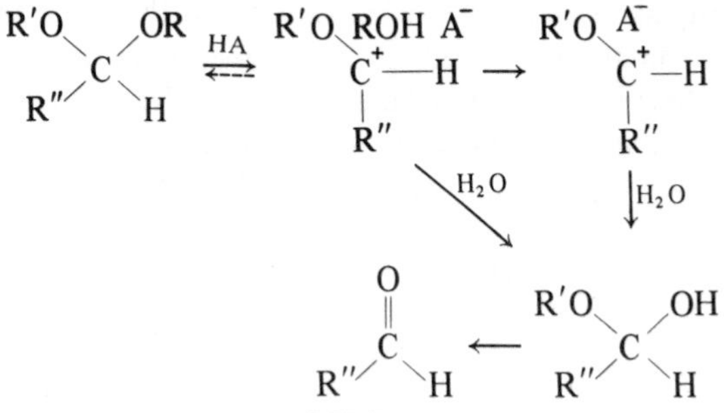

SCHEME 12

The transition state is close in structure to the first intermediate in Scheme 12, which can be seen to be closely associated with two nucleophilic species, the leaving alcohol and a counterion. This may, Jensen suggests, be a clue to the importance of the carboxylate of Asp-52, which assists in preventing reversion of the intermediate to reactant (the dotted arrow in Scheme 12) and so aids the enzymic catalysis.

Young and Jencks[72] have also drawn attention to the fact that lysozyme can catalyze the transfer of a glycosyl group to mono- and disaccharides. This they interpret to mean that in the reaction an intermediate is generated from the reactant glycoside that is sufficiently long-lived so that the alcohol formed with the alkoxycarbenium ion can diffuse away and be replaced by the glycosyl acceptor. An estimated value of the reactivity of the acceptor, relative to water, is 1.5×10^5, so that, even assuming that reaction of the carbenium ion with the acceptor is diffusion controlled, the stabilization of carbenium ion must be at least 5-7 kcal/mol. Both Jensen's and Jencks's interpretations still leave unanswered the question of the precise nature of the stabilization provided for glycosyl carbenium ions by lysozyme, but both envisage a carbenium ion-aspartate-52 ion pair with little if any covalent interaction.

Recent work by Fife and Przystas[75] on the hydrolysis of methyl 3,5-dichlorophenyl acetals of phthalaldehyde (79) and terephthalaldehyde (80) has provided results that lend some support to the notion of electrostatic stabilization of carbenium ions, at least in a nonenzymic situation.

79   80

Unlike hydrolysis of acetals of less acidic phenols, 79 and 80 both show a pH-independent hydrolytic pathway above pH 8, 79 reacting some 100 times as fast as 80. This pathway is taken to be a unimolecular heterolysis of the acetal in its carboxylate form, since the solvent isotope effect for 79 is 1.0. The higher reactivity of 79 is then attributed to electrostatic catalysis of alkoxycarbenium ion formation, although the authors' representation of the charge distribution in the transition state (81) might also be regarded as indicating an intramolecular displacement of 3,5-dichlorophenoxide ion by the neighboring carboxylate group to form the observed product, 3-methoxyphthalide.

81

Another model of lysozyme action, which is more convincing for a number of reasons, not least because the leaving group is an aliphatic alcohol rather than a phenol, has recently been reported by the same authors.[76] Although compounds 82 and 83 show little evidence for carboxyl participation (intramolecular general acid catalysis) at low pH or carboxylate participation (electrostatic or nucleophilic catalysis) at higher pH, 84 shows a bell-shaped pH–rate profile indicating maximal reactivity for the monoanionic form, which, at pH 7.5, is 10^3 times more reactive than 82 and benzaldehyde dicyclohexyl acetal and 4 X 10^4 times more reactive than its own diethyl ester. Both carboxyl groups of 84 appear to be necessary for rapid hydrolysis, one being ionized and one not, and since benzaldehyde diethyl acetal does not show catalysis by buffer acid, the observation of intramolecular general acid catalysis in 84 suggests stabilization of the developing alkoxycarbenium center by the other carboxylate group. The results still do not distinguish electrostatic stabilization from nucleophilic displacement, and it seems doubtful that the distinction can be made unambiguously on the basis of kinetic studies alone.

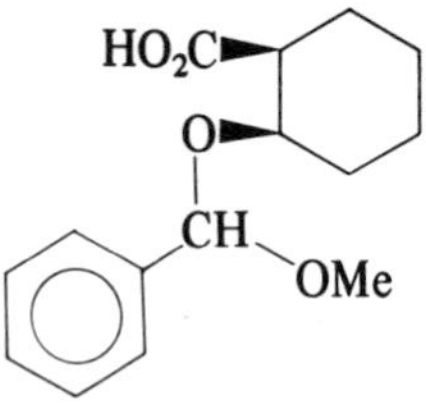

82 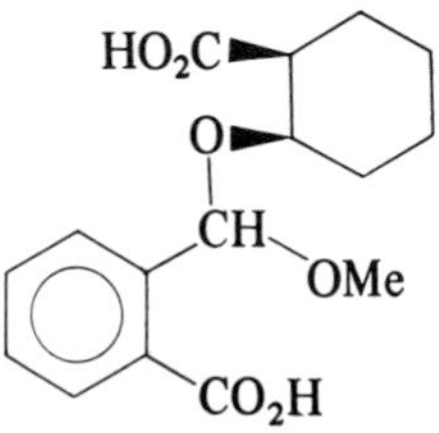   83 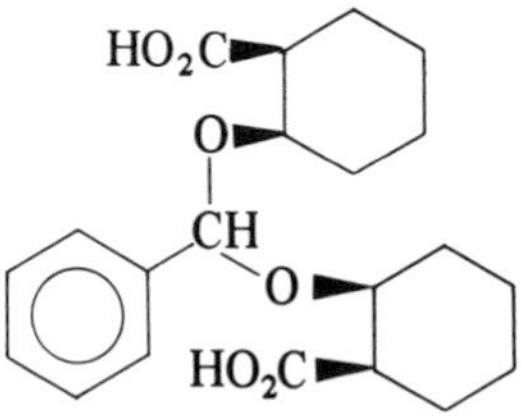   84

Direct evidence that nucleophilic displacement can take place during hydrolysis of acetals has been presented by Kirby and co-workers.[77] They used acetal 85, chosen to maximize the possibility of bimolecular displacement, and by examination of the kinetics of hydrolysis showed the occurrence of a spontaneous process and a second pathway involving catalysis by added nucleophiles. Since many of the nucleophiles used were anions, a crucial feature of the investigation was the disentanglement of kinetic salt effects (i.e., medium effects) from effects due to nucleophilic catalysis. The following observations are of special interest:

1. The sensitivity of 85 to the nucleophilic reactivity of the catalyst was low. A Swain-Scott analysis of the reactivity gave a value of the substrate sensitivity parameter s_N of 0.2–0.3 (cf. 1.0 for MeBr). Application of the Brønsted equation gave $\beta = 0.05 \pm 0.01$, but hydroperoxide ion, by virtue of its α effect, showed high reactivity; $k_{HO_2^-}/k_{HO^-}$ was approximately 100, a value intermediate between that found for highly stabilized cations such as Malachite Green (10^4) and those for typical S_N2 substrates (13–45).

2. Kinetic α-deuterium isotope effects were in the range 5–16% per deuterium atom, values embracing the magnitude normally associated with S_N1 processes and the value observed in reactions catalyzed by lysozyme.

$$MeOCH_2O-\underset{85}{\overset{NO_2}{\underset{}{\bigcirc}}}-NO_2$$

These results have been interpreted in terms of an S_N2 displacement of 2,4-dinitrophenoxide ion by added nucleophiles involving a very loose transition state. This accommodates the evidence of a weak interaction of the nucleophile with the reaction center and kinetic isotope-effect evidence of a high degree of bond breaking in the transition state. While the results seem to be consistent also with nucleophilic attack on a preformed carbenium phenoxide ion pair, Kirby dismisses this on a number of grounds, the more convincing of which are the poor correlation of second-order rate coefficients with Ritchie's N_+ values, and the failure of the kinetic data to fit the rate law derived by Sneen for the ion-pair mechanism.

Of particular relevance for the mechanism of action of lysozyme is the observation that acetate ion possesses about 50% of the nucleophilic reactivity of hydroxide ion toward 85, lending substance to the idea that the carboxylate group of Asp-52 could be an adequate nucleophile to assist in glycosidic cleavage and so generate an acylal intermediate.

Further examination of this question is needed to establish whether, given the steric constraints of the active site of lysozyme, a true acylal can be generated. If such an intermediate cannot be formed because the carboxylate oxygen atom on Asp-52 and the appropriate carbon atom on the N-acetylglucosamine residue bound at subsite D cannot be brought to a bonding distance, then it must be concluded that, notwithstanding Kirby's observations, electrostatic acceleration of alkoxycarbenium ion formation is the crucial factor in the hydrolytic process.

Model systems such as 85 may lend themselves to examination, using the methodology described by Albery and Kreevoy[78] for methyl transfer processes, with a view to developing ideas about reactions in the mechanistic border region between S_N1, S_N2, and nucleophilic substitution on a preformed carbenium ion pair.

VI. REFERENCES

1.* E. M. Arnett and C. Petro, *J. Am. Chem. Soc.*, 100, 2563, 5402, 5408 (1978); see also E. M. Arnett and J. Larsen, *J. Am. Chem. Soc.*, 99, 1438 (1969).

2.   For a discussion of this problem see G. A. Olah and D. J. Donovan, *J. Am. Chem. Soc.*, **100**, 5163 (1978).

3.   J. J. Solomon and F. H. Field, *J. Am. Chem. Soc.*, **98**, 1567 (1976).

4.* E. M. Arnett, C. Petro, and P. von R. Schleyer, *J. Am. Chem. Soc.*, **101**, 523 (1979).

5.   M. H. Abraham, *Prog. Phys. Org. Chem.*, **11**, 1 (1974).

6.   T. W. Bentley and P. von R. Schleyer, *Adv. Phys. Org. Chem.*, **14**, 1 (1977).

7.* P. Kebarle and K. Hiraoka, *J. Am. Chem. Soc.*, **99**, 360 (1977).

8.* R. P. Kirchen, T. S. Sorensen, and K. M. Wagstaff, *J. Am. Chem. Soc.*, **100**, 5134 (1978).

9.   M. Saunders and J. Rosenfeld, *J. Am. Chem. Soc.*, **91**, 7756 (1969).

10. D. Fărcaşiu, *J. Chem. Soc., Chem. Commun.*, 394 (1977).

11. G. A. Olah, G. Liang, and Y. K. Mo, *J. Org. Chem.*, **39**, 2394 (1974).

12. R. P. Kirchen and T. S. Sorensen, *J. Am. Chem. Soc.*, **100**, 1487 (1978).

13.* G. A. Olah, G. K. S. Prakash, T. N. Rawdah, D. Whittaker, and J. C. Rees, *J. Am. Chem. Soc.*, **101**, 3935 (1979).

14. H. C. Brown (with comments by P. von R. Schleyer), *The Nonclassical Ion Problem*, Plenum Press, New York and London, 1977.

15. R. H. Staley, R. D. Wieting, and J. L. Beauchamp, *J. Am. Chem. Soc.*, **99**, 5964 (1977).

16.* P. P. S. Saluja and P. Kebarle, *J. Am. Chem. Soc.*, **101**, 1567 (1979).

17. Using a recently revised value of ΔH_f for the *t*-butyl cation: F. A. Houle and J. L. Beauchamp, *J. Am. Chem. Soc.*, **101**, 4067 (1979); R. G. McLoughlin and J. C. Traeger, *J. Am. Chem. Soc.*, **101**, 5791 (1979).

18. (a) M. J. S. Dewar, R. C. Haddon, A. Komornicki, and H. Rzepa, *J. Am. Chem. Soc.*, **99**, 377 (1977).

     (b) D. W. Goetz, H. B. Schlegel, and L. C. Allen, *J. Am. Chem. Soc.*, **99**, 8118 (1977).

     (c) W. L. Jorgensen and J. E. Munroe, *J. Am. Chem. Soc.*, **100**, 1511 (1978).

     (d) G. Wenke and D. Lenoir, *Tetrahedron*, **35**, 489 (1979).

19.* M. Saunders and M. R. Kates, *J. Am. Chem. Soc.*, **99**, 8071 (1977); the principle was described originally by M. Saunders, M. H. Jaffe, and P. Vogel, *J. Am. Chem. Soc.*, **93**, 2558 (1971); M. Saunders and P. Vogel, *J. Am. Chem. Soc.*, **93**, 2561 (1971).

20.* M. Saunders, L. Telkowski, and M. R. Kates, *J. Am. Chem. Soc.*, **99**, 8070 (1977).

21.* M. Saunders, M. R. Kates, K. B. Wiberg, and W. Pratt, *J. Am. Chem. Soc.*, **99**, 8072 (1977).

22. P. G. Gassman and A. F. Fentiman, *J. Am. Chem. Soc.*, **91**, 1549 (1969).

23. See *Reactive Intermdiates*, Vol. 1, M. Jones and R. Moss, Eds., Wiley-Interscience, New York, 1978.

24.* D. G. Farnum, R. E. Botto, W. T. Chambers, and B. Lam, *J. Am. Chem. Soc.*, **100**, 3847 (1978).

25.* G. A. Olah, G. K. S. Prakash, and G. Liang, *J. Am. Chem. Soc.*, **99**, 5683 (1977).

26. H. C. Brown and E. N. Peters, *J. Am. Chem. Soc.*, **97**, 7442 (1975).

27. H. C. Brown, E. N. Peters, and M. Ravindranathan, *J. Am. Chem. Soc.*, **97**, 7449 (1975).

28. H. C. Brown, S. Ikegami, K.-T. Liu, and G. L. Tritle, *J. Am. Chem. Soc.*, **98**, 2531 (1976).

29. H. C. Brown and K.-T. Liu, *J. Am. Chem. Soc.*, 91, 5909 (1969).

30. H. C. Brown, K. Takeuchi, and M. Ravindranathan, *J. Am. Chem. Soc.*, 99, 2684 (1977).

31. H. C. Brown and K. Takeuchi, *J. Am. Chem. Soc.*, 90, 5268 (1968); H. C. Brown, E. N. Peters, and M. Ravindranathan, *J. Am. Chem. Soc.*, 97, 2900 (1975).

32. H. C. Brown and E. N. Peters, *J. Am. Chem. Soc.*, 97, 1927 (1975).

33. H. C. Brown and M. Ravindranathan, *J. Am. Chem. Soc.*, 99, 299 (1977).

34.* W. L. Jorgensen, *J. Am. Chem. Soc.*, 99, 3840 (1977).

35. E. N. Peters, *J. Am. Chem. Soc.*, 98, 5627 (1976).

36. H. Hogeveen and P. W. Kwant, *Acc. Chem. Res.*, 8, 413 (1975).

37. W. D. Stohrer and R. Hoffmann, *J. Am. Chem. Soc.*, 94, 1661 (1972).

38. S. Masamune, M. Sakai, and H. Ona, *J. Am. Chem. Soc.*, 94, 8955 (1972).

39. S. Masamune, M. Sakai, H. Ona, and A. J. Jones, *J. Am. Chem. Soc.*, 94, 8956 (1972).

40. S. Masamune, M. Sakai, A. V. Kemp-Jones, H. Ona, A. Venot, and T. Nakashima, *Angew. Chem., Int. Ed. Engl.*, 12, 769 (1973).

41. H. Hart and M. Kuzuya, *J. Am. Chem. Soc.*, 96, 6436 (1974).

42. A. V. Kemp-Jones, N. Nakamura, and S. Masamune, *J. Chem. Soc., Chem. Commun.*, 109 (1974).

43. H. Hart and M. Kuzuya, *J. Am. Chem. Soc.*, 97, 2450 (1975).

44. G. A. Olah and A. M. White, *J. Am. Chem. Soc.*, 91, 5801 (1969).

45. S. Masamune, M. Sakai, A. V. Kemp-Jones, and T. Nakashima, *Can. J. Chem.*, 52, 855 (1974).

46. P. W. Kwant, Thesis, University of Groningen, 1974.

47. H. C. Longuet-Higgins, *Chem. Soc. (Lond.) Spec. Publ.*, 21, 109 (1969).

48. H. Hart and R. Willer, *Tetrahedron Lett.*, 4189 (1978).

49.* R. M. Coates and E. R. Fretz, *Tetrahedron Lett.*, 1955 (1977).

50. H. Hogeveen and P. W. Kwant, *J. Am. Chem. Soc.*, 96, 2208 (1974).

51.* C. Giordano, R. F. Heldeweg, and H. Hogeveen, *J. Am. Chem. Soc.*, 99, 5181 (1977).

52.* J. S. Staral, I. Yavari, J. D. Roberts, G. K. S. Prakash, D. J. Donovan, and G. A. Olah, *J. Am. Chem. Soc.*, 100, 8016 (1978).

53.* J. S. Staral and J. D. Roberts, *J. Am. Chem. Soc.*, 100, 8018 (1978).

54. G. A. Olah, R. J. Spear, P. C. Hiberty, and W. J. Hehre, *J. Am. Chem. Soc.*, 98, 7470 (1976).

55.* R. P. Kirchen and T. S. Sorensen, *J. Am. Chem. Soc.*, 99, 6687 (1977).

56.* G. A. Olah, G. K. S. Prakash, D. J. Donovan, and I. Yavari, *J. Am. Chem. Soc.*, 100, 7085 (1978).

57. G. A. Olah, G. Liang, and Gh. D. Mateescu, *J. Org. Chem.*, 39, 3750 (1974).

58. G. A. Olah, G. Liang, P. von R. Schleyer, W. Parker, and C. I. F. Watt, *J. Am. Chem. Soc.*, 99, 966 (1977).

59. B. A. Levi, E. S. Blurock, and W. J. Hehre, *J. Am. Chem. Soc.*, 101, 5537 (1979).

60. S. Winstein, E. C. Friedrich, R. Baker, and T. Lin, *Tetrahedron, Suppl.* 8, *Part II*, 621 (1966).

61. R. M. Coates and E. R. Fretz, *J. Am. Chem. Soc.*, 97, 2538 (1975).

62. D. G. Morris, P. Murray-Rust, and J. Murray-Rust, *J. Chem. Soc., Perkin II*, 1577 (1977).

63. W. L. Jorgensen, *Tetrahedron Lett.*, 3029, 3033 (1976).

64. S. Forsen and R. Hoffman, *Acta Chem. Scand.*, 17, 1787 (1963); *J. Chem. Phys.*, 39, 2892 (1963); *J. Chem. Phys.*, 40, 1189 (1964).

65.* C. Engdahl and P. Ahlberg, *J. Am. Chem. Soc.*, 101, 3940 (1979).

66. S. Winstein, *Q. Rev.*, 23, 141 (1969).

67.* G. A. Olah, G. Asensio and H. Mayr, *J. Org. Chem.*, 43, 1518 (1978).

68. D. V. Banthorpe, P. A. Boullier, and W. D. Fordham, *J. Chem. Soc., Perkin I*, 1637 (1974).

69.* D. Fărcaşiu and L. Craine, *J. Chem. Soc., Chem. Commun.*, 687 (1976).

70.* D. Fărcaşiu, *J. Am. Chem. Soc.*, 100, 1015 (1978).

71. For a recent summary see A. Fersht, *Enzyme Structure and Mechanism*, W. H. Freeman, New York, 1979.

72.* P. R. Young and W. P. Jencks, *J. Am. Chem. Soc.*, 99, 8238 (1977).

73. R. Eliason and M. M. Kreevoy, *J. Am. Chem. Soc.*, 100, 7037 (1978).

74.* J. L. Jensen, L. R. Herold, P. A. Lenz, S. Trusty, V. Sergi, K. Bell, and P. Rogers, *J. Am. Chem. Soc.*, 101, 4672 (1979).

75. T. H. Fife and T. J. Przystas, *J. Am. Chem. Soc.*, 99, 6693 (1977).

76.* T. H. Fife and T. J. Przystas, *J. Am. Chem. Soc.*, 101, 1202 (1979).

77.* G.-A. Craze, A. J. Kirby, and R. Osborne, *J. Chem. Soc., Perkin II*, 357 (1978).

78. W. J. Albery and M. M. Kreevoy, *Adv. Phys. Org. Chem.*, 16, 87 (1978).

7

FREE RADICALS

LEONARD KAPLAN

Union Carbide Corp., South Charleston, West Virginia 25303

251

I. STRUCTURE OF PROTOTYPAL RADICALS

A. Isopropyl

A technique described earlier[1a, 2] has been applied to the isopropyl radical.[3] $Me_2CHCO_2O_2CCHMe_2$ was irradiated ($\lambda > 2900$ Å) in an argon matrix at 6 K to give a transient molecule, reasonably considered to be $Me_2CH\cdot$, whose infrared absorptions at 3045 and 375 cm^{-1} (assigned to an out-of-plane bending mode) synchronously disappeared with the appearance of bands due to $Me_2CHCHMe_2$, $MeCH_2Me$, and $MeCH=CH_2$ when the sample was warmed to $\sim$35 K. By application of earlier resoning,[1a] it was concluded from the band at 3045 cm^{-1} that the isopropyl radical is "planar, or nearly planar."

B. *t*-Butyl

a. *Vacuum UV photoelectron spectroscopy*

From the photoelectron spectrum of the molecule produced by heating $Me_3CN=NCMe_3$ diluted with helium to $\sim$1000 K, an attempt has been made to distinguish a planar from a pyramidal ground-state structure on the basis of whether reasonable assignments could be made of the vibrational structure of the first band of the spectrum on both the high and low ionization potential sides of the band.[4] Because of difficulty in reconciling what was felt to be "the only reasonable assignment for the structure observed on the low ionization potential side of the band" with expectations for a planar structure based on a very simple pictorial model of the bonding and on the results of INDO-CI calculations, such a structure was rejected; in contrast, the low ionization potential structure could be assigned, in a Procrustean manner, for a pyramidal radical. Next, what we believe to be their best argument was presented: ". . . one knows from experience . . . that broad bands are usually indicative of big changes in geometry [on going from the radical to the cation]." Interpretation was then extended even further and a barrier to inversion and an out-of-plane distance of the central carbon were estimated. This may be premature in view of the uncertainties surrounding the assignments.

Additional results are available.[5a] Also, largely on the very plausible basis of work not yet published,[5b, c] the following general opinion was presented:

> "In contrast to methyl radical, which undergoes no significant geometry change on ionization, the methyl-substituted radicals incur complex geometry changes, with excitation of several vibrational modes. This situation is partly the result of very low barriers to internal rotation and out of plane bending in both the ions and the neutrals. Consequently, quantitative analysis . . . of the photoelectron bands of these radicals is very difficult . . . . In view of this, photoelectron data cannot yet be cited as evidence for the planarity or nonplanarity of alkyl radicals having low symmetry."

b. Electron Spin Resonance Spectroscopy

Earlier,[1a] we discussed reports that relied heavily on the sign of the temperature dependence of the H-coupling constant (a_H) and, especially, of the C-coupling constant (a_C) to provide substantial support for the proposition that the structure of the t-butyl radical is not planar at the radical site when time-averaged over all rotamers. The key feature of the temperature dependence of a coupling constant, insofar as it relates to the structure of a radical, is the sign of the dependence. The temperature dependence of a_H of $Me_3C\cdot$ has now been examined in four matrices—t-butyl chloride, neopentane, isobutane, and dilute neopentane in xenon.[6a] Although matrix effects on the weak temperature dependence of a_H were found, the dependence was positive in all matrices; the differences were in slope and limiting value. Arguments were then advanced as if the demonstration of a matrix effect on characteristics of the temperature dependence exclusive of its sign is tantamount to a demonstration of a matrix effect on the sign itself. There were also arguments that appear to be to the effect that the demonstration of a matrix effect on a_H is tantamount to a demonstration of one on a_C, on which earlier[1a] conclusions were principally based.

Above our comments have concerned what can be concluded from that which has been demonstrated experimentally. The discussion[6a] also covers possible explanations of the sign of the temperature dependence of the a_H and a_C observed earlier. The less contrived and *ad hoc* explanation basically says that the matrix is responsible for making the radical nonplanar[7]; its implication is that the earlier analysis of the observations may be inappropriate because the origin of the nonplanarity may not be what it was thought to be. Such an argument addresses why the radical is nonplanar, not whether it is nonplanar. The position is that the structure determined in a matrix is not the same as that which would have been determined in, say, the gas phase. Perhaps. But implicit in a discussion of *any* measurement is reference to a particular medium; the results may or may not be extendable.

As a consequence of such suggestions of medium effects, the temperature dependence of a_C in solution was investigated.[6b] $Me_3^{13}C\cdot$ was generated from $Me_3^{13}CBr$ by means of its photolysis with $Me_3SnSnMe_3$ and with $(t\text{-BuO})_2/Et_3SiH$ in propane and in isooctane. The results were the same in both solvents and for both methods. The value of a_C varied much less with temperature than it did in the solid state and it passed through a minimum, an observation consistent with and suggestive of a pyramidal structure. The possibility was indicated that such a structure *"may have been induced by the solvent"*; then again, it may not have. The broader implications of such a possibility are discussed above.

We note that Symons has reaffirmed his position that "the e.s.r. data for the t-butyl radical strongly indicate effective planarity."[9]

II. CHARACTERISTICS OF PROTOTYPAL RADICALS

A. Ethyl and Isopropyl

Normally,[1a, 10a] ESR spectra have been used to infer something about the structure of a free radical, the inference being strengthened as a result of a vibrational analysis. The vibrational analysis has generally served to provide further support for arguments that rested primarily on determinations of isotropic coupling constants and qualitative trends involving them. Since the principal line of reasoning leading to a conclusion about the radical's structure is usually not definitive,[10a] subsidiary support such as a vibrational analysis, whose relevance rests on the validity of the principal physical model and line of reasoning, is necessarily even less secure even though it may be valid within the framework of the model to which it is applied.[11]

An ambitious program, the detailed, serious vibrational analysis of radicals by use of ESR spectroscopy, has been undertaken.[12] We agree that "the EPR approach to vibrational analysis has not been critically tested."[12a]

$Me_2^{13}CH\cdot$ was generated from $Me_2^{13}CHBr$ by means of its photolysis with $Bu_3SnSnBu_3/(t\text{-BuO})_2$ in isooctane or propane.[12a] The value of a_C was determined as a function of temperature between 131 and 352 K; its variation was considered to be a consequence of the temperature dependence of the out-of-plane vibration of a planar radical, successfully describable quite remarkably as a simple harmonic oscillator. The vibrational frequency calculated for $Me_2^{12}CH\cdot$, based on such a model, was 380 cm^{-1}, as compared to a frequency of 375 cm^{-1} observed in the infrared spectrum and "tentatively assigned to the out-of-plane bending mode of the radical center."[2, 13]

Next,[12b] $Me^{13}CH_2\cdot$ was generated from $Me^{13}CH_2Br$ in the same manner. Then a_C was determined as a function of temperature between 112 and 313 K; its variation was treated on the same basis, but semiempirically, as was that of $Me_2CH\cdot$, and was consistent with an out-of-plane vibrational frequency of ~ 500

cm^{-1}, as compared to a frequency of 541 cm^{-1} observed in the infrared spectrum and assigned to that mode.[2]

B. Dipole Moment of Trifluoromethyl

The dipole moment of $CF_3 \cdot$, generated by reaction of $F \cdot$ and $CF_2 = CF_2$, has been determined from the effect of electric field strength on the "focusing"[14] of $CF_3 \cdot$ versus NF_3 in an inhomogeneous electric field.[15] From this effect, because of the similarity of the structure of NF_3, of known dipole moment, and that estimated for $CF_3 \cdot$, the relative dipole moments could be determined. That of $CF_3 \cdot$ was calculated to be 0.43 D.

C. Alkoxy Radicals

The ESR spectrum of an alkoxy radical in solution has not been observed. The usual explanation[16] is based on an approximate, pictorial model and proceeds along the following lines: The degeneracy, or near-degeneracy, of the orbitals of the unpaired electron and a lone pair on oxygen leads, as a consequence of the dependence of the g factor on the spin–orbit coupling, to a highly anisotropic g factor that, in the fluctuating environment of a fluid medium, produces short relaxation times and consequent line broadening. Such an argument also has been used to explain the independent observation that CIDNP effects were not observed during reactions believed to involve radical pairs containing t-butoxy radicals.[17]

Based on such a model, it was anticipated that a spectrum could be observed if the radical were in an environment that interacted strongly and asymmetrically with it, so that the relevant orbitals would be of significantly different energy.[16a] Hydrogen bonding would be such a perturbation.[16b] Unfortunately, the perturbing factors that would permit observation of $RO \cdot$ also introduce the problem that the observed spectral parameters need not be characteristic of free $RO \cdot$.[18]

The anticipation has been realized in single-crystal, powder, and polycrystalline matrices. A variety of such radicals, having constitutionally complex precursors,[19] has been reported.[20] The spectra of the primary alkoxy radicals were characterized generally in terms of a large hyperfine coupling of $\sim$70–80 G to each of the β-hydrogens and a large maximum g value of 2.05–2.09.[21a, 21b]

Spectra of simple alkoxy radicals also have been described. The ESR spectrum of the product of X-irradiation of polycrystalline CH_3OH at 4.2 K was assigned reasonably to $CH_3O \cdot$.[22, 23] A g_{max} of 2.088, typical of $RO \cdot$, was observed. It is large, as expected for a radical in which the unpaired electron density is centered primarily on an atom (oxygen) with a higher spin–orbit coupling constant than that of carbon. What was considered to be a pseudo-seven-line multiplicity of the splitting has been discussed in terms of tunneling rotation[24] of the methyl group. Might Jahn-Teller[16a, 18, 25a, 25b] or site-induced distortion from

C_{3v} symmetry account for the spectrum that was actually observed, without any requirement that tunneling by invoked? Also, note that, even in the absence of such distortion, almost all the $RO\cdot$'s that have been studied cannot have C_{3v} symmetry. The "origin" of the characteristics of the spectrum in this particular case[22] and the more general question of the "reason" for the lack of observation of $RO\cdot$ under "normal" conditions are quantitative, not qualitative, questions to be answered through quantitative, not qualitative, argument. This is so even within the framework of the simple approximate model used so far. What should be considered is how close the orbitals are in energy and what the quantitative consequences are of that separation and the other relevant factors on observability.

It was argued[22] from the similarity of the spectral parameters of $CH_3O\cdot$ in methanol and those of substituted alkoxy radicals[19] in matrices very different from methanol that the spectra are indeed characteristic of free $RO\cdot$ and that the concern expressed above[18] does not materialize. It has also been suggested,[21a] however, that the large β-couplings reported are a result of hydrogen bonding of the alkoxy radical and that, in the absence of hydrogen bonding, the couplings would not be unusually large. Since the characteristics of free and of strongly matrix-perturbed species are, in principle, different, the burden of proof must remain on a proponent of the idea that the magnitude of that difference is small, that is, that spectra characteristic of pure $RO\cdot$ have been observed.

The spectrum of $CD_3O\cdot$ may have been observed upon irradiation of CD_3-OD.[21c, 22] The spectra observed upon γ-irradiation of ROH (R = t-butyl, t-amyl) ($g_{max} = 2.11$, a_H unresolved) and β-irradiation of c-$C_6H_{11}OH$ ($g_{max} = 2.09$, $a_H \approx 70$ G, doublet) have also been attributed to the corresponding alkoxy radicals.[21c, 26]

In general, how strong are the above assignments to $RO\cdot$? "There can be no doubt of the identification [Ref. 21d] as an $RCH_2O\cdot$ radical, since the form of the g-tensor, the large coupling to two inequivalent β-type protons and the directions of the g-tensor components relative to the hydrogen-bond framework in the crystal are all as expected for such radicals."[27a] Although we believe the assignments to be quite reasonable, more work with isotopically labeled materials would strengthen them. There has been, however, a report[28] of great similarity between spectral parameters reported for $RCH_2O\cdot$ and those attributed[28] to an acyl radical, $ClCH_2\dot{C}O$, and there has been a reference[21e] to "distinguishing primary alkoxy radicals from other as yet unidentified free radical products . . . that give rise to somewhat similar ESR spectra."

Work on the vibrational transitions of methoxyl has resumed[29] and its laser magnetic resonance spectrum has been reported (Section IV.B).

D. Alkyl Thiyl Radicals

Alkyl thiyl radicals are in the same situation with respect to ESR observation as are alkoxy radicals (Section II.C), presumably for the same reason.[16a, 27b, 30a]

ESR spectra generated from a variety of precursors,[31] in single crystal and glassy matrices and adsorbed on surfaces, and attributed to alkyl thiyl radicals have been reported[20]; some of those incorrectly assigned are not cited.

E. Radical Anions Comprised of Tetracoordinated Carbon Only

The title compounds constitute a subset of a class of molecules known as σ^*-radicals, σ^*-ions, or σ^* radical anions.[20,32]

We described earlier[1b] a report of the observation by use of ESR of the anion radical of perfluorocyclobutane. That work has since been described in more detail and expanded upon (Sections II.E.a–II.E.d). The title class of compounds has also been observed in other systems (Sections II.E.e and II.E.f).

We merely list[20] uncritically another subset, σ^*-ions that contain conjugated systems,[32] and exclude from discussion any system that could have been, for example, π-delocalized.[32] Reports in this category are of molecules that contain aromatic, amide, acetylenic, and olefinic groups.[33,34a] These are cases in which the electron could have gone into an extended π-system, but instead the resulting radical anion, according to the "best description" has the electron in a σ^*-orbital.

a. Perfluorocyclopropane (c-C_3F_6) Radical Anion[34b]

Methods of generation. (1) γ-Irradiation of c-C_3F_6 in neopentane and tetramethylsilane matrices. (2) Addition of tetramethyl-p-phenylenediamine to the above mixture, followed by UV irradiation.

Basis of identification of carrier. (1) Methods of generation. (2) ESR spectrum consistent with expectation. (3) Spectrum can be completely photobleached.

b. Perfluorocyclobutane (c-C_4F_8) Radical Anion[34]

Methods of generation. (1) γ-Irradiation of c-C_4F_8 in neopentane and tetramethylsilane matrices.[34b] (2) As 2 above for c-C_3F_6[34b] and also in 2-methyltetrahydrofuran.[34b] (3) γ-Irradiation of $CF_2{=}CF_2$ in tetramethylsilane-d_{12} or in 2-methyltetrahydrofuran or in 3-methylhexane matrices.[34c] (4) γ-Irradiation of c-C_4F_8 in 2-methyltetrahydrofuran.[34a]

Basis of identification of carrier. (1) As in 1–3 above for c-C_3F_6.[34b] (2) $CF_2{=}CF_2$ in 2-methyltetrahydrofuran was γ-irradiated at 77 K. The ESR spectrum recorded at 83 K "clearly resembles the powder spectrum which has been analyzed in terms of $C_2F_4^-$." "On annealing the sample to 95 K, these features belonging to $C_2F_4^-$ gradually disappeared and were replaced by . . . [those] of $[c$-$C_4F_8^{-\cdot}]$".[34c] (3) "The cycloaddition reaction was also monitored by optical absorption, the structured band of $C_2F_4^-$ . . . being replaced by a broad band at 390 nm which can be assigned to c-$C_4F_8^-$ . . . ."[34c]

c. Perfluorocyclopentane (c-C_5F_{10}) Radical Anion[34b]

c-C_5F_{10} of unspecified purity in tetramethylsilane matrix was γ-irradiated. The ESR spectrum was consistent with expectation and could be completely photobleached.

d. General Comments on Aspects of the Above

The results with tetramethyl-p-phenylenediamine constitute reasonable evidence that a step in the formation of the carrier is electron attachment to the substrate.

The photobleaching results are consistent with photodetachment and are therefore consistent with the carrier being an anion; "electrons can be transferred from c-$C_4F_8^-$ to . . . SF_6 or methyl halides by photobleaching when one of the latter compounds is also incorporated in the tetramethylsilane matrix."[34d]

"The isotropic e.s.r. spectra of $[c$-$C_nF_{2n}(n = 3$-$5)]$ . . . show the second-order structure characteristic of 6, 8 and 10 equivalent fluorines, respectively, the total ^{19}F coupling being approximately the same value (1170 ± 20 G) in each case."[34b] In the Abstract of Ref. 34b it is stated that "the equivalence of the fluorines *indicates* (italics ours) that the unpaired electron is delocalized over the entire molecular framework in an orbital of high symmetry." However, in the Discussion,[34b] the phrase "suggests that" is used and it is emphasized "that this observation of equivalent fluorines does not require stereochemically rigid (D_{nh}) ring-structures. . . ." Elsewhere,[34a] it is stated that the spectrum assigned to $C_4F_8^-$ "suggests some real inequivalence between the fluorine atoms" and "it seems probable that . . . there is a major deviation from planarity."

The evidence is strong, particularly in the C_4 system, that the spectrum of *a* radical anion is being observed. However, is it the spectrum of *the* radical anion, that is, that of the substrate? This determination rests on the breadth of the consistency of the spectrum with expectation which, as indicated earlier,[1b] could be increased by examination of systems containing isotopically labeled substrates.

We also expressed interest in the study of the radical anion of a tetrahedrally symmetric fluorocarbon.[1b] 1,3,5,7-Tetrafluoroadamantane was then a known compound. Perfluoro-1,3,5,7-tetramethyladamantane has since been reported.[35]

e. Halomethane Radical Anions

Our earlier statement[1b] that there had been no previous knowledge of the existence of the title species may have been incorrect: γ-irradiation at 77 K of CBr_4 in CD_3OD led to the observation of an ESR spectrum assigned to CBr_4^- on the basis of, at most, suggestive evidence.[36a] There have recently been several reports of the observation of radical anions of halomethanes by use of ESR and IR spectroscopy.

γ-Irradiation of CF_3X (X = Cl, Br, I) of undescribed purity in tetramethyl-silane, neopentane, and 2-methyltetrahydrofuran matrices and of CF_2Cl_2 of undescribed purity in tetramethylsilane led to the observation of ESR spectra assigned to the corresponding radical anions. The spectrum of CF_3Cl^- also was generated by UV irradiation of tetramethyl-p-phenylenediamine in a CF_3Cl-tetramethylsilane matrix.[34b, 36b] We assess the assignments as follows:

CF_3Cl^- Assignment is very reasonable and probably correct.

CF_3Br^- Assignment is reasonable.

CF_3I^- Information provided is insufficient for assessment of the assignment.

$CF_2Cl_2^-$ Spectrum is not inconsistent with the assignment.

Two methods have been reported to permit observation of the infrared spectra of halomethane radical anions: (1) photoionization-photoirradiation-proton irradiation of the halomethane with excess argon at 15 K, and (2) photoirradiation of the halomethane with sodium.[37] We assess the assignments as follows:

$$CHCl_3^-, CDCl_3^-, CHCl_2Br^-, CHClBr_2^-, CHBr_3^-, CDBr_3^-, \text{ and } CF_3Cl^-$$

The assignments are reasonable.

$$CF_2Cl_2^-, CFCl_3^-, CF_3Br^-, CF_2Br_2^-, CFBr_3^-, CF_2ClBr^-, \text{ and } CF_3I^-$$

The assignments are tentative.

From a study of the vibrational progressions of the low-energy electron transmission spectra of CF_4 and CF_3Cl conclusions have been drawn regarding the electronic structure of the corresponding radical anions formed by capture of the electrons.[38] It is argued as follows:[38]

	CF_4	CF_4^-	CF_3Cl	CF_3Cl^-
Vibrational energy (cm^{-1})	1283	1283 ± 15	1212	1146 ± 30

The vibrational energy of CF_4^- [predominantly antisymmetric C—F stretch] ... is exactly the same as that of the neutral molecule .... So the extra electron is captured in an orbital that does not affect the C—F bond. In CF_4 there are no low lying valence orbitals and, moreover, any valence orbital would be C—F antibonding [so what?]. Therefore the extra electron must be in a Rydberg type orbital ... the similarity between the structure in the CF_3Cl transmission spectrum and the Rydberg type resonance in CF_4 is striking. So we believe that there ... [is a] Rydberg [resonance of CF_3Cl] .... EPR studies in a low temperature matrix show that . . . the unpaired electron [in CF_3Cl^{-}] is in an orbital localized in the C—Cl bond. However, since state with an electron in a diffuse Rydberg orbital will be strongly disturbed in the condensed phase, and the valence type states are stabilized in such a matrix, this does not affect our conclusions. Summarizing, our conclusion is that there is a relatively stable state in CF_4^{-} and CF_3Cl^{-} with an electron in a Rydberg orbital.

However, in a matrix (see above) the infrared frequency for CF_3Cl^{-} has been reported to be close to that of CF_3Cl.[37a] The vibrational frequency just is not that sensitive an indicator of electronic structure.

f. Ethylene oxide

The ESR spectrum observed upon adsorption of N_2O on tungsten supported on silica gel was assigned to O^{-}. Introduction of ethylene at 77 K led to a decrease in the spectrum assigned to O^{-} with "the simultaneous appearance" of a new spectrum, characterized by $g = 2.003$ and five hyperfine lines of $a = 25.4$ G and in the ratio $1:4:6:4:1$, whose carrier "may be considered as" a species whose "structure . . . resembles that of . . ." the radical anion of ethylene oxide.[39] ESR spectra attributed to the isomer $\cdot CH_2CH_2O^{-}$ have also been reported.[40]

F. Radical $\cdots$ Anion (R·/X⁻), Radical $\cdots$ Ion Pair (R·/X⁻M⁺) and (Carbene $\cdots$ HX)⁻ Complexes

The title compounds, closely related to each other and to the class of σ^*-ions (Section II.E), constitute an area of ongoing significance.[9, 20, 27c, 33a, 37a, 37b, 37c, 41]

G. What Are We Talking about in Sections II.E and II.F?

What is the difference between the title compounds of Sections II.E and II.F? For example, when would a molecule be called a σ^*-ion (RX⁻) and when would it be called a radical–anion complex (R·/X⁻)? Dichotomous distinctions based on continuous variables (structural parameters) can be made arbitrarily at best. Although a structure-based meaningful distinction between the two classes of compounds could conceivably be possible if, for example, there were a bimodal distribution of bond lengths in such species, it is much more likely that we are faced with the common question: When does a compound go from having a long bond to being a mere complex? It may be best to rely on common usage, which is divided between (1) operational definitions and (2) those based on approximate theoretical models:[32]

1. ... these species, R · · · X⁻, are not properly described as the radical anions, RX⁻, but are genuine alkyl radicals which, formed in a solid-state "cavity," are unable to leave the site of the anions X⁻ and exhibit a weak charge-transfer interaction which does not modify their shape or reactivity appreciably, but only their e.s.r. spectra.[41a,41b]

   ... alkyl radical–halide ion complexes that have effectively reached the limit of complete dissociation, but which remain together because of the influence of the rigid solvent cage.[42]

   [R·/X⁻] display ESR parameters for the alkyl radicals which are only slightly reduced from those for the "free" alkyl radicals, and hyperfine coupling to the halogen nuclei is small.[33b]

2. [Distinction between] a methyl radical undergoing weak charge-transfer interaction with a bromide ion [and] MeBr⁻ in its limiting form in which the unpaired electron is almost entirely confined to the carbon $2p(\sigma)$ orbital.[43a]

   ... if any residual covalent bonding is present [in R·/X⁻], it must be small.[33b]

   ... for the σ^*-radicals ... the unpaired electron is evenly distributed between the alkyl group and the halogen atom, whereas for the alkyl radical–halide ion complexes, the spin is essentially confined to the planar alkyl radical, the interaction with the halide ion being very small, and probably indicative of slight charge transfer.[43b]

An argument has been put forward to the effect that

σ^*-anions [of simple monoalkyl halides] cannot exist for longer than a few vibrational periods,[9] [that is] ... monoalkyl halides do not have an energy minimum of the type clearly displayed by ... F_3C-hal⁻ ions [σ^* ions] .... My reason ... is based upon the detection of alkyl radical–halide ion complexes ....[42] It seems to me very unlikely that alkyl halide anions can exist both as the established radical–halide ion adducts and also as genuine σ^*-anions ... a unique double minimum reaction coordinate is required if σ^* anions have a finite lifetime, as well as the R·hal⁻ adducts. The key point is that an extra barrier is needed ... to prevent the R·hal⁻ adduct ... from moving back to the σ^* complex. Without such a barrier, either the σ^* complex or the R·hal⁻ adduct can be favored, but not both.[43b]

We do not believe that the facts require the conclusion. Regardless of whether the system is in a solid matrix or not, the existence of a double-minimum potential is not prohibited *a priori*.

Why do some molecules form σ^*-radicals more easily than others? Note at the outset that this question is different from, although not unrelated to, Why is the free energy of formation of some σ*-radicals from their parents more favorable than that of others?

> Why . . . do CH_3—Br and CF_3Br, for example, give such different products with electrons [dissociative electron capture versus σ*-radical, respectively]? I suggest that the key to understanding the difference lies in the fact that methyl radicals are planar, whilst ·CF_3 radicals are markedly pyramidal . . . . As the C·Br bond stretches in CH_3Br on electron addition, the CH_3 group flattens. This increases the $2p_z$ character at carbon, and concurrently reduces the overlap, thus weakening the bond. This continuous weakening leads ultimately to planar methyl and an effectively broken bond. No such flattening occurs for F_3C—Br and hence a balance is achieved and the weakened bond is retained.[42,44]

While this argument does have a certain internal consistency to it, when viewed in more general terms it does not appear to us to follow uniquely from the evidence currently available. Its central point seems to be that the species that requires only a little structural reorganization to go to product is consequently the more stable. We do not defend the very common antithetical thinking; however, regardless of the validity of such thinking, the absence of a need for large structural change does not, in itself, require that the barrier be high. How about a simple electronegativity argument? ("The CF_3X^- radical anions have . . . been produced in several types of experiments, indicating possible stabilization of the negative charge by the inductive effect of the fluorine atoms."[37d])

H. Olefin Radical Cations

The radical cation of adamantylidineadamantane (1) has been assigned as the carrier of the ESR spectrum (broadened multiplet for many hydrogens with a 3.0 G splitting constant, $g = 2.0031$) generated by electrolytic oxidation of the olefin in acetonitrile.[45a] It was considered to be "the first example of a long-lived monoolefin cation radical," it being so as a result of "Bredt's rule stabilization," although no evidence for an unusual lifetime other than the fact of its observation was presented. The magnitude of the coupling was attributed to "hyperconjugation" involving the γ-hydrogens; the magnitude of the g value (see

1

also Ref. 46) "points to unusually important σ,π interaction." Other radical cations of olefins[46] have been reported.[20]

γ-Irradiation of $Me_2C{=}CMe_2$ in an alkane glass in the presence of electron acceptors resulted in the observation of an ESR spectrum that was assigned reasonably to $Me_2C{=}CMe_2^{\ddagger}$.[45b] As the concentration of olefin was increased, that spectrum was replaced by one attributed to $(Me_2C{=}CMe_2)_2^{\ddagger}$. At these conditions of high concentration, a broad optical absorption band was observed at $\sim$860 mμ. That band became weaker as the concentration of olefin was decreased and disappeared when the concentration was such that the spectrum assigned to $Me_2C{=}CMe_2^{\ddagger}$ was present. "$Me_2C{=}CMe_2$ does not give a distinct band at wavelengths longer than 500 nm" (cf. Refs. 47a–47e). A band at $\sim$290 mμ was tentatively assigned to $Me_2C{=}CMe_2^{\ddagger}$. These observations cast doubt on other optical absorption assignments to olefin radical cations.[47a,47f–47i] The products of γ-irradiation of cyclohexene in alkane/alkyl chloride solutions also have been studied by use of optical spectroscopy.[45c] It was concluded that ". . . both monomer and dimer cations of cyclohexene . . . have been found and it is argued that many of the cationic olefin species found by other workers are dimeric [and that] the long wavelength bands of [cyclohexene and] other olefins are probably also due to dimers."

ESR spectra have been assigned to the isoelectronic $>B{-}\dot{C}<$. For example, γ-irradiation of Me_3B at 77 K gave a spectrum ($a = 4.3, 18\ G, g \sim 2.003$) assigned to $Me_2BCH_2^{\ddagger}$.[48]

I. Comparative Chemistry of Radicals in Different Electronic States

Several workers have addressed experimentally the question of the comparative chemistry of radicals, as exemplified by amidyl and imidyl, in different electronic states[49] following comments on such radicals by Koenig and Wielesek,[51a]

> . . . all three electronic structures [of succinimidyl] [Π, $\Sigma_{nitrogen}$, Σ_{oxygen}] are potential reaction intermediates. Different precursors or methods of observation could involve different electronic states. . . . The Σ_0 structure could undergo ring opening (β-scission) to the electronic ground state of an acyl isocyanate. The same nuclear motions for the Π structure leads to an excited state of the isocyanate. The β-scission of the Π radical would thus be expected to be slow. β-scission of the Σ_N configuration correlates with ground state isocyanate.

and by Hedaya et al.,[51b]

> Conjugated Π's may react at more than one site, whereas . . . [many] Σ's . . . are . . . characterized by predominant location of spin at one site, at which reaction should always (or almost always) occur. In general, Σ's should be more reactive and less selective than Π's.

Some of this work has involved observing the chemistry undergone by a particular presumed radical intermediate and concluding from the nature of that chemistry whether the radical was Σ or Π.[52] Goosen and Skell and co-workers, however, have reported observations that they interpret in terms of the behavior of the same radical in more than one electronic state.

Intramolecular cyclizations of amido-radicals generated by homolysis of N-iodo-amides onto aromatic systems gives γ- and δ-lactams . . . . Kinetic studies of the cyclization of N-methylbiphenyl-2-carboxamide with t-butyl hypoiodite under irradiation have shown that γ- and δ-cyclization products are formed in parallel and that the reactions have different activation energies and entropies. Reactions of N-methyl derivatives of 1,2,3,4-tetrahydro- and 1,2-dihydro-phenanthrene-4-carboxamides with t-butyl hypoiodite under irradiation have provided evidence that π-conjugation is a prerequisite for δ-lactam formation. From a consideration of the molecular orbitals involved it is suggested that γ- and δ-cyclization reactions occur with amidyls in the Σ- and Π-electronic states.[54a,54b]

Skell et al. have reported the following:[54e–54i]

Imidoyl radicals as the chain-carrying hydrogen abstracting intermediates can be attained under a variety of conditions with . . . substrates such as neopentane . . . and methylene chloride. . . . Identical selectivities [competitive halogenation] are obtained with (bromo imides + Br$_2$) or (iodo imides + I$_2$) reagents . . . . Also, in the case of succinimidoyl, in the presence of Br$_2$ *there is no opening of the ring* to give the product BrCH$_2$CH$_2$-C(O)N=C=O. . . . The reactions carried out in the presence of small amounts of bromine-scavenging olefins show a strikingly different selectivity . . . [Also], a rearrangement product, β-halopropionyl isocyanate, can be a major product . . . when *N*-bromosuccinimide or *N*-iodosuccinimide is used.

What is concluded?

Our results *require* [emphasis added] the presence of two radicals in these systems, a ground state . . . and an excited state radical . . . . It is impossible to explain [the results] . . . with a single hydrogen-abstracting imidyl radical . . . . [and] we are *forced to the conclusion* [emphasis added] that a *different* imidyl radical is involved . . . [Ref. 54h, received by journal 9/26/77].

 . . . . the chemistry . . . *is attributed to* [emphasis added] . . . [Ref. 54g, received by journal 9/26/77].

The kinetic experiments *require* [emphasis added] two succinimidoyl radical intermediates [Ref. 54f, received by journal 10/27/77].

The two methods for generating succinimidyl radicals . . . produce different succinimidyl radicals, ground state Π . . . and excited state (σ_N) . . . [stated as fact (Ref. 54e, received by journal 12/23/77)].

[With reference to] "the assemblage of experimental results which are to be accommodated by the hypothesis . . . these new results add force to the earlier *suggestion* [emphasis added] that thermal chain reactions involving succinimidyl radicals can involve either π or σ states of the radical [Ref. 54i, received by journal 3/29/78)].

It is unclear what the conclusions are; there are great variations, in no temporal sequence, in their strength. But regardless of these variations, any conclusions cannot be assessed because of the absence of experimental details. This is interesting work, but acceptance of the conclusions must await publication of a full paper with complete experimental details.

III.   ENERGETICS OF PROTOTYPAL RADICALS

A.   Thermochemistry of t-Butyl $\cdot$ $[\Delta H_f(t\text{-Bu}\cdot), D(\text{Me}_3\text{C-H})]$

Attempts to deal with the mutual interdependence of structure, thermodynamics, and kinetics in free radical chemistry and to reconcile the various arguments and conclusions[1c] continue to lead to the questioning of "known" thermochemical quantities.

The experimental data on . . . [Bu^tBu^t] decomposition from shock-tube and radical buffer studies and radical combination from very-low-pressure pyrolysis and modulation spectroscopy are shown to be consistent. They lead to . . . [a heat] of formation for . . . t-butyl . . . [$\sim$5 kcal] higher than currently accepted . . . from methathesis reactions.[55a]

The equations describing Benson's radical buffer systems were formulated originally to permit calculation of rates of radical combination, for example, $2\text{Bu}^t\cdot \rightarrow \text{Bu}^t\text{Bu}^t\cdot$ They were restructured[55a] so as to permit calculation of rates of dissociation, for example, $\text{Bu}^t\text{Bu}^t \rightarrow 2\text{Bu}^t\cdot$, which can be done more accurately from the same experimental data because the calculation no longer involves thermochemical properties of the alkyl radical. Note that this is not an "improvement," however, as different information is being obtained. The rate was compared with that extrapolated from the results of high-temperature shock-tube experiments; the agreement was considered acceptable and would provide a "check" on the shock-tube results. The high-temperature shock-tube results were combined with results obtained by extrapolation of data obtained for the reaction in the opposite direction, $2\text{Bu}^t\cdot \rightarrow \text{Bu}^t\text{Bu}^t$,[1c] to give entropies and enthalpies of reaction at high temperature. After modeling of t-butyl · and "best judgment" calculations, $\Delta H_{f,298}(\mathrm{Bu}^t \cdot)$ was calculated to be ~ 12.5 kcal/mol, which is ~ 5 kcal/mol higher than the "traditional" value.

The rate of decomposition of $\mathrm{Bu}^t \mathrm{Bu}^t$ has since been studied[55b] at 700–900 K and the results are in excellent agreement with those presented earlier,[55a] an agreement that strengthens both sets of results. As before, a high-temperature ΔH_f (6.9 kcal/mol at 700 K) was determined and converted to $\Delta H_{f,300}(\sim 12.5$ kcal/mol).

The kinetics of oxidation of $\mathrm{Bu}^t \mathrm{Bu}^t$ have been studied.[55c] After a complex sorting out of and correction for a chain component of the reaction, Arrhenius parameters for $\mathrm{Bu}^t \mathrm{Bu}^t \rightarrow 2\mathrm{Bu}^t \cdot$ were calculated; they are strengthened by comparison with values calculated from the results of an earlier study of the competitive decomposition of $\mathrm{Bu}^t \mathrm{Bu}^t$ and decyclization of cyclohexene. As was done independently with the "old" decomposition data,[55a] combination of these results with literature values of the recombination rate constant led to a high-temperature value of $\Delta H_f(\mathrm{Bu}^t \cdot)$, which then had to undergo significant and uncertain correction over the more than 400° to standard conditions. The results were a $\Delta H_{f,298}(\mathrm{Bu}^t \cdot)$ (10.5 ± 1 kcal) and a $D(\mathrm{Me}_3\mathrm{C}{-}\mathrm{H})$ ~ 3 kcal higher than the "traditional" values.

The above study[55c] of the oxidation of $\mathrm{Bu}^t \mathrm{Bu}^t$ has been chemically elaborated upon by conducting it in the presence of H_2.[55d] After a complex, frequently approximate analysis, during which corrections were often made for departures from the initially presumed simplicity of the chemistry, the relative Arrhenius parameters for the reactions of $\mathrm{Bu}^t \cdot$ with O_2 and H_2 to give isobutylene and isobutane, respectively, were calculated. By use of (1) reported values for the Arrhenius parameters for the reverse of the reaction with H_2, and (2) thermochemical properties of $\mathrm{Bu}^t \cdot$,[55c] the equilibrium constant for $\mathrm{Bu}^t \cdot + \mathrm{H}_2 \rightleftarrows \mathrm{H} \cdot + i\text{-}\mathrm{C}_4\mathrm{H}_{10}$ was calculated. From this equilibrium constant and an estimated rate constant for $\mathrm{H} \cdot + i\text{-}\mathrm{C}_4\mathrm{H}_{10} \rightarrow \mathrm{H}_2 + \mathrm{Bu}^t \cdot$, $k(\mathrm{Bu}^t \cdot + \mathrm{H}_2)$ was calculated; it was then combined with the relative rate constants for reaction of $\mathrm{Bu}^t \cdot$ with O_2 and with H_2 to give $k(\mathrm{Bu}^t \cdot + \mathrm{O}_2 \rightarrow$ isobutylene). The rate constants for reaction with H_2 and with O_2 were then compared with the corresponding rate constants for other simple alkyl radicals and it was concluded that those of $\mathrm{Bu}^t \cdot$ were out of line. The thermochemical parameters of $\mathrm{Bu}^t \cdot$,[55c] were chosen as the culprit and the $\mathrm{Bu}^t \cdot$ rate constants were brought into line by moving the values[55c] of $\Delta H_f(\mathrm{Bu}^t \cdot)$ and $D(\mathrm{Me}_3\mathrm{C}{-}\mathrm{H})$ ~ 1 kcal in the direction of the "traditional" values.

An overall assessment has been presented[55a]:

The present results have important But implications on the use of iodination and bromination [metathesis reactions] as a method of determining bond [dissociation] energies or heats of formation of radicals . . . . There may well be a need to critically reexamine the assumptions and procedures employed. A most obvious possibility is that the halogenation results are based on kinetic measurements in one direction with the assumption of zero activation energy for the reverse process. Direct determination of the latter would be extremely informative [see also Section VIII.A].

We believe that in both studies the conceptual and computational pathway was long and involved enough so that the "new" should be accepted only cautiously, but reasonable enough so that the "old" should be viewed with suspicion.

In contrast, the decomposition of neopentane was studied at 1000–1260 K by use of the very-low-pressure pyrolysis technique.[55e] These high-temperature results fit well into a procedure that (1) combined them with a value of the rate constant of the reaction in the opposite direction estimated from a t-Bu· recombination rate constant extrapolated from earlier literature values; (2) incorporated the traditional value of ΔH_f; (3) modeled t-Bu· and the transition state leading to it; and (4) involved correction over more than 700° to standard conditions. It was judged that a ΔH_f 4 kcal from the traditional value would have given an unsatisfactory fit, but that one 2 kcal away would not have.

The reaction of $Bu^tN{=}NBu^t$ with DI to give Bu^tD was studied at 644–722 K, also by use of the very-low-pressure pyrolysis technique.[55f] The system was dominated by uncharacterized chemistry hypothesized on the basis of inference to be irrelevant to the determination of the desired $k(Bu^t \cdot + DI \rightarrow Bu^tD + I\cdot)$. Typically, the rate constants were determined from the slopes of what are essentially three-point straight lines; they were assigned an uncertainty of ±20%. From these and from (1) an estimate of the isotope effect on k, (2) data for the reverse reaction, (3) a typical assortment of assumptions and thermochemical estimates regarding species involved, and (4) the usual extrapolation of data, a value of $\Delta H_f^\circ(Bu^t\cdot) = 8.4$ kcal/mol emerged; use of a different reasonable value for the out-of-plane bending frequency of $Bu^t\cdot$, which enters into one of the thermochemical estimates, alone would have increased ΔH_f° by 1.1 kcal/mol.

"We feel that the balance of evidence still favors ... 'accepted' values. ..."[55e]

The photoelectron spectrum of $Bu^t\cdot$ generated by pyrolysis of neopentyl nitrite at 500–600°C was determined.[5a] An adiabatic ionization potential was assigned and was then combined with $\Delta H_f(t\text{-}Bu^+)$ [determined from ion cyclotron resonance determination of ΔH ("benzyl"$^+$ + $Bu^tX \rightleftarrows t\text{-}Bu^+ + PhCH_2X$) $(X = Cl, Br)$, and values of the heats of formation of $PhCH_2^+$ and $PhCH_2X$] to give $\Delta H_f(Bu^t\cdot)$. Based on those results, a more "traditional" $\Delta H_{f,298}(Bu^t\cdot)$ of 8.4 kcal was recommended and it was stated that "it seems likely that the higher radical heats of formation [Ref. 55a] should be viewed with caution."

The uncertainties and inaccuracies possibly associated with the above[5a] calculation of $\Delta H_f(Bu^t\cdot)$ have been stressed.[55g] Results of the photoionization mass spectrometric determination of the appearance energies of the $C_4H_9^+$ formed from isobutane and t-butyl halides were then presented.[55g] A value of $\Delta H_f(Bu^t\cdot)$

in good agreement with that recommended earlier[5a] was calculated and it was concluded that "the validity of the proposed higher radical heat of formation [Ref. 55a] must be questioned."

As things stand now, the futures market in $\Delta H_f(\text{Bu}^t\cdot)$ is as dangerous as any. And the outcome is hardly inconsequential (see Section III.I).

B. D(Cyclopropyl-H)

The reaction of $Cl\cdot$ with cyclopropane was studied at $25°C$ in a "very low-pressure reactor" (see Section IV.A) and the concentrations of $Cl\cdot$, HCl, cyclopropane, and the cyclopropyl radical were measured.[56,57a] The rate constant measured was in good agreement with that found earlier.[57b] Upon introduction of HCl into the reaction system, $Cl\cdot$ was produced.[57e] Its concentration was followed as a function of that of HCl and the results appropriately plotted so as to yield the equilibrium constant K(cyclopropyl$\cdot$ + HCl $\rightleftarrows$ $Cl\cdot$ + cyclopropane); data were sparse and their graphical treatment was quite imaginative. From K, D(cyclopropyl-H) was estimated to be 106 kcal/mol, higher even than that for methane.

A thorough characterization of the chemistry of a system is a necessary part of a study of its kinetics.

C. The Phenoxy Radical. D(PhO−H), ΔH_f(PhO$\cdot$)

The very-low-pressure pyrolyses of PhOEt and of $PhOCH_2CH=CH_2$ have been studied at 950-1220 and 720-983 K, respectively, by following their disappearance by mass spectrometry.[58] Based on inferential interpretation of limited and suggestive circumstantial evidence, the results were analyzed in terms of the following decomposition reactions only:

$$PhOCH_2CH=CH_2 \rightarrow PhO\cdot + CH_2=CHCH_2\cdot$$

$$PhOEt \rightarrow PhO\cdot + Et\cdot$$

To derive high-pressure activation energies from the data, RRK theory was used and the number of active oscillators was estimated. Then, to obtain $\Delta H_{f,298}^{\circ}$(PhO$\cdot$), estimates were made of A factors and $\Delta C_p(T)$ as well as of an activation energy for the reverse (recombination) reaction. The reactions of $PhOCH_2CH=CH_2$ and PhOEt gave ΔH_f = 10.5 and 12.2 kcal/mol, respectively, which, even though many of the procedures and types of estimates were common to both, are in good agreement for two different reactions, each with its own ambiguities. D(PhO-H) = 86.5 ± 2 kcal/mol was obtained, which agreed well with the value obtained earlier[1d] by use of a different method in solution.

D. The Anilino Radical. D(PhNH-H), ΔH_f(PhNH·)

The very-low-pressure pyrolysis (VLPP) of PhNHMe has been studied at 952–1257 K by following its disappearance mass spectrometrically.[59] As in the work discussed in Section III.C, arguments were assembled to support the proposition that the decomposition proceeds solely as PhNHMe → PhNH· + Me·. To go from the measured low-pressure to the limiting high-pressure activation energies, RRK theory was used approximately and the number of active oscillators were estimated empirically. To calculate $\Delta H^\circ_{f,298}$(PhNH·), the A factor was estimated by approximated analogy and ΔC_p (T) was estimated to permit extrapolation of the high-temperature activation energy to standard conditions; the long extrapolation is a problem with the VLPP method. Also, the activation energy of the reverse reaction was assumed to be zero. D(PhNH-H) was calculated to be 86.4 ± 2 kcal/mol.

E. $D(R_2\overset{+}{O}\text{-H})$

The proton affinities of ethers and alcohols ($H^+ + R_2O \rightarrow R_2\overset{+}{O}H$), determined by use of ion cyclotron resonance (ICR) techniques, were combined with their known photoionization potentials ($R_2O \rightarrow R_2O^{\ddagger} + e^-$) to yield $D(R_2\overset{+}{O}\text{-H})$, the oxygen-hydrogen bond dissociation energies ($R_2\overset{+}{O}H \rightarrow R_2O^{\ddagger} + H\cdot$) of several aliphatic and alicyclic oxonium ions.[60a,60b]

F. The Electron Affinity of Methyl

The photoelectron spectrum of the species of m/e = 15 produced by subjecting ketene to an electrical discharge was considered to be that of the methyl anion on the basis of its consistency "with that which we expect for CH_3^-, and not that of other species (e.g., NH^-, CHD^-) with m/e 15."[61] Its photodetachment energy was determined relative to that of O^-, the electron affinity of the oxygen atom being well known.

G. The Electron Affinity of Allyl

The electron affinity of allyl has been determined to be 12.7 ± 1.2 kcal/mol from a highly judgmental analysis of measured cross section as a function of wavelength for electron photodetachment from the species, presumed to be $CH_2{=}CHCH_2^-$, produced by proton abstraction from propene by OH^- in an ICR spectrometer.[62a]

Rate and equilibrium constants have been measured for the reaction $CH_3CH{=}CH_2 + OH^- \rightleftarrows C_3H_5^- + H_2O$ at 23°C. The equilibrium constant was combined with an estimate of ΔS to yield ΔH, from which $\Delta H_f(CH_2=CHCH_2^-)$ and thus the electron affinity of allyl (12.4 ± 1.9 kcal/mol were calculated.[62b]

H. Electron Affinity of Alkoxyl Radicals

a. Methoxyl

The electron affinity of methoxyl was determined, via that of oxygen atom as a reference, to be 36.2 ± 0.5 kcal/mol from the photoelectron spectrum of CH_3O^-, which had been produced by subjecting CH_3OH, CH_3OH/N_2O, CH_3OD, and dimethyl oxalate to an electrical discharge. That of $CD_3O\cdot$ was determined to be 35.8 ± 0.5 kcal/mol. The difference of 0.42 ± 0.14 kcal/mol was considered to be statistically significant.[18]

An ICR spectrometric study of the cross section for electron photodetachment from CH_3O^- generated by electron impact on MeOOMe yielded a value of 36.7 ± 0.9 kcal/mol.[25b]

b. t-Butoxyl

An ICR spectrometric study of the cross section for electron photodetachment from $t\text{-}BuO^-$ generated by electron impact on Bu^tOOBu^t and by reaction of F^- with Bu^tOH yielded a value of 43.1–43.8 kcal/mol for the electron affinity of t-butoxyl.[25b,63]

I. It's a Small World

We indicated earlier how the mutual interdependence of arguments and conclusions regarding structure, thermodynamics, and kinetics in free radical chemistry can lead to the revelation of what would otherwise be unapparent problems or errors in one area as a result of attempts to reconcile differences in another.[1c] Such interdependence can extend far beyond free radical chemistry. For example, would it have been anticipated that new values for the ionization potential and heat of formation (accepted hastily?–see Section III.A) of $Bu^t\cdot$ would end up causing revisions of the gas-phase proton affinities of a wide variety of bases? It so happens that the absolute proton affinity of ammonia is the reference used to convert relative gas-phase proton affinities to absolute ones and that that proton affinity is in turn determined from the heat of formation and ionization potential of $Bu^t\cdot$ via the proton affinity of isobutylene and heat of formation of Bu^{t+}. The values determined for the gas-phase basicities of a wide variety of molecules, such as amino acids, amines, amides, carboxylic acids, ketones, arenes, and olefins, do indeed hinge on the energetics of a hydrocarbon free radical. The situation has been described by several workers.[5a,64a–64c] However, see also references 64d,e.

IV. EXPERIMENTAL TECHNIQUES

A. Equilibria Involving Radicals

A new method for measuring rate constants and equilibrium constants of molecule–radical reactions has been reported.[65a,b] It uses a very low pressure reactor and is a variation of the very low pressure pyrolysis technique; however, it does not suffer two deficiencies of the latter: (1) the necessity of operating at high temperature in order to generate radicals rapidly and pyrolytically, and a consequent long extrapolation of results to ambient temperature; (2) problems associated with reactions on the walls of the reactor. In the present method, radicals were generated at low temperature via microwave discharge and the walls of the reactor were coated with a fluorocarbon. "The apparatus described lends itself quite generally to the quantitative study of atom–molecule, radical–molecule, and radical–radical reactions in the temperature range – 80 to 150°. . .".

This first report describes work that "calibrates" the method, a study of the equilibrium

$$Cl \cdot + CH_4 \underset{k'}{\overset{k}{\rightleftharpoons}} HCl + CH_3 \cdot$$

"Gas phase equilibrium data for radical reactions at room temperature represent the most direct route to the thermochemistry of free radicals."

From steady-state equations for the forward reaction and a plot of $[Cl \cdot]_0 / [Cl \cdot]$ versus $[CH_4]$, k was obtained and found to be in agreement with that resulting from earlier determinations. From steady-state equations that describe a system to which HCl had been added and a plot of $\frac{[Cl \cdot]}{[Cl \cdot]_0} - [Cl \cdot]$ versus $\frac{[HCl]}{[CH_4]}$, k/k', that is K was found to be 1.3 ± 0.3 at 25°. When combined with thermochemical data for the other species involved in the reaction, this value led to $\Delta H_f^{\circ}(CH_3 \cdot) = 34.9 \pm 0.15$ and $D(CH_3\text{-}H) = 104.9 \pm 0.15$ kcal/mol, in good agreement with "known" values.

"To the best of our knowledge," "report Baghal-Vayjooee, Colussi, and Benson,[65a] "this represents the first time the equilibrium constant for a chemical reaction involving a very reactive organic free radical has ever been measured directly."

The steady-state absorbances of $CH_3 \cdot$ generated by decomposition in a shock tube at 1600–1800 K of dilute mixtures of Me_4Sn and of ethane in argon, were determined spectrophotometrically at 216 nm.[65c] They were converted into steady-state concentrations by use of an extinction coefficient calculated from the maximum absorbance of $CH_3 \cdot$ resulting from the decomposition assumed with experimental support to be "instantaneous," of Me_4Sn. Equilibrium constants for $2CH_3 \cdot \rightleftharpoons C_2H_6$ were then calculated based on the experimentally supported assumption that the only significant fate of $CH_3 \cdot$ was dimerization to ethane. These equilibrium constants agreed acceptably well with each other and with values calculated from reported thermodynamic data. "The reaching of a single equilibrium ('from above' through recombination of CH_3 and 'from below' through decomposition of ethane) in itself [i.e., in the absence of additional experimental support] practically excludes the possibility of any significant influence of secondary reactions." Evaluation of this work is difficult because of insufficient experimental and computational details.

In an attempt to determine the thermodynamic parameters of R· by establishing the "equilibrium" $CH_3 \cdot$ + olefin $\rightleftarrows$ R· and studying the recombinative draining of the radicals from it, the reactions of azomethane with ethylene[66a] and with propylene[66b] have been studied. Unfortunately, the systems were complex and the approach became a labyrinth of reactions and of estimates and assumptions regarding them.

B. Use of Radicals/ESR to Study Quite Another Phenomenon

The interaction of 2 with salts in ethanol was studied by use of ESR spectroscopy with the following results[67]:

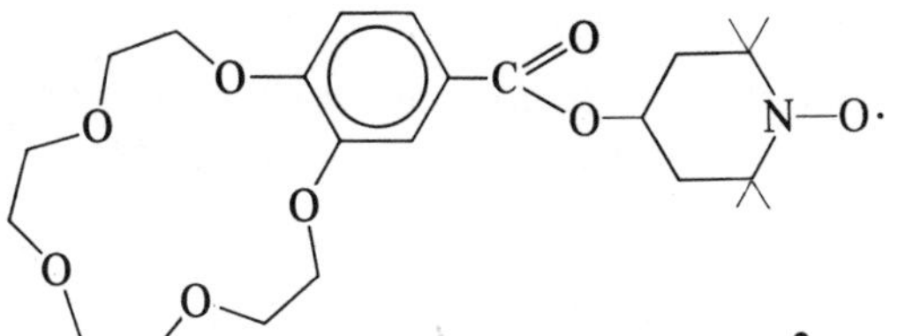

2

Compound	ESR Spectrum
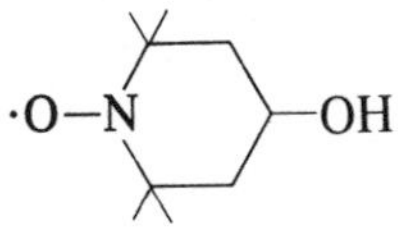	Doublet spectrum. $a_N = 15.99$ G, $g_0 = 2.0058$
2	Doublet spectrum. $a_N = 15.86$ G, $g_0 = 2.0058$
2· NaSCN	Doublet spectrum. $a_N = 15.81$ G, $g_0 = 2.0058$
2· $\frac{1}{2}$ KSCN (room temperature)	$a_N = 15.86$ G; $a_N' = 7.92$ (i.e., $\frac{1}{2} a_N$)
2· $\frac{1}{2}$ KSCN (matrix, 77 K)	Triplet spectrum

Although Na^+ would be expected to "fit" into the crown of 2, K^+ would be "too big." It is generally true that sandwiches can form when the cation is too big for the crown; it is known specifically for combination of K^+ with a 15-crown-5.[68]

The spectra of $2 \cdot \frac{1}{2}$ KSCN were attributed reasonably to a complex containing two nitroxyl groups and having the K^+ sandwiched between two crown ether rings.

Thus ESR spectroscopy has been used to study crown–cation complexation.

C. Magnetic Resonance Spectroscopy

The work of groups at the Free University of Berlin illustrates and collectively advances a significant methodological area of free radical chemistry. The reader can get the flavor of this work by scanning the titles[69] and, made easy by a fixation on "firsts" and the commendable[70a] absence of any reluctance to use that label, the following list:

First measurements of deuterium quadrupole coupling constants of a polyatomic doublet state radical . . . .[69a]

For the first time alkali metal cation [ENDOR and triple-] resonance lines could be observed at frequenies below 4 MHz belonging to counterion hyperfine constants as small as 100 kHz."[69b]

Deuterium ENDOR resonances in solution have been detected for the first time.[69c]

For the first time the fluid-solution and rigid-media ENDOR spectra of triradicals in the quartet spin state could be obtained.[69d]

. . . . the first successful ^{13}C ENDOR experiment in a liquid crystal.[69e]

Chlorine ENDOR signals could not be detected up to now . . .[69f]

For the first time natural abundance ^{13}C-ENDOR measurements have been successful.[69g]

Another characteristic is the ability to refer "with a straight face" to "double resonance methods . . . like INDOR, ENDOR, ELDOR, ODMR, and MODOR . . . ,"[69h] accompanied by the gesture of sparing us CRENDOR.[69i]

V. MECHANISTIC TECHNIQUES

A. Generation and Trapping of Radicals[70b]

The reaction in Eq. 1 has been discovered and the mechanism of Eq. 2

$$(Ph_3P)_2N^+ \, CpV(CO)_3H^- + RX \xrightarrow{\text{THF}} (Ph_3P)_2N^+ \, CpV(CO)_3X^- + RH \quad (1)$$

$(X = Br, I; R = n\text{-alkyl, cycloalkyl, }t\text{-alkyl, benzyl, alkenyl, phenyl})$

$$CpV(CO)_3^{\cdot-} + RX \longrightarrow CpV(CO)_3X^- + R\cdot$$

$$R\cdot + CpV(CO)_3H^- \longrightarrow RH + CpV(CO)_3^{\cdot-}$$

(2)

proposed[71]; the radical nature of the mechanism is based on strong preponderance of circumstantial evidence.

B. Trapping of Radicals. Horsemanship

We have discussed earlier[1f] the general question of trapping of radicals and the utility of organotin hydrides as trapping agents. R_3SnH's are excellent radical traps. They are actually too good, as the radical may be trapped before it can undergo the desired chemistry. While, in principle, such a difficulty can be overcome by proper choice of concentrations and direction of addition, life in practice is not so simple; various practical difficulties usually leave one with only a few orders of magnitude in rate to play with. With R_3SnH, this is often not enough.

Reactions similar to the stannane reductions occur with the corresponding silicon and germanium hydrides, but with differences in reactivity in the order $Sn > Ge > Si$. We have pointed out the potential that the silicon and germanium hydrides, when used complementarily with the tin hydrides, will overcome the above difficulty remaining with the use of tin hydrides as radical generating and trapping agents, while retaining the advantages.[72] By choice of one or a series of such hydrides, it may be possible (a) to identify kinetically radicals along a reaction path and establish their intermediacy and (b) to study configurational and structural isomerization of radicals. Trapping ability varies greatly from one class of R_3MH to another; "fine control" may be achieved by changing R. It may be possible, in a wide variety of cases and by use of the same method, to implicate radicals as intermediates along a reaction path in as rigorous and convenient a manner as can be done in carbonium ion chemistry. Present standards in the field of free radical chemistry are quite inferior, as the intermediacy of a given species is usually claimed as a result of a showing that it is present during the reaction and that its conversion to product can be formulated easily. In a slogan, radicals of different lifetime may be "kinetically titrated" with Group IV hydrides.

The above approach involves the use of a variety of external trapping agents (a "stable" in Ingold's terminology; see below) that show similar chemistry and are adaptable, through the establishment of appropriate competitions, to the study, in the same way, of reactions whose rates vary significantly. A conceptually different approach has been described by Ingold.[73a] It is directed toward the determination of the rates of bimolecular radical reactions by competing them against a variety of appropriately chosen "standard" internal rearrangements. He presents the availability of "an extremely versatile STABLE of primary alkyl radical rearrangements. Within this stable we can find radicals whose rearrangement rates will enable them to compete in a wide variety of radical–molecule races. That is, amongst these radicals we can find entrants suitable for competition against reactions whose speed varies from the gallop of, $R \cdot + O_2 \rightarrow ROO \cdot$ to the walk of, $R \cdot + RH \rightarrow RH + R \cdot$."

The mainstay of Ingold's stable is the 5-hexenyl radical, $CH_2{=}CHCH_2CH_2\cdot$ $CH_2CH_2 \cdot$. Its cyclization to the cyclopentylcarbinyl radical has become a standard basis for the demonstration of the intermediacy of radicals and for the determination of the rates of their bimolecular reactions. Its rate of cyclization has been reexamined.[73b] Di-(6-heptenoyl) peroxide was photolyzed in cyclopropane between 183 and 232 K. The intensities of the ESR signals of 5-hexenyl and cyclopentylcarbinyl radicals were determined, calibrated against diphenylpicrylhydrazyl (DPPH), and converted to "effective" radical concentrations in two steps:

1. Empirically determining the relationship between the signal intensity and the effective concentration of t-Bu·. The rate constant for self-termination of t-Bu· was measured by use of signal intensities calibrated against DPPH and then compared to the "known" rate constant. The "effective" radical concentration was calculated from the calibrated signal intensity by bringing the calculated rate constant into line with the known one.

2. Assuming that the effective concentrations of 5-hexenyl and cyclopentylcarbinyl bear the same relationship to their calibrated signal intensities as does that of $Bu^t \cdot$ to its calibrated signal intensity. From the steady-state concentrations, k(5-hexenyl → cyclopentylcarbinyl)/k(bimolecular termination) can be calculated if it is assumed that (1) the only reactions undergone by the radicals are rearrangement of 5-hexenyl to cyclopentylcarbinyl, as well as their symmetric and unsymmetric bimolecular terminations, and (2) the various rate constants for bimolecular termination are equal. If it is further assumed that the various rate constants for bimolecular termination of 5-hexenyl and cyclopentylcarbinyl are equal to that determined for n-hexyl, then k(5-hexenyl → cyclopentylcarbinyl) can be calculated: $\log k \, (\sec^{-1}) = (9.5 \pm 1.1) - (6.1 \pm 1.1)/RT$.

The kinetics of the cyclization that involves the analogous secondary radical, that is, the conversion of the 6-hepten-2-yl radical to the 2-methylcyclopentylcarbinyl radical (Eq. 3), has been studied.[73c] $CH_2{=}CH(CH_2)_3\dot{C}HCH_3$ was generated between 183 and 232 K by reaction of $CH_2{=}CH(CH_2)_3CHBrCH_3$ with photochemically generated $Bu_3Sn \cdot$ in n-pentane. The concentrations of the open and closed radicals were determined by calibration against DPPH. From the

$$CH_2=CHCH_2CH_2CH_2\dot{C}HCH_3 \longrightarrow \overset{\text{CH}_2^{\cdot}}{\bigsqcup} \quad (cis \text{ and } trans) \quad (\text{Eq. 3})$$

steady-state concentrations, k(6-hepten-2-yl → 2-methylcyclopentylcarbinyl)/ k(bimolecular termination) can be calculated as in the parent 5-hexenyl cases above. The second-order rate constant for bimolecular self-reaction of the closed radical in n-pentane at 240 K was measured by use of ESR and was found to be near the "diffusion-controlled limit." It was therefore assumed that its variation with temperature could be estimated from the temperature coefficient of the viscosity of the solvent. This permitted the calculation of log k(cyclization) (sec^{-1}) = $(9.8 \pm 0.3) - (6.4 \pm 0.3)/RT$, which is essentially the same as that calculated for cyclization of the 5-hexenyl radical.

C. Spin Trapping of Radicals

Although "spin trapping" of radicals by nitroso compounds and nitrones has long been a common technique, it has not been possible to use it quantitatively for lack of reliable rate constants for trapping of ordinary alkyl radicals.

a. Trapping of Primary Radicals

The competition between spin trapping and cyclization of the 5-hexenyl radical has been studied.[74a-74d] It was generated from the thermal decomposition in benzene of impure di(6-heptenoyl) peroxide, ^{13}C-labeled in the 2-position, in the presence of 2-methyl-2-nitrosopropane, 2,4,6-tri-$tert$-butylnitrosobenzene, and phenyl-N-$tert$-butyl nitrone at 23–71°C.

From (1) knowledge of the concentration of the trap, tacitly considered to be monomeric; (2) ESR measurement of the ratio of initial rates of formation[75] of the spin adducts of 5-hexenyl and cyclopentylcarbinyl; (3) assumption that the only reactions undergone by the radicals are rearrangement of 5-hexenyl to cyclopentylcarbinyl and their trapping by the spin trap; (4) knowledge of the rate constant of rearrangement; and (5) assumption[76] that the only route for formation of the spin adducts is trapping of the radicals, rate constants of trapping of 5-hexenyl were calculated. They were reported and taken to be representative of those of primary alkyl radicals.

b. Trapping of Secondary Radicals

The competition between spin trapping and cyclization of the 6-hepten-2-yl radical, $CH_2=CH(CH_2)_3\dot{C}HCH_3$, has been studied. It was generated by decomposition of impure $CH_2=CH(CH_2)_3CHMeCO_2)_2$ in benzene at 40°C in the presence of 2-methyl-2-nitrosopropane,[77a] nitrosodurene,[77b] and 2,4,6-tri-t-

butylnitrosobenzene. From procedures and assumptions similar to those described for 5-hexenyl in the preceding section, rate constants of trapping of 6-hepten-2-yl were calculated. They are reported and taken to be characteristic of secondary radicals.[73c]

The γ-radiolysis of pentamethylnitrosobenzene in cyclohexane has been studied at 299 K by following the formation and decay of an ESR signal attributed to the spin adduct of the cyclohexyl radical. Based on a variety of liberally made assumptions regarding what was and was not occurring, what was being observed, and the form of the corresponding descriptive kinetic equations, a rate constant of spin trapping of 1.6×10^7 M^{-1} sec^{-1} was extracted from a study of the ability of added Bu$_3$SnH to influence the effect of radiation dose on the concentration of spin adduct.[74e]

c. Trapping of Tertiary Radicals

The photochemistry of 2-methyl-2-nitrosopropane in benzene has been studied by ESR at 299 K as was the above (Section V.C.a) chemistry of the cyclohexyl radical.[74f, 74g] From a study of the effect of added Bu$_3$SnH on the rate of formation of Bu$_2^t$NO· the number, considered to be the rate constant of trapping of Bu^t·, 3×10^6 M^{-1} sec^{-1} was obtained.[78a]

d. Immobilized Spin Traps

Some very preliminary, encouraging experiments directed toward development of an immobilized spin trap insoluble in the reaction medium have been described. Use of such a spin trap could alleviate mechanistic ambiguities and experimental difficulties in some situations.[79]

D. Use of Behavioral Differences to Distinguish Transient Intermediates

A method commonly used to distinguish the possible intermediacies of two transient species in a reaction, say, radical versus cation or radical versus anion, is to run the reaction on a substrate chosen such that different intermediates would be "expected to" give different products. For example, a choice between an anion and a radical could be made on the following basis:

R_1. "is known to rearrange to" R_2., that is, it does so in a reference reaction. R_1^- "is known to give products of retained structure," that is, it does so in a reference reaction. The reaction of unestablished mechanism would be run starting with an R_1 skeleton. If the products contained an R_2 skeleton, it would be concluded that a radical is an intermediate. If the products contained an R_1 skeleton, it would be concluded that an anion is an intermediate.

The problem here is that a preferred reaction course is being considered to be a characteristic of the reacting molecule, rather than of the reacting molecule and its chemical environment.

The observation that R_1 · rearranges to R_2 · means that $k > \sum_i k_i'$ [trapping agent]$_i$ in Eq. 4. If different trapping agents, with different values of k', or different concentrations of the same trapping agent were employed, that is, if the chemical environment were different, the inequality might have come out in the opposite direction. The same is true when it is observed that $k'' < \sum_i k_i'''$ [trapping agent]$_i$ in Eq. 5, that is, that R_1^- "is known to give products of retained structure."

$$R_1 \text{ products} \xleftarrow[\text{trapping}]{k'} R_1 \cdot \xrightarrow{k} R_2 \cdot \qquad (4)$$

$$R_1 \text{ products} \xleftarrow[\text{trapping}]{k'''} R_1^- \xrightarrow{k''} R_2^- \qquad (5)$$

Even though $R_1 \cdot$ may rearrange to $R_2 \cdot$ in one reaction (the reference reaction) in which $R_1 \cdot$ is generated (one particular $\sum_i k_i'$ [trapping agent]$_i$), it cannot be concluded that, were $R_1 \cdot$ an intermediate in another reaction of unestablished mechanism, it also would rearrange to $R_2 \cdot$. In this other reaction $\sum_i k_i'$ [trapping agent]$_i$ might be greater than in the reference reaction. Similarly, even though R_1^- is observed to give products of retained structure in one reaction [the reference reaction) in which R_1^- is generated (one particular $\sum_i k_i'''$ [trapping agent]$_i$), it cannot be concluded that, were R_1^- an intermediate in another reaction of unestablished mechanism, the products would contain the R_1 skeleton. In this other reaction, $\sum_i k_i'''$ [trapping agent]$_i$ might be less than in the reference reaction.

Preferred reaction pathway is not a characteristic of a particular species. It depends, in each circumstance, on the relative "availabilities" to that species of the competing pathways. Of course, if one were dealing, for example, with the above $R_1 \cdot$ versus R_1^- situation and saw, say, rearrangement, and if a convincing argument could be made that $\sum_i k_i'''$ [trapping agent]$_i$ in the reaction of interest would be greater than in the correspoinding reference case, that is, if the observed rearrangement could not have been of R_1^- since it could be shown that it was not "more able" to rearrange (weaker trapping) in the case of interest than in the reference case, the technique could validly be used.

A paper on the "Formation of the Totally Degenerate Bicyclo[3.2.2]nonatrienyl Radical by Electron Transfer"[80] is a good subject for application of the above principles.

E. A Chemical Amplifier of Radical Pair Leakage

The photolysis of $BrCCl_3$ in benzene has been studied. An accurate measure of small changes in the quantity of radicals that escaped the initial radical pair was desired.[81]

To overcome the problem of quantitative analysis of small amounts of materials . . . a chemical amplifier in the form of a chain reaction, was built into the system. By this means a small quantity of radical escaping from the initial cage produced a much larger quantity of product. The reaction employed was the $CCl_3 \cdot$-initiated radical bromination of toluene in which typically 150 bromotoluene molecules may result from a single initial radical. Whether the chain-propagating species is $CCl_3 \cdot$ or $Br \cdot$ is uncertain but in either reaction scheme simple kinetic analysis shows that the logarithm of the observed rate of disappearance of toluene . . . should vary linearly with time and with the concentration of the initiating radical. . . .

As a further check a competitive scavenger for radicals, phenyl thiol, was added. This has a hydrogen abstraction rate that appears to be $\approx 10^3$ times that of toluene and so it is initially consumed before the toluene reacts at the concentrations used. Thus a plot of log (toluene concentration) against time is constant until sufficient radical has escaped from the cage to consume essentially all the thiol, when it falls linearly as before. Any change in the cage escape efficiency . . . should consequently be reflected in the time . . . when the plot departs from constancy; this time should be definable quite accurately thanks to the amplifying effect of the chain reaction.

VI. PERSISTENT RADICALS

A. The Other Side of Persistence

We mentioned earlier that Ingold's work on persistent radicals "illustrates the important role of F-strain in the reference compound in determining 'stability' . . . ."[1g] On this basis, it would be reasonable for "dimers" of persistent radicals to homolyze to those radicals unusually rapidly via relief of strain. Such behavior of alkanes and cycloalkanes, molecules that would yield radicals "which hardly differ in their electronic stabilization,"[83a] is presented here.[84]

"To study the steric acceleration of the homolysis of carbon–carbon bonds . . . we have prepared a series of 'hexasubstituted ethanes' and examined their thermolysis."[83a] The results of Rüchardt et al.[83], expressed on a common basis in Table 7.1[83], were discussed within the context of Ingold's work. They show "distinctly the importance of the steric ground state reciprocal effect for the thermal stability, because in all examples unconjugated alkyl radicals form, which hardly differ in their electronic stabilization."[83a]

"[Tetra-t-butylethane] is the most thermolabile alkane hydrocarbon known."[83b]

What are the substrates?[90] Et_3CCEt_3,[91a] $(c\text{-}C_6H_{11})Me_2CCMe_2(c\text{-}C_6H_{11})$,[91b-91d] $(c\text{-}C_6H_{11})MeEtCCMeEt(c\text{-}C_6H_{11})$,[91b-91d] $Me_2EtCCEtMe_2$,[91b, 91d] $Bu_2^tCHCHBu_2^t$,[91d-91h] meso- and d,1-$(c\text{-}C_6H_{11})Bu^tCHCHBu^t(c\text{-}C_6H_{11})$.[91a]

What reactions are being observed?[83,90]

$Et_3CCEt_3 \rightarrow 77\% \ Et_3CH$[91i] $ + 8\% \ MeCH=CEt_2$[91i] $ + 10\%$?

$(c\text{-}C_6H_{11})CMe_2CMe_2(c\text{-}C_6H_{11}) \rightarrow$ "the corresponding main products"[91j]

$(c\text{-}C_6H_{11})CMeEtCMeEt(c\text{-}C_6H_{11}) \rightarrow$ "the corresponding main products"[91j]

$Bu_2^tCHCHBu_2^t \xrightarrow[\text{mesitylene}]{} 72\% \ Bu^tCH_2Bu^t$ [91k,91l] $ + 18\% \ Bu_2^tCHCH_2-$ [91k]

$meso\text{-}(c\text{-}C_6H_{11})CHBu^tCHBu^t(c\text{-}C_6H_{11}) \xrightarrow{\text{Ref. 91m}} 152\% \ (c\text{-}C_6H_{11})CH_2Bu^t$

$+ 15\% \ (c\text{-}C_6H_{11})CH=CH(c\text{-}C_6H_{11}) + 3\% \ (c\text{-}C_6H_{11})CH_2CHBu^t(c\text{-}C_6H_{11})$

$+ 3\% \ (c\text{-}C_6H_{11})CH=CBu^t(c\text{-}C_6H_{11})$

What is the process whose rate is being measured? Good first-order kinetics were observed over more than three half-lives for those compounds decomposed in octane, dodecane, and octadecane (see also footnote e to Table 7.1). However, might there be, for example, radical recombination in the primary radical pair, causing the observed rate of decomposition not to be the rate of homolysis? No relevant evidence is reported, although it should be less likely than for "ordinary" geminate alkyl radicals.

What are the bond lengths? It would be reasonable for such strained molecules to have unusually long C—C bonds. Rüchardt[83a] pointed out the following work:

$$Me_3C-CMe_3 \quad 1.58 \ \text{Å}$$ [92a]

$$(Me_3C)_3CH \quad 1.61 \ \text{Å}$$ [92b]

$\quad 1.60 \ \text{Å}$ [92c]

The "longest known unbridged C—C bond" has since been reported.[92d]

$$Bu_2PhC-CPhBu_2 \quad 1.64 \ \text{Å}$$

B. Persistent Radicals. An Update

a. A Caveat

$(Me_3Si)_2CHCH(SiMe_3)_2$ (2) $(Me_3Si)_2\dot{C}CH(SiMe_3)_2$ (7)

"We wish to take this opportunity to point out the difficulty of determin-

TABLE 7.1 Thermolysis of $R_1R_2R_3C-CR_1R_2R_3$

R_1	R_2	R_3	Solvent	Comments	Temperature at which $t_{1/2} = 1$ hr (°C)	$\Delta H^{\ddagger}$ (kcal/mol)
H	H	H	Gas phase	a	695	89
H	H	CH_3	Gas phase	a,b,c	590	87
H	CH_3	CH_3	Gas phase	a,c,d,e	565	~77
$(CH_3)_3CCH_2C(CH_3)_3$				f	502	
CH_3	CH_3	CH_3	Gas phase	a,c,e,g,h	475,[83a] 490[83c]	70
CH_3	CH_3	CH_3CH_2	Octadecane	e,i,j	420	63
$(CH_3)_3CCH(CH_3)C(CH_3)_3$				f	415	
CH_3	CH_3	$CH_3CH_2CH_2CH_2$			412	
CH_3	CH_3	$CH_3CH_2CH_2$			411	
CH_3	CH_3	$CH_2CH(CH_3)_2$		Ref. 83d	384	
H	$c\text{-}C_6H_{11}$	$c\text{-}C_6H_{11}$			384	
$(CH_3)_3CC(CH_3)_2C(CH_3)_3$				f	350	
$d,l\text{-}$H	$C(CH_3)_3$	$c\text{-}C_6H_{11}$	Tetralin	e,k	330	
CH_3	CH_3	$CH(CH_3)_2$		e	329	
CH_3	CH_3	$CH_2C(CH_3)_3$		e	321	
CH_3	CH_3	$c\text{-}C_6H_{11}$	Dodecane	e,j	315	61
$meso\text{-}$H	$C(CH_3)_3$	$c\text{-}C_6H_{11}$	Tetralin	e	285	53
CH_3CH_2	CH_3CH_2	CH_3CH_2	Octadecane	e,j	285	51
CH_3	CH_3CH_2	$c\text{-}C_6H_{11}$	Octane	e,j	245,[83a] 250[83c]	50
$[(CH_3)_3C]_3CH$			Mesitylene	f,l	243	47
CH_3	CH_3	$C(CH_3)_3$		e	195	
H	$C(CH_3)_3$	$C(CH_3)_3$	Mesitylene	e,m	141	36

[a]Earlier work by others; cited by Rüchardt et al.
[b]T extrapolated ~45°C.
[c]See also Ref. 55a and 55b.
[d]T extrapolated ~160°C.
[e]"Product analyses . . . show that bond cleavage occurs mainly at the central $C-C$ bond."[83c]
[f]Based on statistically corrected rate constants.
[g]See also Refs. 55c and 89.
[h]T extrapolated ~240°C.
[i]T extrapolated ~15°C.
[j]0.06–0.1M in substrate, 1.5M in PhSH.
[k]T extrapolated 30°C, based on an estimated $\Delta H^{\ddagger}$.
[l]0.05M in substrate.
[m]0.4–0.6M in substrate.

282

ing a true lifetime for persistent radicals because of their tendency to decay by reacting with impurities in the medium or on (or diffusing from) the surface of the containing vessel."

The radical 7 was generated in concentrations of 10^{-5}–$10^{-7}M$ by photolysis of di-*tert*-butyl peroxide solutions of 2 in carefully degassed quartz EPR tubes. All samples of 7 decayed with "clean" first-order kinetics. After decay the radical could be regenerated by a brief irradiation of the sample. The half-lives for decay were found to increase dramatically with each succeeding measurement. For example, a sample prepared in a quartz tube that had been heated with a flame for 30 min under vacuum had an initial half-life at 23°C of ca. 3 days. The rate of decay decreased continuously, reaching a value which corresponded to a half-life of 300 days after 1 year. It would appear that 7 (and probably many other persistent radicals) is not subject to decay at ambient temperature by unimolecular or bimolecular self-reactions. Decay probably occurs only when 7 encounter other reactive species."[93a]

b. A Review

"Long-lived Free Radicals"[93b]

c. Recent Papers

"Persistent Cyclopropyl Radicals with Novel Configurations."[93c]

"The First Reversible Thermal Dissociation of Distannanes, $R_3 Sn-SnR_3$."[93d]

"Syntheses with Compounds $R_3M-Hg-MR_3$, XVI. (Silylamino)methyl Radicals and their Dimers."[93e]

"An Electron Spin Resonance Study of the Group IVB Organometallic Adducts of 2,6-Di-*t*-butylbenzoquinone."[93f]

"Reactions of Silyl and Stannyl Radicals: Persistent Metalated Methyl, Oxymethyl and Aminomethyl Radicals."[93g]

VII.   CHARACTERISTICS OF PROTOTYPAL REACTIONS

A.   Polar Effects on Radical Reactions. Bromination of Arylmethanes by N-Bromosuccinimide

The relative reactivities of the ring-substituted toluenes in benzyl bromination are frequently correlated through linear Hammett-type equations. Preference seems to be given to correlation through . . . σ^+. . . . The argument advanced in favor of such treatment is that the data on the rapid bromination of *p*-methoxytoluene are best correlated by . . . σ^+ plots.

The anomalously high rate of bromination of *p*-methoxytoluene is explained by the fact that the . . . strong electron-donor substituent . . . can couple directly with the electron-deficient reaction center to form transition states with . . . polar structure and radically different values of the σ and σ^+ constant.[94a]

Is this textbook case real?

> . . . ring bromination [of p-methoxytoluene by N-bromosuccinimide] affects the mechanism of the overall bromination process. . . . The data obtained here suggest that correlations based on relative values of the rate constants for benzyl bromination of p-methoxytoluene are untenable since the . . . mechanism of bromination of p-methoxytoluene is quite different from the mechanism of bromination of the other substituted toluenes.[94a]

One body of work in this area may not be invalidated by such problems because what were considered to be the relative rates of appearance of the $ArCH_2Br$ were determined, rather than the relative rates of disappearance of the $ArCH_3$[94b]; however, the work necessary to establish that the relative amounts of $ArCH_2Br$ present at the end of the reaction were equal to the relative rates of formation of the $ArCH_2\cdot$ was not done.

For other potential problems with the common picture, see Refs. 95.

VIII. RATES OF PROTOTYPAL REACTIONS

A. Rate Constants for the Scavenging of Radicals by Iodine

Although iodine has long been used as a radical scavenger and the activation energy of Eq. 6 is a key element in the determination of much of what is known about the thermochemistry of radicals,[96] "no detailed studies on the kinetics of these reactions have been carried out."[97]

$$R\cdot + I_2 \longrightarrow RI + I\cdot \qquad (6)$$

$$I\cdot + I\cdot \longrightarrow I_2 \qquad (7)$$

The rate of Eq. 6 has been determined for cyclopentane and cyclohexane through an optical absorption study on the nanosecond time scale of the buildup and decay of iodine atoms. The data were fitted to equations corresponding to Eqs. 6 and 7. The value of k_6 was found to be $1.2 \times 10^{10} \; M^{-1} \; sec^{-1}$ (23°C) for cyclohexane and 1.9×10^{10} (room temperature), with $E_a = 1.6$ kcal/mol, for cyclopentane.

B. Rate Constants for Reaction of Bu^{t}O·

Alkoxy radicals enter into all the usual types of radical reactions. That is, they react with radicals (combination and disproportionation with themselves and with other radicals), they react with molecules (intermolecular hydrogen abstraction, addition to unsaturated systems, and substitution at multivalent atoms . . .), and they undergo unimolecular reactions (intramolecular hydrogen abstractions, additions, etc., and β-scission . . .).[98]

The importance of alkoxy radicals in organic, biological, and atmospheric chemistry has motivated numerous studies of their reactions. These studies have frequently centered on the *tert*-butoxy radical, reflecting both the availability of several thermal and photochemical sources and the fact that the corresponding peroxide can be handled safely without any special precautions. . . . In spite of the large number of relative hydrogen abstraction rate constants that have been determined . . . , the measurement of absolute values has so far been unsuccessful. Attempts to use ESR spectroscopy also have been unsuccessful.[99a]

We have called attention[1h] to a discussion of the advantages and limitations of optical absorption detection as compared to ESR detection in a flash-photolysis system. Both of these techniques have since been used in an attempt to determine the absolute rate constants of some reactions involving *t*-butoxy radicals.

In this study we have used a nanosecond laser flash photolysis technique and optical absorption spectroscopy to examine the reaction of *tert*-butoxy radicals produced in the photolysis of di-*tert*-butyl peroxide with a number of substrates. We have been able to obtain absolute abstraction rate constants which provide a solid base allowing the conversion of all the relative rate constants into absolute values. Several substrates have been carefully chosen so that absolute rates from previous competitive studies can be obtained in a "one-step" calculation rather than using numerous intermediate ratios.[99a, b]

A solution containing Ph$_2$CHOH and the substrate in $1:2$ benzene/(Bu^{t}O)$_2$ was photolyzed at 19–23°C; the substrates (RH) included a variety of alkylbenzenes, cycloalkanes, cycloalkenes, cycloalkadienes, alcohols, ethers, and fatty acids. The absorption of Ph$_2$ĊOH at 535 nm was monitored, the rate of increase of its concentration being dependent on the rate of reaction of Bu^{t}O· with RH. From knowledge of the kinetics of the RH-free system and a study of the rate of increase of [Ph$_2$ĊOH] as a function of [RH] it was possible to obtain k(Bu^{t}O · + RH → R·) if it was assumed that (1) k(Bu^{t}O· + RH → products) = k(Bu^{t}O· + RH → R·), and that (2) Ph$_2$ĊOH is produced only via Bu^{t}O· + Ph$_2$CHOH.[100a]

$k(\text{Bu}^t\text{O}\cdot + \text{Ph}_2\text{CHOH} \rightarrow \text{Ph}_2\dot{\text{C}}\text{OH})$ and the kinetics of the decay of $\text{Bu}^t\text{O}\cdot$ were determined by following the buildup of $\text{Ph}_2\dot{\text{C}}\text{OH}$ in an RH-free system as a function of $[\text{Ph}_2\text{CHOH}]$ and making the following assumptions: (1) $\text{Bu}^t\text{O}\cdot$ reacts only in first-order processes[100b]; (2) $\text{Ph}_2\dot{\text{C}}\text{OH}$ is stable under the reaction conditions; and (3) $\text{Bu}^t\text{O}\cdot$ is generated much faster than it reacts. These assumptions were supported in part by reasonable, and in part by suggestive, arguments. Among others, the rate constants in Table 7.2 were reported for ($\text{Bu}^t\text{O}\cdot + \text{RH} \rightarrow \text{R}\cdot$):

TABLE 7.2. Rate Constants for Abstraction by $\text{Bu}^t\text{O}\cdot$

RH	$k(M^{-1}\ \text{sec}^{-1})$	RH	$k(M^{-1}\ \text{sec}^{-1})$
PhCH_3	2.3×10^5	CH_3OH	2.9×10^5
$\text{PhCH(CH}_3)_2$	8.7	$\text{CH}_3\text{CH}_2\text{OH}$	11.
Cyclopentane	8.8	$(\text{CH}_3)_2\text{CHOH}$	18.
Cyclohexene	57.	Tetrahydrofuran	83.

"We have used ESR techniques in order to measure several ratios of reactivities. They . . . provide more accurate values in the case of the less reactive substrates like methanol."[99a] A pair of substrates (RH and R'H) in neat $(\text{Bu}^t\text{O})_2$ was UV irradiated as the solution passed through an ESR cavity. From the intensities of the signals corresponding to the radicals ($\text{R}\cdot$ and $\text{R}'\cdot$), the absolute steady-state concentrations of those radicals were determined. These in turn were used to calculate a number, considered to be the relative rate constant for abstraction of hydrogen from the substrates by $\text{Bu}^t\text{O}\cdot$. The equations used were based on an assumed simplified model of the chemistry of the system wherein the sole origin of $\text{R}\cdot$ and $\text{R}'\cdot$ was reaction of $\text{Bu}^t\text{O}\cdot$ with RH and R'H, respectively, and their sole fate was either self- or mutual cross-termination. Those relative rate constants were combined with the absolute rate constant reported for ethanol (see above), and the absolute rate constants in Table 7.3 were reported for ($\text{Bu}^t\text{O}\cdot + \text{RH} \rightarrow \text{R}\cdot$). The agreement between these results and those obtained directly by use of optical absorption detection (see above) is very good and increases the likelihood that they are correct.

TABLE 7.3. Rate Constants for Abstraction by $\text{Bu}^t\text{O}\cdot$

RH	$k(M^{-1}\ \text{sec}^{-1})$
CH_3OH	3.4×10^5
Me_2CHOH	15.5
$\text{Me}_2\text{CHOCHMe}_2$	11.8

The growth of radicals resulting from abstraction by $Bu^tO\cdot$ from substrates such as cyclopentane, $PhOCH_3$, $MeOBu^t$, and MeOH has been followed by use of flash photolysis/ESR spectroscopy in neat $(Bu^tO)_2$ or 1:4 $(Bu^tO)_2$/benzene solvent.[99c] It was desired to follow the rise of [R·] and, from those kinetics, extract a rate constant $k(Bu^tO\cdot + RH \rightarrow R\cdot)$. There is a multitude of problems, in addition to "the rather large errors involved in the present measurements." Although it is necessary to show that R· was produced only by reaction of $Bu^tO\cdot$ with RH, what was done was to present a case to the effect that there are conditions under which $Bu^tO\cdot$ is consumed only by reaction with RH, that is, the $Bu^tO\cdot$/RH reaction can "dominate over [all other] reactions . . . for the decay of $Bu^tO\cdot$ radicals." This, of course, is a *different* matter: $Bu^tO\cdot$ reacting only with RH does not necessarily mean that RH is reacting only with $Bu^tO\cdot$. Also, when the argument which *should have been* advanced is set aside, the kinetic analysis is still confused, even within the context of the argument which *was* advanced.[101] From this procedure, rate constants (20°C) of 3.4 × 10⁵ and 1.3 × 10⁵ M^{-1} sec^{-1}, as well as Arrhenius parameters, emerged for reaction with cyclopentane and methanol, respectively.

C. Dimerization of Radicals

We discussed earlier the rate constants of dimerization of prototypal alkyl radicals and characterized the situation as confusing, but approaching stabilization.[1i] Where are things 3 years later? Recent reports of rate constants for ethyl,[102a] isopropyl,[102b] and *t*-butyl[102c, 102d] are in fair agreement with the range cited earlier, especially if that portion in the direction of diffusion-controlled rates is emphasized.

If you have ever completed a difficult climb only to find a little girl with a red pocketbook at the top of the mountain, you can appreciate how workers in this area must have felt when a paper entitled "A Student Experiment for Measuring Rate Constants of Radical Recombination" appeared in the *Journal of Chemical Education*.[103a]

Based on a very superficial kinetic analysis of the photosensitized reaction of PhI with 1-hexene, and the weakly supported and implausible presumption that the sole termination step was 2Ph· → PhPh, a rate constant for dimerization of phenyl of 8 × 10⁹ M^{-1} sec^{-1} has been calculated.[103b]

IX. SYNTHETIC APPLICATIONS

A. Asymmetric Induction

We have pointed out previously the potential utility of optically active Group IV hydrides in the conversion of racemic halides (RX) to the corresponding optically active compound (RH), the activity being asymmetrically induced at the stage of abstraction of hydrogen from the Group IV hydride by R· during the process of reduction of RX to RH by the Group IV hydride.[72] At that time, optically active organogermanium and organosilicon, but not the more useful organotin, hydrides were known. Optically active organotin hydrides have now become available, and a different example of the general reaction, required in principle, has been reported to have possibly occurred.

Gielen and Tondeur, having been wise enough to ignore what must have been a strongly discouraging climate of opinion, have reported the preparation of chiral triorganotin hydrides[104a]:

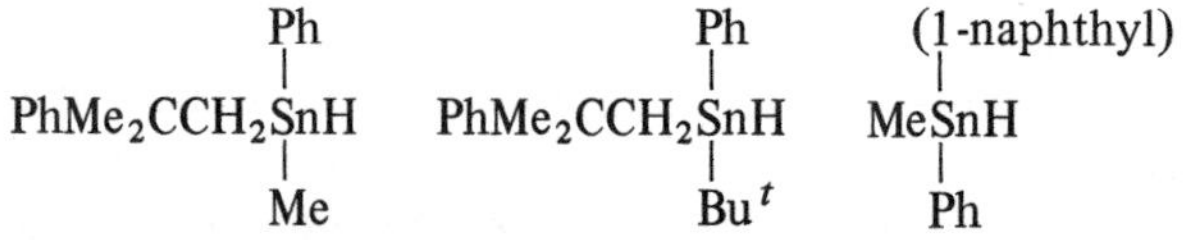

$$
\begin{array}{ccc}
\text{Ph} & \text{Ph} & \text{(1-naphthyl)} \\
| & | & | \\
\text{PhMe}_2\text{CCH}_2\text{SnH} & \text{PhMe}_2\text{CCH}_2\text{SnH} & \text{MeSnH} \\
| & | & | \\
\text{Me} & \text{Bu}^t & \text{Ph}
\end{array}
$$

Fujihara and his co-workers have made the first report "that deals with the radical addition of an optically inactive substrate with the chain transfer reagent giving an optically active product."

The reaction[s] of cis-2-Octene with (a) cyclohexanone and (b) thiolacetic acid in the presence of l-menthol or deoxycholic acid (DCA) were carried out using benzoylperoxide in the former case and 2,2'-azobisisobutryonitrile in the latter case. Addition products obtained were found to be optically active, as shown below. This may be caused by the interaction of a growing radical with the chiral substance at the transition state.[104b, 104c]

Product(a)   Product(b)

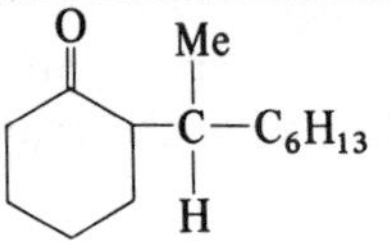

Product(a):

- cyclohexanone ring with $\overset{\text{Me}}{\underset{\text{H}}{\text{C}}}-\text{C}_6\text{H}_{13}$ $[\alpha]_D^{25} - 1.5°(\text{MeOH})$ 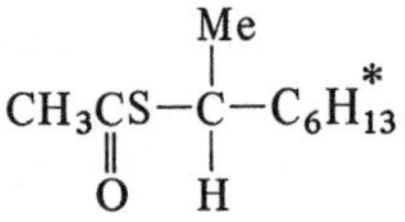 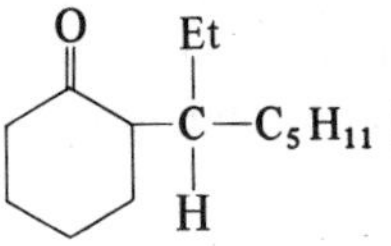
- cyclohexanone ring with $\overset{\text{Et}}{\underset{\text{H}}{\text{C}}}-\text{C}_5\text{H}_{11}$ $[\alpha]_D^{25} - 3.1°(\text{MeOH})$

Product(b):

$$\text{CH}_3\overset{}{\underset{\text{O}}{\text{C}}}\text{S}-\overset{\text{Me}}{\underset{\text{H}}{\text{C}}}-\overset{*}{\text{C}_6\text{H}_{13}}$$

*in the presence of l-menthol

$$[\alpha]_D^{25} - 3.0°(\text{Ph}-\text{H})$$

*in the presence of DCA

$$[\alpha]_D^{25} + 6.1°(\text{Ph}-\text{H})$$

B. Interconversions of Functional Groups

We have discussed the replacement of hydroxyl by another group, as well as the unified and generalized chemistry centered around $RO\overset{\cdot}{C}=O$ and its thio analogues.[1j] Activity in this area has continued. Additional reactions that can be formulated in terms of the fragmentation of $RO\overset{\cdot}{C}=O$, formed from either $ROCOX$ or $RO\cdot + CO$, to $R\cdot + CO_2$ have been reported,[105] as has another reaction describable as $Ar\cdot + CS_2 \rightarrow ArS\overset{\cdot}{C}=S \rightarrow ArS\cdot + CS.$[106]

Although the stannane-induced conversion of an alcohol, as its chloroformate ester, to the corresponding hydrocarbon, that is, $ClCO_2R \rightarrow RH$, has long been known, yields are low; it was presumed that the tin hydride trapped an intermediate alkoxycarbonyl radical before it could decarboxylate.[107] We discuss the obviation of such a problem by use of silanes in Section V.B.

The silane-induced conversion of an alcohol, as its ester, to the corresponding hydrocarbon, that is, $R'CO_2R \rightarrow RH$, has been reported; however, the reaction conditions were harsh.[72, 109] The conversion in Eq. 8 has now been reported.[110]

$$ROH \xrightarrow{COCl_2} ROCOCl \xrightarrow[(Bu^tO)_2]{Pr_3SiH,\ 140^\circ C} RH \qquad (8)$$

$R = n$-octyl, cyclohexyl, Et_3C, 3β-cholestanyl, benzyl and $MeCO(CH_2)_3$

Based on inference and analogy, it was proposed that the reaction proceeds via the mechanism of Eq. 9. Also mentioned, in another context, is a scheme similar

$$ROCOCl + Pr_3Si\cdot \longrightarrow RO\overset{\cdot}{C}=O \xrightarrow{-CO_2} R\cdot \xrightarrow{Pr_3SiH} RH + Pr_3Si\cdot \qquad (9)$$

to Eq. 10.

$$\overset{\displaystyle OSiPr_3}{\underset{\displaystyle |}{ROCOCl + Pr_3Si\cdot \longrightarrow RO\overset{\cdot}{C}Cl}} \longrightarrow Pr_3SiOCOCl + R\cdot \xrightarrow{Pr_3SiH} RH \qquad (10)$$

This family of reactions may be expanded by inclusion of the reports of $F_5SO\cdot + CO \rightarrow F_5S + CO_2.$[111]

C. Host–Guest Interactions. Macrocyclically Complexed Radicals and Radical Transfer Agents

We believe the title subject to be potentially a fruitful area of free radical chemistry, the rather obvious conceptual relationship to the corresponding chemistry of ions notwithstanding. Presentation of two recent sets of work provides for us the opportunity to get into a broader discussion of possible free radical host–guest chemistry.

In an extension of earlier work, "the e.s.r. spectra of . . . organic radicals . . . generated from organic 'guests' in cycloamylose matrices by γ-irradiation at 77 K have been investigated." The radicals reported included cyclohexadienyl, $Me_2\dot{C}OH$, and benzyl.[112]

Radicals have long been known to have an affinity for π-systems and for non-bonded electron pairs centered on other-than-first-row atoms. An optical absorption spectrum has been assigned to $Et_2\dot{S}Br$; "similar results are obtained for other organic sulfides, e.g. $(CH_3)_2S$, $(t\text{-butyl})_2S$ or 1,4-dithiane, and other halides, e.g., . . . [Cl]."[113a] ESR spectra have been attributed to a variety of $RR'\dot{S}X$,[113b,113d,113e] $R_2R'\dot{P}X$,[113b,113d–113f] $Ph_2\dot{S}eX$,[113g,113h] and $Ph_3\dot{A}sCl$,[113i] radicals (X = Cl, Br, I).

The chemical behavior of radicals could be modified significantly as a result of multicenter complexation by appropriate additives and solvents, for example, the well-known sulfur and phosphorus analogues of materials such as cryptands, crown ethers, and their acyclic counterparts.[114] By such complexation, radicals could have altered chemical behavior and could be made more persistent; stable complexes might be isolable.

Complementarily, the behavior of common radical transfer agents may be modified significantly as a result of the complexation with crown ethers.[115]

D. Isotope Enrichment via Magnetic Field Effects

This field was opened conceptually by Lawler and Evans:

> . . . no measurements of isotope effects on cage combination efficiencies per se appear to have been reported. The anticipated changes in cage effect of up to 25% for ^1H vs. ^2H are, however, encouraging and appear to be worth seeking. Isotope effects with heavy atoms where one isotope has nuclear spin of zero, such as the $^{12}C-^{13}C$, $^{16}O-^{17}O$, and $^{32}S-^{33}S$ pairs, seem to be doubly attractive for study. The mass effects should be much smaller than with hydrogen and nuclear-spin-induced intersystem crossing could be ignored for the isotope with zero spin.[117a]

Its practical potential was indicated by Atkins and Lambert,

> . . . magnetic field effects . . . may have considerable industrial significance. We shall have to wait in order to see if this will emerge through their application to . . . the separation of nuclear isotopes . . . .[117b]

by Evans and Lawler,

For a singlet-born pair, in zero magnetic field the combination product in a 50:50 ^{235}U—^{238}U mixture would be enriched by 1.3 per cent in ^{238}U (with scavenging product correspondingly enriched in ^{235}U). For a triplet-born pair, the enrichment would be larger .... [117c]

and by Buchachenko,

The principle of the enrichment of magnetic isotopes in chemical reactions ... is quite general and applicable to the sorting of ... carbon-13-12 ... oxygen 17-16, uranium 235-238, boron-10-11, *etc.*[117d]

The "origin" of the effect has been elaborated upon by Buchachenko,[117d,117e]

Nuclear spins influence through hyperfine interaction the rate of singlet-triplet transitions in ... [a radical] pair, and hence the probability of chemical reaction (*e.g.* recombination) in the pair ....

Thus the probability of singlet triplet conversion of a pair and the probability of chemical reaction depend on the hyperfine interaction energy, *i.e.* on the sign of the nuclei and their magnetic moments.

This idea is readily illustrated by an example in which triplet radical pairs are generated ... from the molecule X—Y: We suppose that one of the radicals (*e.g.* X·) contains a carbon nucleus (^{12}C or ^{13}C). Radical pairs containing the magnetic carbon-13 isotope undergo more rapid triplet-singlet conversion than do pairs containing the non-magnetic carbon-12 isotope. Therefore, pairs containing the former isotope become singlet more rapidly and recombine more rapidly to form the original molecules X—Y, whereas triplet-singlet conversion of pairs containing carbon-12 is retarded, and such non-magnetic pairs have less chance of recombining. Thus molecules of the initial substance X—Y are enriched with the magnetic carbon-13 isotope, while conversion products are depleted in this isotope. If singlet pairs are generated from X—Y, magnetic pairs will clearly be converted more rapidly into triplet pairs than the reverse singlet-triplet evolution and dissociate, whereas the singlet-triplet evolution of non-magnetic pairs is retarded; and such pairs have a greater chance of recombining. In this case the initial X—Y molecules are impoverished in carbon-13, while the conversion products are enriched.

Molecules that have passed through the first stage [in which each X—Y molecule has undergone only one chemical change, only one act of decomposition] and are enriched with the magnetic isotope can obviously undergo fresh chemical change and become still more enriched in a second stage; and so on. In fact, with increase in the degree of chemical conversion the quantity of substance X—Y remaining will diminish, but its isotope content will continually increase ....

.... the degree of enrichment depends on the hyperfine interaction constants in the radicals .... The largest magnetic effects must be expected in radical pairs involving σ-electron radicals [high a_C] ..., CF_3..., and also hetero-organic radicals containing isotopes of tin, germanium, thallium, phosphorus, silicon, etc.

[In summary,] the difference in rates of singlet-triplet conversion between magnetic and non-magnetic pairs leads to a difference in the probabilities of their recombination. This is the magnetic isotope effect, a consequence of which is the enrichment of magnetic isotopes in chemical reactions.

and by Sagdeev and Molin.[117f,117g]

This field was opened experimentally by Buchachenko and by Sagdeev and Molin, and their co-workers.

The photolysis of $(PhCH_2)_2CO$ was studied in benzene and hexane and the unreacted $(PhCH_2)_2CO$ was found to be enriched in ^{13}C.[117h] Although the isotopic selectivity could have resulted from a "magnetic isotope effect," the results are not necessarily attributable uniquely to it:

1. " ... a kinetic isotope effect should also cause an enrichment of the compound in ^{13}C. However, ... [it] should not exceed ~2%. Furthermore, ... [it] should lead to an enrichment of dibenzyl in ^{12}C." $^{13}C/^{12}C$ in dibenzyl was found to be 0.01094_6 as compared to 0.01093_3 in the starting $(PhCH_2)_2CO$.

2. "In contrast to kinetic isotope effects, isotope effects caused by a hyperfine interaction should depend on the strength of the magnetic field. Experiments conducted in different magnetic fields showed that enrichment drops, in fact, with an increase in field strength" (Table 7.5). Any statement regarding the dependence of the enrichment on the field strength rests on the

TABLE 7.4. ^{13}C Content of $(PhCH_2)_2CO$ for Various Conversions[a]

Conversion (%)	$^{13}C/^{12}C$ per carbon in $(PhCH_2)_2CO$	Enrichment[b] in ^{13}C (%)
0	0.01093	—
75	0.01098	6.5
80	0.01101	10.6
94	0.01105	15.9
98	0.01106	17.0

[a] Photolysis in Earth's magnetic field.

[b] Assuming it all occurred at the central carbon; this assumption should be verified.

TABLE 7.5.   Dependence of Enrichment on Magnetic Field Strength

Field Strength (Oe)	$^{13}C/^{12}C$ per carbon in $(PhCH_2)_2CO$	Enrichment in ^{13}C (%)
Earth	0.01100	9.5
180	0.01097	5.0
350	0.01098	6.0
530	0.01098	6.0

[a] Conversion ~80%.

measurement in the earth's field; that measurement is very sensitive to the percent conversion (see Table 7.4). With what confidence can it be stated that the conversion was 80% at each field strength? (It is referred to as "~80%".) How was it measured? An error of ~5% would nullify a statement that there is a dependence of percent enrichment on field strength.

Our subjective feeling is that the accuracy of the measurements may be inadequate; the effect has probably been observed, but the data do not really show it.

Turro and Kraeutler also[118] have studied the photolysis of $(PhCH_2)_2CO$ and found that[117i]

1. a . . . small $^{13}C/^{12}C$ isotope separation occurs in homogeneous (benzene) solution;
2. the efficiency of . . . separation is greatly enhanced in soap solution . . . ;
3. the carbonyl carbon of $(PhCH_2)_2CO$ is specifically and exponentially enriched as photolysis proceeds;
4. an external magnetic field significantly influences the efficiency . . . .

The results obtained may be expressed as in Table 7.6.

The ^{13}C content in the 1-position of the phenyl ring of phenyl benzoate produced by triplet (acetophenone) sensitized photolysis of Bz_2O_2 in CCl_4 has been reported to have increased by $23 \pm 5\%$ while that in the other positions remained unchanged.[117k]

An experiment more definitive with respect to distinguishing between magnetic and "usual" kinetic isotope effects, in that it relies on qualitative, not quantitative, results has been reported.[117l] Among the common isotopes of oxygen (^{18}O, ^{17}O, ^{16}O) it is only ^{17}O, the isotope *intermediate* in mass, which has a magnetic moment. Therefore, the relative patterns of enrichment should be *qualitatively* different for a magnetic versus a "classical" kinetic isotope effect. The relative amounts of $^{16}O^{16}O$, $^{16}O^{17}O$, and $^{16}O^{18}O$ in the initial O_2 reagent and in the O_2 remaining after the dicyclohexyl peroxydicarbonate-initiated autoxidation of ethylbenzene at 50°C in PhEt/benzene solvent were determined. It was

TABLE 7.6. Carbon Isotope Ratios

	$\dfrac{^{13}C/^{12}C \text{ in Residual } (PhCH_2)_2CO}{^{13}C/^{12}C \text{ in Initial } (PhCH_2)_2CO}$		
Conversion (%)	System a[a]	System b[b,d]	System c[c,e]
0	(1.00)	(1.00)	(1.00)
10	1.00	1.03	ND
50	1.02	1.25	ND
90	1.06	2.08	1.28

[a] Benzene, room temperature, Earth's magnetic field.
[b] Soap (hexadecyltrimethylammonium$^+$Cl$^-$) solution, room temperature, Earth's magnetic field.
[c] Same as system b, except ~15000 G.
[d] Position of enrichment determined by NMR.
[e] It has more recently been reported that system b at 100,000 G behaves as system a.[117j]

found[118] that

$$\frac{(^{16}O^{17}O/^{16}O^{16}O)_{residual}}{(^{16}O^{17}O/^{16}O^{16}O)_{initial}} = 1.13 \quad \text{and} \quad \frac{(^{16}O^{18}O/^{16}O^{16}O)_{residual}}{(^{16}O^{18}O/^{16}O^{16}O)_{initial}} = 0.998$$

This is an elegantly simple experiment. It proves that a "usual" kinetic isotope effect is not being observed. Similar results obtained in a photochemical experiment would be reconcilable with difficulty to an explanation involving an isotope effect on intersystem crossing (via an effect on vibronic interactions) versus deactivation versus chemical reaction.

Turro also has studied a case of ^{17}O enrichment in O_2 (Eq. 11).[117m]

The AIBN-induced decomposition of Me_3SnH, containing ^{117}Sn and ^{119}Sn in natural abundances of 7.6 and 8.6%, respectively, led to $Me_3SnSnMe_3$ and Me_3SnH, which contained 17.8 and 13.5%, respectively, of $^{117}Sn + ^{119}Sn$ at 50% reaction.[117r]

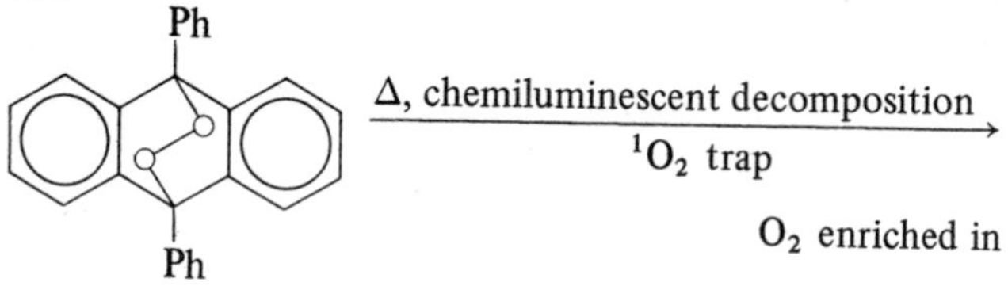

$$\xrightarrow[\text{$^1O_2$ trap}]{\Delta, \text{ chemiluminescent decomposition}}$$

O_2 enriched in ^{17}O (11)

The analysis,[117e,117n]

Measurements of $\alpha = k_3/(k_2 + k_3)$ in the thermal decomposition of two isotopic forms of azoisobutyronitrile

$$RN_2R \rightarrow |\dot{R}\ \dot{R}| \begin{cases} \xrightarrow{k_2} RR \\ \xrightarrow{k_3} \dot{R} + \dot{R} \end{cases}$$

where $R = (CH_3)_2C(CN)\cdot$ and $(CD_3)_2C(CN)\cdot$, gave [Ref. 119] $\alpha_H = 0.56$ and $\alpha_D = 0.51$. On the assumption that k_3 is independent of the isotopic composition of the radicals, the isotope effect in the recombination is easily found to be $(k_2)_H/(k_2)_D \simeq 0.8$. There is little doubt that this isotope effect is magnetic in origin. The hyperfine interaction with protons in the $(CH_3)_2C(CN)\cdot$ radical is sixfold that in the $(CD_3)_2C(CN)\cdot$ radical. Thermal decomposition takes place via singlet radical paris, so that pairs containing protons pass into the triplet state more rapidly and have a greater probability of dissociating (than pairs containing deuterated radicals). Hence, the probability of recombination of protonated pairs should be less than that of deuterated pairs (in agreement with experiment).

provides two lessons:

1. In the flush of enthusiasm and success, it is possible to get carried away. The observed isotope effect is, after all, of a magnitude and direction compatible with a "usual" kinetic origin.[120]

2. On the other hand, let us not be too sure that the usual kinetic explanations that we have been giving for some of the isotope effects observed in free radical chemistry are "correct." ". . . the possibility of the manifestation of 'magnetic' effects must be borne in mind in the analysis of the nature of the experimental kinetic isotope effects in radical recombination reactions."[117f] "In analyzing experimental kinetic isotope effects in radical reactions one should be cognizant of possible isotope effects of a magnetic nature."[117o]

In summary,

A magnetic isotope effect appears only in radical reactions, so that isotope enrichment is a reliable means for detecting radicals and radical stages of chemical reactions. The new method of investigating reaction mechanisms is similar in this respect to the well known method of the chemical polarization of nuclei . . . . The difference is merely that the polarization of nuclei is detected only during a reaction, whereas change in the isotopic composition of the molecules . . . can be measured at any time after reaction has terminated. The reaction is conducted . . . in the Earth's magnetic field, without a nuclear magnetic resonance spectrometer, ... It is necessary merely to isolate the reaction products and make an isotopic analysis on them . . . .

The new method is only beginning to be developed, for its application is not restricted to the range of classical organic reactions in which radicals and radical pairs may appear. A magnetic isotope effect can occur in electron-transfer reactions involving ions of variable valency, in biochemical processes (redox reactions, respiratory processes, phosphorylation, certain enzymatic reactions, *etc*.). A necessary condition for this effect is the establishment of sufficiently long lived states, in which transitions are possible between levels differing in spin multiplicity. In these cases changes in electron spin lead to changes in the nuclear spin system.[117e]

It would be tempting to use the magnetic isotope effects to explain the anomalous abundance of magnetic isotopes (C^{13}, N^{15}, and others) in objects of terrestrial and cosmic origin [Ref. 117p].[117q]

See also Refs. 121.

X. ADDENDA

A nice reaction scheme, illustrative of several free radical reactions and portraying a complex overall transformation, has been presented.[122a] Similarly, reaction schemes have been presented that incorporate sequences of different, common radical processes[122b]; we suggest that the reader go through this to get a kaleidoscopic view of many of the things radicals do.

A. Awards

For the "good sense" treatment of data and the avoidance of cloaking what are basically arbitrary choices with "respectability" resting on handwaving arguments or conclusion-by-decree:

R.G. Marshall and L. Rahman, "Radical Equilibrium Studies; The Thermodynamic Parameters of *n*-Propyl." *Int. J. Chem. Kinet.*, **9**, 705 (1977).

For the application of basic principles to research in free radical chemistry:

Fluorocarbon Group (Shanghai Institute of Organic Chemistry, Academia Sinica), "Structure-Property Relationships of Fluoroolefins. I. The Polar and Solvent Effects in Free-Radical Additions," *Sci. Sin.* **20**, 353 (1977).

XI. POSTSCRIPTS TO VOLUME 1

I. Structure of Prototypal Radicals

a. Methyl

"Matrix Isolation Study of the Infrared Spectrum and Structure of the CH_3 Free Radical."[123a]

"Theoretical Studies of CH_3, CH_3^+, and CH_3^- Using Correlated Wavefunctions."[123b]

b. Ethyl

The full paper corresponding to the previously described communications of the infrared spectra of ethyl and phenyl has appeared: "Matrix Isolation Studies of Alkyl Radicals. The Characteristic Infrared Spectra of Primary Alkyl Radicals."[2,124a]

II. Characteristics of Prototypal Radicals

b. Infrared Spectrum of Phenyl

See Section I.B.

e. The Isoelectronic Series $H_2CN\cdot$, $H_2CO^{+\cdot}$, $H_2BO\cdot$

An ESR spectrum has been assigned to $H_2C=C^{\cdot}$, another member of the series.[36,40a]

III. Energetics of Protypal Radicals

b. $D(R_3\overset{+}{N}\text{-}H)$

"Effects of Alkyl and Fluoroalkyl Substitution on the Heterolytic and Homolytic Bond Dissociation Energies of Protonated Amines."[125]

c. The 3-Cyclopropenyl Radical

"Stabilities of Trivalent Carbon Species. 4. Electrochemical Reduction of Carbocations in Sulfuric Acid."[126]

d. Stability and Selectivity

A full paper that covers essentially the same ground and leads us to the same conclusions as did the communication discussed earlier, has appeared.[127]

IV. Experimental Techniques

Rotational tunneling in free radicals and its interaction with other processes.[128]

V. Mechanistic Techniques

c. Recent Examples of the Generation and Trapping of Radicals

"Free Radical Addition of 3-Chloro-4,4-dimethyl-2-oxazolidinone to Allylstannane and Allylsilane"[129a]; "Reaction of α-Chloro Ketones with Allyltri-n-

butyltin. Control of Reaction Site by the Variation of Reaction Conditions."[129b]
"Photochemistry of Alkyl Halides. 5. 2,4-Dehydroadamantane and Proto-adamantene from 2-Bromo- and 2-Iodoadamantane"[129c]; "Photochemistry of Alkyl Halides. 6. *gem*-Diiodides. A Convenient Method for the Cyclopropanation of Olefins"[129d,129e]; "4-Phenyl-1-iodobutane Photochemistry."[129f]

"A Radical-Chain Mechanism for some Sulfur Dioxide 'Insertion' Reactions of Organocobaloximes"[129g]; "Homolytic Displacements at Carbon Centers. Part 1. Reaction of Allyl and Allenyl cobaloximes with Polyhalogenomethanes"[129h]; "Reactions of Allyl- and Propadienyl-rhodium(III) and -iridium(III) Complexes with Polyhalogenomethanes. Rhodium(II) and Iridium(II) Species as Reactive Intermediates"[129i]; "Homolytic Displacements at Carbon. Part 3. Regiospecific Syntheses of Allyl Sulphones in the Reaction of Allylcobaloximes with Organo-sulphonyl Chlorides."[129j]

VII. New Radical Processes

b. Disproportionation of Radicals via α-Abstraction

Chemistry of ·CCl_3 in radical pairs.[130a,130b] Reports of the title reaction.[130c-130f]

VIII. Characteristics of Prototypal Reactions

The decomposition of *meso*-1,1'-diphenylazoethane in the presence of an optically active cobalt complex was studied in the presence and absence of the trap, nitrosobenzene. Based on a highly conjectural view of the chemistry and a tacitly assumed purity of the product, the signs of the very low optical rotation of the 2,3-diphenylbutane produced under the two circumstances was taken as indicative of its formation via an S_H2 reaction, a bimolecular carbon radical displacement on carbon, proceeding with net inversion.[131]

XII. ERRATA FOR VOLUME 1

Page 180	Equation 10	Superscript should be *f*
Page 182	Section D, footnote *f*	Delete "see footnote *c*"
Page 185	Paragraph after Eq. 21	Replace "75" by "75s"
Page 196	Reference 75s	Replace "75a" by "75q"

XIII. REFERENCES AND NOTES

1.* L. Kaplan, in *Reactive Intermediates. A Serial Publication*, Vol. 1, M. Jones, Jr., and R. A. Moss, Eds., Wiley Interscience, New York, 1978, (a) p. 165; (b) p. 166; (c) p. 184; (d) p. 167; (e) pp. 171–174; (f) pp. 173–174; (g) p. 176; (h) p. 171, Ref. 27; (i) p. 183; (j) pp. 184–186.

2.   J. Pacansky, D. E. Horne, G. P. Gardini, and J. Bargon, *J. Phys. Chem.*, 81, 2149 (1977).

3.* J. Pacansky, G. P. Gardini, and J. Bargon, *Ber. Bunsenges. Phys. Chem.*, 82, 19 (1978).

4.* J. Dyke, N. Jonathan, E. Lee, A. Morris, and M. Winter, *Phys. Scr.*, 16, 197 (1977).

5.* (a) F. A. Houle and J. L. Beauchamp, *J. Am. Chem. Soc.*, 101, 4067 (1979).
     (b) Cited in Ref. 5a.
     (c) F. A. Houle and L. B. Harding, submitted for publication in *Proc. Natl. Acad. Sci. U.S.*.
     (d) F. A. Houle, J. L. Beauchamp, G. Prakash, and G. A. Olah, *J. Am. Chem. Soc.*, to be submitted.
     (e) F. A. Houle and J. L. Beauchamp, *J. Am. Chem. Soc.*, to be submitted.

6.* (a) L. Bonazzola, N. Leray, and J. Roncin, *J. Am. Chem. Soc.*, 99, 8348 (1977).
     (b) D. Griller, K. U. Ingold, P. J. Krusic, and H. Fischer, *J. Am. Chem. Soc.*, 100, 6750 (1978).

7.   It is unclear which of two arguments is being advanced: (1) that there is a dependence on temperature of the equilibrium geometry in the matrix, a pyramidal structure being favored at lower temperature, or (2) that the equilibrium geometry at all temperatures is planar, but the radical is trapped by the matrix in a pyramidal structure at the time of its formation at low temperature. Since the second possibility can be addressed rather simply through a determination of whether the coupling constants observed upon initial formation of the radical at lower temperature are reproduced upon returning the radical to lower temperature after its having been at higher temperature, and since the argument advanced in an earlier discussion[8a] of $Me_3C\cdot$ resembles the first possibility more closely, we assume that it is the first possibility that is being advanced, even though, in a recent discussion[8b] of Me_3N^{+}, isoelectronic with $Me_3C\cdot$, somewhat less ambiguous arguments resembling the second possibility more closely were advanced.

8.   (a) C. Hesse and J. Roncin, *Mol. Phys.*, 19, 803 (1970).
     (b) J. P. Michaut and J. Roncin, *Can. J. Chem.*, 55, 3554 (1977).

9.   M. C. R. Symons and I. G. Smith, *J. Chem. Soc. Perkin II*, 1362 (1979).

10. L. Kaplan, in *Free Radicals*, Vol. 2, J. K. Kochi, Ed., Wiley, New York, 1973: (a) p. 361; (b) p. 421, footnote g.

11. Analogy: An exact mathematical solution to equations that describe approximately an approximate model of reality.

12.* (a) D. Griller and K. F. Preston, *J. Am. Chem. Soc.*, 101, 1975 (1979).
     (b) D. Griller, P. R. Marriott, and K. F. Preston, *J. Chem. Phys.*, 71, 3703 (1979). A preprint and helpful correspondence from Dr. Griller are gratefully acknowledged.

13. See also Section I.A.

14.   (a) P. R. Brooks, E. M. Jones, and K. Smith, *J. Chem. Phys.*, **51**, 3073 (1969).
      (b) N. I. Butkovskaya, M. N. Larichev, I. O. Leipunskii, I. I. Morozov, and V. L. Tal'rose, *Chem. Phys.*, **12**, 267 (1976).

15.* N. I. Butkovskaya, M. N. Larichev, I. O. Leipunskii, I. I. Morozov, and V. L. Talrose, *Chem. Phys. Lett.*, **63**, 375 (1979).

16.* (a) M. C. R. Symons, *J. Chem. Soc., Perkin II*, 1618 (1974).
      (b) M. C. R. Symons, *J. Am. Chem. Soc.*, **91**, 5924 (1969).
      (c) E. Sagstruen, *J. Chem. Phys.*, **69**, 3206 (1978).
      (c) J. Y. Lee and H. C. Box, *J. Chem. Phys.*, **59**, 2509 (1973).

17.   J. A. Den Hollander and R. Kaptein, *Chem. Phys. Lett.*, **41**, 257 (1976). See also N. C. Verma and R. W. Fessenden, *J. Chem. Phys.*, **65**, 2139 (1976). However, see also B. H. Bakker, Th. R. Bok, H. Steinberg, and Th. J. de Boer, *Rec. Trav. Chim.*, **96**, 31 (1977) and V. M. Kuznets, A. Z. Yankelevich, B. D. Sviridov, G. A. Nikiforov, C. de Jonge, Kh. I. Khageman, and V. V. Ershov, *Izv. Akad. Nauk SSSR*, 1251 (1979).

18.* P. C. Engelking, G. B. Ellison, and W. C. Lineberger, *J. Chem. Phys.*, **69**, 1926 (1978).

19.   D,L-Serine,[16d,21e] $DOCH_2CH(\overset{+}{N}D_3)CO_2^-$,[16d] thymidine,[21d,21e,21n,21p] 5-bromodeoxyuridine,[21e,21g,21o,21p] dulcitol,[21f] 3'-cytidylic acid,[21g] 5-chlorodeoxyuridine,[21g] adenosine ·HCl,[21g] uracil-β-D-arabinofuranoside,[21q] deoxycytidine 5'-phosphate,[21a] deoxyadenosine monohydrate,[21g] α-methyl-D-glucopyranoside,[21b] myo-inositol,[21h] per(OD)-myo-inositol,[21i] L(+)arabinose,[21j] rhamnose,[21k,21s] deoxycytidine 5'-phosphate-H_2O,[21l] α,D-glucopyranose,[21r] 6-methylmercaptopurine riboside,[21m] sucrose,[21k,21s] thymidine,[21c] α,D-glucose,[21c] polyvinyl alcohol,[21t] $R_2NCH_2OOR_1$.[21w]

20.   Citations are extensive because we have tried to assemble a body of work not previously brought together.

21.* (a) D. Krilov, A. Velenik, and J. N. Herak, *J. Chem. Phys.*, **69**, 2420 (1978).
      (b) K. P. Madden and W. A. Bernhard, *J. Chem. Phys.*, **70**, 2431 (1979).
      (c) M. C. R. Symons and G. W. Eastland, *J. Chem. Res.*(M)*, 2901 (1977).
      (d) H. C. Box and E. E. Budzinski, *J. Chem. Phys.*, **62**, 197 (1975).
      (e) H. C. Box, E. E. Budzinski, and G. Potienko, *J. Chem. Phys.*, **69**, 1966 (1978).
      (f) H. C. Box, E. E. Budzinski, H. G. Freund, and W. R. Potter, *J. Chem. Phys.*, **70**, 1320 (1979).
      (g) W. A. Bernhard, D. M. Close, J. Hüttermann, and H. Zehner, *J. Chem. Phys.* **67**, 1211 (1977).
      (h) H. C. Box and E. E. Budzinski, *J. Chem. Phys.*, **67**, 4726 (1977).
      (i) E. E. Budzinski, and H. C. Box, *J. Chem. Phys.*, **68**, 5296 (1978).
      (j) H. C. Box, E. E. Budzinski, and H. G. Freund, *J. Chem. Phys.*, **69**, 1309 (1978).
      (k) E. E. Budzinski, W. R. Potter, G. Potienko, and H. C. Box, *J. Chem. Phys.*, **70**, 5040 (1979).
      (l) D. M. Close and W. A. Bernhard, *J. Chem. Phys.*, **70**, 210 (1979).
      (m) E. Sagstuen and C. Alexander, Jr., *J. Chem. Phys.*, **68**, 762 (1978).
      (n) H. C. Box and H. G. Freund, *Ann. N. Y. Acad. Sci.*, **222**, 446 (1973).
      (o) H. C. Box and E. E. Budzinski, *J. Chem. Soc., Perkin II*, 553 (1976) (see also Ref. 21d for a passing reference to this work).
      (p) H. C. Box, *Disc. Faraday Soc.*, **63**, 264 (1977).
      (q) R. Bergene and R. A. Vaughan, *Int. J. Radiat. Biol.*, **29**, 145 (1976).
      (r) K. P. Madden and W. A. Bernhard, *J. Phys. Chem.*, **83**, 2643 (1979).

(s)   Assignment made only inferentially.

(t)   D. N. Aleni and G. P. Shakhovskoi, *Khim. Vys. Energ.*, **12**, 397 (1978) [assignment probably wrong: (1) spectrum is a singlet; (2) g value is "almost identical" to that of a spectrum assigned to $-CH_2\dot{C}HOH$].

(u)   R. V. Kucher, A. A. Turovskii, L. V. Luk'yanenko, and N. V. Dzumedzei, *Teor. Eksp. Khim.*, **12**, 41 (1976) (assignment very speculative).

22.* K. Toriyama and M. Iwasaki, *J. Am. Chem. Soc.*, **101**, 2516 (1979).

23.   See also C. Chachaty and A. Forchioni, *C. R.* **268C**, 300 (1969).

24.   For example, see Section X.IV of this chapter.

25.   (a) D. R. Yarkony, H. F. Schaefer III, and S. Rothenberg, *J. Am. Chem. Soc.*, **96**, 656 (1974); *(b) B. K. Janousek, A. H. Zimmerman, K. J. Reed, and J. I. Brauman, *J. Am. Chem. Soc.*, **100**, 6142 (1978).

26.   See also M. C. R. Symons, *J. Chem. Res. (M)*, 3565 (1978).

27.   M. Symons, *Chemical and Biochemical Aspects of Electron-Spin Resonance Spectroscopy*, Wiley, New York, 1978: (a) p. 153; (b) p. 78; (c) p. 76; (d) p. 102.

28.   B. Kalyanaraman and L. D. Kispert, *J. Chem. Phys.*, **68**, 5331 (1978).

29.   (a) D. W. G. Style and J. C. Ward, *Trans. Faraday Soc.*, **49**, 999 (1953).

      (b) H. W. Brown and G. C. Pimentel, *J. Chem. Phys.*, **29**, 883 (1958).

      (c) H. R. Wendt and H. E. Hunziker, *J. Chem. Phys.*, **71**, 5202 (1979).

      (d) K. Ohbayashi, H. Akimoto, and I. Tanaka, *J. Phys. Chem.*, **81**, 798 (1977).

      (e) G. Inoue, H. Akimoto, and M. Okuda, *Chem. Phys. Lett.*, **63**, 213 (1979).

30.* (a) D. J. Nelson, R. L. Petersen, and M. C. R. Symons, *J. Chem. Soc. Perkin II*, 2005 (1977).

      (b) D. Nelson and M. C. R. Symons, *Chem. Phys. Lett.*, **36**, 340 (1975).

      (c) T. Gillbro, *Chem. Phys.*, **4**, 476 (1974).

      (d) P. S. H. Bolman, I. Safarik, D. A. Stiles, W. J. R. Tyerman, and O. P. Strausz, *Can. J. Chem.*, **48**, 3872 (1970).

      (e) A. Torikai, S. Sawada, K. Fueki, and Z.-i. Kuri, *Bull. Chem. Soc. Jap.*, **43**, 1617 (1970).

      (f) R. F. Wheaton and M. G. Ormerod, *Trans. Faraday Soc.*, **65**, 1638 (1969).

      (g) V. G. Krivenko, L. P. Kayushin, and M. K. Pulatova, *Biofizika*, **14**, 615 (1969).

      (h) K. Akasaka, *J. Chem. Phys.*, **43**, 1182 (1965).

      (i) H. C. Box, H. G. Freund, and E. E. Budzinski, *J. Chem. Phys.*, **45**, 809 (1966).

      (j) W. W. H. Kou and H. C. Box, *J. Chem. Phys.*, **64**, 3060 (1976).

      (k) E. E. Budzinski and H. C. Box, *J. Phys. Chem.*, **75**, 2564 (1971).

      (l) G. Saxebøl and O. Herskedal, *Radiat. Res.*, **62**, 395 (1975).

      (m) L. P. Kayushin, V. G. Krivenko, and M. K. Pulatova, *Stud. Biophys.*, **33**, 59 (1972).

31. CH_3SH,[30a-30d] CH_3SSCH_3,[30c,30d] $EtSH$,[30a,30b,30d,30e] $EtSSEt$,[30d] $PrSH$,[30d] $PrSSPr$,[30d] $n\text{-}C_5H_{11}SH$,[30d] $n\text{-}C_6H_{13}SH$,[30d] Bu^tSH,[30d] CF_3SH,[30d] CF_3SSCF_3,[30d] $(CF_3S)_2Hg$,[30d] $HSCH_2COOH$,[30a,b] $HSCH_2CH_2COOH$,[30a,30b] $HSCH_2CH(OH)\text{-}CH_2Cl$,[30a,30b] $HSCH_2CH(OH)CH(OH)CH_2SH$,[30a,30b] $HSCHMeCOOH$,[30a,30b] $HSCH_2\text{-}CH(NH_2)COOH$,[30a,30b,30f] $HSCH_2CH(NH_3Cl)COOH$,[30g-30j,30m] $DSCH_2CH(ND_3Cl)\text{-}COOD$,[30j] $HSCMe_2CH(NH_3Cl)COOH$,[30k] $HSCMe_2CH(NH_2)COOH$,[30a,30b] $HSCH_2\text{-}CH(NHAc)COOH$.[30l]

32.   Although it is not completely meaningful to define and classify experimental observations in terms of an approximate theoretical model (e.g., σ^*orbitals, conjugated systems, π-delocalization), we use common jargon here and elsewhere in this section.

33. (a) S. P. Mishra and M. C. R. Symons, *J. Chem. Soc., Perkin II*, 391 (1973).
    (b) D. J. Nelson and M. C. R. Symons, *Chem. Phys. Lett.*, 47, 436 (1977).
    (c) S. Nagal and T. Gillbro, *J. Phys. Chem.*, 81, 1793 (1977);
    (d) L. D. Kispert, R. Reeves, and T. C. S. Chen, *J. Chem. Soc. Faraday II*, 871 (1978).
    (e) G. W. Neilson and M. C. R. Symons, *J. Chem. Soc. Faraday II*, 1582 (1972).
    (f) S. P. Mishra, G. W. Neilson, and M. C. R. Symons, *J. Chem. Soc. Faraday II*, 1280 (1974).
    (g) H. Riederer, J. Hüttermann, and M. C. R. Symons, *J. Chem. Soc., Chem. Commun.*, 313 (1978).
    (h) M. C. R. Symons, *J. Chem. Soc. Chem. Commun.*, 408 (1977);
    (i) G. W. Neilson and M. C. R. Symons, *Mol. Phys.*, 27, 1613 (1974);
    (j) M. B. Yim and D. E. Wood, *J. Am. Chem. Soc.*, 98, 2053 (1976).
    (k) T. Higashimura, *Int. J. Radiat. Phys. Chem.*, 6, 393 (1974).
34.* (a) M. C. R. Symons, R. C. Selby, I. G. Smith, and S. W. Bratt, *Chem. Phys. Lett.*, 48, 100 (1977).
    (b) A. Hasegawa, M. Shiotani, and F. Williams, *Disc. Faraday Soc.*, 63, 157 (1977).
    (c) R. I. McNeil, M. Shiotani, F. Williams, and M. B. Yim, *Chem. Phys. Lett.*, 51, 438 (1977).
    (d) R. McNeil, M. Shiotani, and F. Williams, unpublished results cited in Ref. 34b.
35. (a) R. E. Moore and G. L. Driscoll, *J. Org. Chem.*, 43, 4978 (1978);
    (b) G. Robertson, E. K. S. Liu, and R. J. Lagow, *J. Org. Chem.*, 43, 4981 (1978).
36.* (a) S. P. Mishra and M. C. R. Symons, *J. Chem. Soc., Chem. Commun.*, 577 (1973).
    (b) A. Hasegawa and F. Williams, *Chem. Phys. Lett.*, 46, 66 (1977).
37.* (a) F. T. Prochaska and L. Andrews, *J. Chem. Phys.*, 68, 5577 (1978).
    (b) F. T. Prochaska and L. Andrews, *J. Phys. Chem.*, 82, 1731 (1978).
    (c) F. T. Prochaska and L. Andrews, *J. Chem. Phys.* 68, 5568 (1978).
    (d) F. T. Prochaska and L. Andrews, *J. Am. Chem. Soc.*, 100, 2102 (1978).
    (e) B. A. Ault and L. Andrews, *J. Chem. Phys.*, 63, 1411 (1975).
38.* G. J. Verhaart, W. J. Van Der Hart, and H. H. Brongersma, *Chem. Phys.*, 34, 161 (1978).
39.* V. A. Shvets, V. B. Sapozhnikov, N. D. Chuvylkin, and V. B. Kazansky, *J. Catal.*, 52, 459 (1978).
40. (a) Y. Ben Taarit, M. C. R. Symons, and A. J. Tench, *J. Chem. Soc., Faraday Trans. I*, 1149 (1977).
    (b) Y. Kirino, *J. Phys. Chem.*, 79, 1296 (1975).
    (c) V. B. Sapozhnikov, V. A. Shvets, N. D. Chuvylkin, and V. B. Kazanskii, *Kinet. Katal.*, 17, 1251 (1976).
    (d) see also N. D. Chyvylkin, G. M. Zhidomirov, and V. B. Kazanskii, *Kinet. Katal.*, 20, 250 (1979).
    (e) See also V. B. Kazanskii, S. L. Kalyagin, G. A. Kozlov, S. A. Surin, and B. N. Shelimov, *Kinet. Katal.*, 19, 1264 (1978).
    We are grateful to Drs. E. M. Thorsteinson and T. P. Wilson for assistance in locating pertinent literature.
41. (a) L. Andrews and F. T. Prochaska, *J. Am. Chem. Soc.*, 101, 1190 (1979).
    (b) L. Andrews, J. H. Miller, and E. S. Prochaska, *J. Am. Chem. Soc.*, 101, 7158 (1979).
    (c) T. Izumida, T. Ichikawa, and H. Yoshida, *J. Phys. Chem.*, 83, 374 (1979).
    (d) E. D. Sprague, *J. Phys. Chem.*, 83, 849 (1979).

(e) L. Andrews and F. T. Prochaska, *J. Phys. Chem.*, **83**, 824 (1979).

(f) Y. J. Chung and F. Williams, *J. Phys. Chem.*, **76**, 1792 (1972).

(g) Y. J. Chung, K. Nishikida, and F. Williams, *J. Phys. Chem.*, **78**, 1882 (1974).

(h) M. Irie, M. Shimizu, and H. Yoshida, *J. Phys. Chem.*, **80**, 2008 (1976).

(i) L. Andrews, C. A. Wight, F. T. Prochaska, S. A. McDonald, and B. S. Ault, *J. Mol. Spectrosc.*, **73**, 120 (1978).

(j) F. T. Prochaska, B. W. Keelan, and L. Andrews, *J. Mol. Spectrosc.*, **76**, 142 (1979).

(k) J. Burdett, *J. Mol. Spectrosc.*, **36**, 365 (1970).

(l) L. Andrews and F. T. Prochaska, *J. Chem. Phys.*, **70**, 4714 (1979).

(m) L. Andrews and G. C. Pimentel, *J. Chem. Phys.*, **47**, 3637 (1967).

(n) L. Y. Tan and G. C. Pimentel, *J. Chem. Phys.*, **48**, 5202 (1968).

(o) T. G. Carver and L. Andrews, *J. Chem. Phys.*, **50**, 4223 (1969).

(p) T. G. Carver and L. Andrews, *J. Chem. Phys.*, **50**, 4235 (1969).

(q) T. G. Carver and L. Andrews, *J. Chem. Phys.*, **54**, 5425 (1971).

(r) E. D. Sprague and F. Williams, *J. Chem. Phys.*, **54**, 5425 (1971).

(s) Y. Fujita, T. Katsu, M. Sato, and K. Takahashi, *J. Chem. Phys.*, **61**, 4307 (1974).

(t) K. Toriyama and M. Iwasaki, *J. Chem. Phys.*, **65**, 2883 (1976).

(u) T. Izumida, Y. Tanabe, T. Ichikawa, and H. Yoshida, *Bull. Chem. Soc. Jap.*, **52**, 235 (1979).

(v) T. Saito and H. Yoshida, *Bull. Chem. Soc. Jap.*, **47**, 3167 (1974).

(w) J. B. Gallivan and W. H. Hamill, *Trans. Faraday Soc.*, **61**, 1960 (1965).

(x) D. Nelson and M. C. R. Symons, *Tetrahedron Lett.*, 2953 (1975).

(y) B. Brocklehurst and M. I. Savadatti, *Nature*, **212**, 1231 (1966).

(z) A. R. Lyons, M. C. R. Symons, and S. P. Mishra, *Nature*, **249**, 341 (1974).

(aa) E. D. Sprague, K. Takeda, J. T. Wang, and F. Williams, *Can. J. Chem.*, **52**, 2840 (1974).

(bb) M. C. R. Symons, *J. Chem. Soc. Perkin II*, 908 (1976).

(cc) M. E. Jacox and D. E. Milligan, *Chem. Phys.*, **16**, 381 (1976).

(dd) M. E. Jacox and D. E. Milligan, *Chem. Phys.*, **16**, 195 (1976).

(ee) M. Iriè, M. Shimizu, and H. Yoshida, *Chem. Phys. Lett.*, **25**, 102 (1974).

42. M. C. R. Symons, in *Radicaux Libres Organiques Colloq. Int. C.N.R.S.*, **278**, Paris 105 (1978).

43. (a) Reference 33e; see also Ref. 33g.

(b) M. C. R. Symons, *J. Chem. Res. (S)*, 360 (1978); see also Ref. 36b.

44. See also Ref. 33b, 33h, and 43b.

45.* (a) S. F. Nelsen and C. R. Kessel, *J. Am. Chem. Soc.*, **101**, 2503 (1979).

(b) T. Ichikawa, N. Ohta, and H. Kajioka, *J. Phys. Chem.*, **83**, 284 (1979).

(c) B. Badger and B. Brocklehurst, *Trans. Faraday Soc.*, **65**, 2576 (1969).

(d) P. L. Corio and S. Shih, *J. Phys. Chem.*, **75**, 3475 (1971).

(e) M. Iwasaki, H. Muto, K. Toriyama, M. Fukaya, and K. Nunome, *J. Phys. Chem.*, **83**, 1590 (1979).

(f) A. E. Hirschler, W. C. Neikam, D. S. Barmby, and R. L. James, *J. Catal.*, **4**, 628 (1965) (spectra not assigned to a cation radical).

(g) L. Toman, J. Pilař, and M. Marek, *J. Polymer Sci., Polymer Chem.*, **16**, 371 (1978).

(h) M. Marek, L. Toman, and J. Pilař, *J. Polymer Sci., Polymer Chem.*, **13**, 1565 (1975).

(i) R. M. Dessau, *J. Am. Chem. Soc.*, **92**, 6356 (1970).

(j) T. Ichikawa and P. K. Ludwig, *J. Am. Chem. Soc.*, 91, 1023 (1969).

(k) F. Nauwelaerts and J. Ceulemans, *J. Mag. Res.*, 25, 141 (1977).

(l) however, see also Ref. 45b.

46. n-$C_6H_{13}CH=CH_2$,[45f] $Me_2C=CH_2$,[45g,45h] $EtCH=CHMe$,[45f] $Me_2C=CHMe$,[45f] Pr^i-$CH=CHMe$,[45f] $Me_2C=CMe_2$,[45b,45d,45e,45i,45j,45k,45l] (g = 2.0028[45d], 2.0025[45e]), $Me_2C=CMeEt$,[45i] $Me_2C=CMePr^i$,[45i] 1,2-dimethylcyclohexene.[45i]

47. (a) A. Owczarczyk and W. Stachowicz, *Radiat. Phys. Chem.*, 10, 319 (1977).

(b) T. Shida and W. H. Hamill, *J. Am. Chem. Soc.*, 88, 5376 (1966).

(c) J. P. Guarino and W. H. Hamill, *J. Am. Chem. Soc.*, 86, 777 (1964).

(d) E. P. Bertin and W. H. Hamill, *J. Am. Chem. Soc.*, 86, 1301 (1964).

(e) J. B. Gallivan and W. H. Hamill, *J. Chem. Phys.*, 44, 2378 (1966).

(f) Gy. Cserép, O. Brede, W. Helmstreit, and R. Mehnert, *Radiochem. Radioanal. Lett.*, 32, 15 (1978).

(g) Gy. Cserép, O. Brede, W. Helmstreit, and R. Mehnert, *Radiochem. Radioanal. Lett.*, 34, 383 (1978).

(h) O. Brede, R. Mehnert, and Gy. Cserép, *Radiochem. Radioanal. Lett.*, 39, 169 (1979).

(i) See also Ref. 47h and 47j, wherein assignments are made to the dimer.

(j) J. Bös, O. Brede, W. Helmstreit, and R. Mehnert, *Proc. 4th Symp. Radiat. Chem.*, 881 (1976).

48. A. R. Lyons and M. C. R. Symons, *J. Chem. Soc., Faraday Trans. II*, 502 (1972).

49. See also Ref. 50.

50. (a) J. M. McBride, M. W. Vary, and B. L. Whitsel, in Organic Free Radicals, W. A. Pryor, Ed. A.C.S. Symp. Ser., Vol. 69, American Chemical Society, Washington, D.C. 1978, p. 208.

(b) J. A. Baban and B. P. Roberts, "Ligand-σ Phosphoranyl Radicals," *J. Chem. Soc., Chem. Commun.*, 537 (1979).

51. (a) T. Koenig and R. A. Wielesek, "INDO Configurations for Succinimidyl" *Tetrahedron Lett.*, 2007 (1975).

(b) E. Hedaya, R. L. Hinman, V. Schomaker, S. Theodoropulos, and L. M. Kyle, "The Succinimidyl Radical Problem. The Ease of Formation of Nitrogen and Oxygen II and Σ Free Radicals," *J. Am. Chem. Soc.*, 89, 4875 (1967).

52. See, for example, Ref. 53.

53. (a) F. L. Lu, Y. M. A. Naguib, M. Kitadani, and Y. L. Chow, *Can. J. Chem.*, 57, 1967 (1979).

(b) T. C. Joseph, J. N. S. Tam, M. Kitadani, and Y. L. Chow, *Can. J. Chem.*, 54, 3517 (1976).

(c) A. R. Forrester, A. S. Ingram, I. Lennox John, and R. H. Thomson, *J. Chem. Soc., Perkin I*, 1115 (1975).

(d) P. Mackiewicz, R. Furstoss, B. Waegell, R. Cote, and J. Lessard, *J. Org. Chem.*, 43, 3746 (1978).

(e) B. Danieli, P. Manitto, and G. Russo, *Chem. Ind.*, 203 (1971).

54.* (a) S. A. Glover and A. Goosen, "N-Iodo-amides: Mechanism of Intramolecular Reactions with Aromatic Rings of Amido-radicals in Σ- and Π-Electronic States," *J. Chem. Soc., Perkin I*, 1348 (1977).

(b) See also Ref. 54c, 54d, and 54j.

(c) S. A. Glover and A. Goosen, "Evidence for the Reaction of Amido Radicals via the Σ- and Π-States," *Prog. Theor. Org. Chem.*, 2 (Appl. MO Theory Org. Chem.), 297 (1977).

(d) S. A. Glover and A. Goosen, *J. Chem. Soc., Perkin I*, 653 (1978).

(e) P. S. Skell and J. C. Day, in *Organic Free Radicals*, A.C.S. Symp. Ser., W. A. Pryor, Ed., Vol. 69, 1978, p. 290 (a brief statement, without details).

(f) P. S. Skell and J. C. Day, *Acc. Chem. Res.*, 11, 381 (1978) (a review, without experimental details).

(g) J. C. Day, M. G. Katsaros, W. D. Kocher, A. E. Scott, and P. S. Skell, *J. Am. Chem. Soc.*, 100, 1950 (1978) (a communication, without experimental details).

(h) P. S. Skell and J. C. Day, *J. Am. Chem. Soc.*, 100, 1951 (1978) (a communication, without experimental details).

(i) P. S. Skell, J. C. Day, and J. P. Slanga, *Angew. Chem., Int. Ed. Engl.*, 17, 515 (1978) (a communication, without experimental details).

(j) A. Goosen, "N-Haloamides and amido–Radicals," *S. Afr. J. Chem.*, 32, 37 (1979).

55.* (a) W. Tsang, *Int. J. Chem. Kinet.*, 10, 821 (1978).

(b) J. A. Walker and W. Tsang, *Int. J. Chem. Kinet.*, 11, 867 (1979).

(c) G. M. Atri, R. R. Baldwin, G. A. Evans, and R. W. Walker, *J. Chem. Soc. Faraday I*, 366 (1978).

(d) G. A. Evans and R. W. Walker, *Int. J. Chem. Kinet.*, 1458 (1979).

(e) A. C. Baldwin, K. E. Lewis, and D. M. Golden, *Int. J. Chem. Kinet.*, 11, 529 (1979).

(f) M. Rossi and D. M. Golden, *Int. J. Chem. Kinet.*, 11, 969 (1979).

(g) R. G. McLaughlin and J. C. Traeger, *J. Am. Chem. Soc.*, 101, 5791 (1979).

56.* M. H. Baghal-Vayjooee and S. W. Benson, *J. Am. Chem. Soc.*, 101, 2838 (1979).

57. (a) Insufficient experimental details are provided in this full paper of just what was measured (for example, what reaction was being observed) when k(Cl· + cyclopropane) was determined. All that is said is that "the products of the equilibrium reaction Cl + cyclopropane $\rightleftarrows$ HCl + cyclopropyl were the only ones detected" and that " . . . we found no evidence for such isomerization [cyclopropyl· → allyl·] . . . ." Just what was done and what was observed? For example, what about the well-known reaction Cl· + cyclopropane → ClCH₂CH₂CH₂·? Complex formation between Cl· and cyclopropane? Concerted Cl· + cyclopropane → HCl + allyl·?

(b) Earlier work consisted of studies of the photochlorination of C_2H_6/cyclopropane to cyclopropyl chloride in unreported yield in which " . . . an appreciable dark reaction took place which did not . . . form cyclopropyl chloride. This product may have been 1:3-dichloropropane."[57c] Also carried out were studies of photochlorination of CF_3CH_3/cyclopropane to cyclopropyl chloride, produced in unreported yield and identified by use of VPC, in which "not enough runs were done to obtain accurate Arrhenius parameters."[57d]

(c) J. H. Knox and R. L. Nelson, *Trans. Faraday Soc.*, 55, 937 (1959).

(d) P. Cadman, A. W. Kirk, and A. F. Trotman-Dickenson, *J. Chem. Soc., Faraday I*, 1027 (1976).

(e) How was it produced? How does HCl react with cyclopropyl? Electrophilically? How good a material balance is obtained?

58.* A. J. Colussi, F. Zabel, and S. W. Benson, *Int. J. Chem. Kinet.*, 9, 161 (1977).

59.* A. J. Colussi and S. W. Benson, *Int. J. Chem. Kinet.*, 10, 1139 (1978).

60.* (a) J. F. Wolf, R. H. Staley, I. Koppel, M. Taagepera, R. T. McIver, Jr., J. L. Beauchamp, and R. W. Taft, *J. Am. Chem. Soc.*, 99, 5417 (1977);

(b) See also Ref. 60c.

(c) D. H. Aue, H. M. Webb, W. R. Davidson, L. D. Betowski, M. C. Vidal, and M. T. Bowers, reported by D. H. Aue and M. T. Bowers, in *Gas Phase Ion Chemistry*, Vol. 2, M. T. Bowers, Ed., Academic, 1979, Table VII, p. 25.

61.* G. B. Ellison, P. C. Engelking, and W. C. Lineberger, *J. Am. Chem. Soc.*, 100, 2556 (1978).

62.* (a) A. H. Zimmerman and J. I. Brauman, *J. Am. Chem. Soc.*, 99, 3565 (1977);
    (b) G. I. Mackay, M. H. Lien, A. C. Hopkinson, and D. K. Bohme, *Can. J. Chem.*, 56, 131 (1978).

63.* K. J. Reed and J. I. Brauman, *J. Am. Chem. Soc.*, 97, 1625 (1975).

64. (a) M. Meot-Ner (Mautner), E. P. Hunter, and F. H. Field, *J. Am. Chem. Soc.*, 101, 686 (1979).
    (b) M. J. Locke, R. L. Hunter, and R. T. McIver, Jr., *J. Am. Chem. Soc.*, 101, 272 (1979).
    (c) Y. K. Lau, P. P. S. Saluja, P. Kebarle, and R. W. Alder, *J. Am. Chem. Soc.*, 100, 7328 (1978).
    (d) S. T. Ceyer, P. W. Tiedemann, B. H. Mahan, and Y. T. Lee, *J. Chem. Phys.*, 70, 14 (1979).
    (e) But, see the comments on Ref. 64d in Ref. 5a.

65.* (a) M. H. Baghal-Vayjooee, A. J. Colussi, and S. W. Benson, *J. Am. Chem. Soc.*, 100, 3214 (1978).
    (b) M. H. Baghal-Vayjooee, A. J. Colussi, and S. W. Benson, *Int. J. Chem. Kinet.*, 11, 147 (1979).
    (c) I. S. Zaslonko and V. N. Smirnov, *Kinet. Katal.*, 20, 575 (1979).

66. (a) R. M. Marshall and L. Rahman, *Int. J. Chem. Kinet.*, 9, 705 (1977).
    (b) R. M. Marshall and N. D. Page, *Int. J. Chem. Kinet.*, 11, 199 (1979).

67.* K. Ishizu, H. Kohama, and K. Mukai, *Chem. Lett.*, 227 (1978).

68. P. R. Mallinson and M. R. Truter, *J. Chem. Soc., Perkin II*, 1818 (1972).

69.* (a) R. Biehl, W. Lubitz, K. Mobius, and M. Plato, "Observation of Deuterium Quadrupole Splittings of Aromatic Free Radicals in Liquid Crystals by ENDOR and TRIPLE Resonance," *J. Chem. Phys.*, 66, 2074 (1977).
    (b) W. Lubitz, R. Biehl and K. Möbius "Sodium and Proton ENDOR and Triple Resonance Experiments on Biphenyl and Fluorenone Ion Pairs in Solution," *J. Mag. Resonance.*, 27, 411 (1977).
    (c) R. Biehl, K. Hinrichs, H. Kurreck, W. Lubitz, U. Mennenga, and K. Roth, "ESR, NMR and ENDOR Studies of Partially Deuterated Phenyl Substituted Anthracenes. $\pi-\sigma$ Delocalization," *J. Am. Chem. Soc.*, 99, 4278 (1977).
    (d) B. Kirste, H. van Willigen, H. Kurreck, K. Möbius, M. Plato, and R. Biehl, "ENDOR of Organic Triplet- and Quartet State Molecules in Liquid Solutions and in Rigid Media," *J. Am. Chem. Soc.*, 100, 7505 (1978).
    (e) B. Kirste, H. Kurreck, H.-J. Fey, Ch. Hass, and G. Schlomp, "^{1}H, ^{2}H, and ^{13}C ENDOR Studies of Phenalenyl Radicals in Nematic and Smectic Mesophases of Liquid Crystals," *J. Am. Chem. Soc.*, 101, 7457 (1979).
    (f) R. Biehl, Ch. Hass, H. Kurreck, W. Lubitz, and S. Oestreich, "ESR and ENDOR Studies of Partially Deuterated and Chlorinated Phenalenyls. New Synthetic Pathways," *Tetrahedron*, 34, 419 (1978).
    (g) W. Lubitz, W. Broser, B. Kirste, H. Kurreck, and K. Schubert "^{13}C- and ^{1}H-ENDOR Studies of a Phenoxyl Type Radical," *Z. Naturforsch.*, 33a, 1072 (1978).

(h) K. Mobius, "ENDOR and ELDOR," in *Electron Spin Resonance* (Specialist Periodical Report), Vol. 4, P. B. Ayscough, Ed., The Chemical Society, 1977, p. 16.

(i) Ref. 69h, p. 21.

(j) K. Hinrichs, B. Kirste, H. Kurreck, and J. Reusch, "On Galvinols and Galvinoxyls-IV. Investigation of Dynamic Effects on Galvinoxyl Doublet Radicals with ENDOR in Solution," *Tetrahedron*, 33, 151 (1977).

(k) H. J. Fey, H. Kurreck, and W. Lubitz, "ENDOR Studies of [6] Helicene Anion Radical," *Tetrahedron*, 35, 905 (1979).

(l) W. Broser, H. Kurreck, S. Oestreich-Janzen, G. Schlömp, H.-J. Fey, and B. Kirste, "ESR, ^{1}H-, D-, ^{13}C-ENDOR and TRIPLE Resonance Studies of Alkyl Substituted Phenalenyl Radicals," *Tetrahedron*, 35, 1159 (1979).

(m) E. Boroske and K. Möbius, "^{1}H, ^{2}H, and ^{13}C Distant-ENDOR and DNP Studies of the Polarization and Depolarization Mechanisms In Partially Deuterated Succinic Acid," *J. Magn. Resonance.*, 28, 325 (1977).

(n) E. Boroske, L. Mayas, and K. Möbius, "A Molecular Structure Study of Partially Deuterated Succinic Acid by ^{1}H and ^{2}H Distant ENDOR," *Tetrahedron*, 35, 231 (1979).

(o) B. Kirste, H. Kurreck, W. Lubitz, and K. Schubert, "^{13}C ENDOR Studies of Organic Doublet and Triplet State Molecules," *J. Am. Chem. Soc.*, 100, 2292 (1978).

(p) B. Kirste, H. Kurreck, W. Harrer, and J. Reusch, "ENDOR Investigation of Internal Dynamics in Cyclopropyl Galvinoxyl Radicals," *J. Am. Chem. Soc.*, 101, 1775 (1979).

(q) H.-J. Fey, W. Lubitz, H. Zimmermann, M. Plato, K. Möbius, and R. Biehl, "^{13}C- and Proton-ENDOR Studies of ^{13}C- labelled Organic Radicals," *Z. Naturforsch.*, 33a, 514 (1978).

(r) S. Oestreich, W. Broser, and H. Kurreck, "Substituted Pentaphenylcyclopentadienyls and Tetraphenylcyclopentadienones, XII. The Conformation of Tetraphenylcyclopentadienones and their Anion Radicals as Studied by Polarography and EPR Spectroscopy" *Z. Naturforsch.*, 32b, 686 (1977).

(s) R. Biehl, M. Plato, and K. Mobius, "The g-Factors of Planar and Non planar Hydrocarbon Radicals. π-σDelocalization," *Mol. Phys.*, 35, 985 (1978).

(t) L. Mayas, M. Plato, C. J. Winscom, and K. Möbius, "Deuterium Quadrupole Coupling Constants in Hydrogen Bonded Dicarboxylic Acids. A Distant ENDOR Study," *Mol. Phys.*, 36, 753 (1978).

(u) B. Kirste, H. Kurreck, and K. Schubert, "Detection of a Biradical with ^{1}H and ^{13}C-ENDOR-in-Solution," *Tetrahedron Lett.*, 777 (1978).

(v) K. Möbius and R. Biehl, "Electron-Nuclear-Nuclear TRIPLE Resonance of Radicals in Solution," in *Multiple Electron Resonance Spectroscopy*, M. M. Dorio and J. H. Freed, Eds., Plenum Press, New York, 1979, p. 475.

(w) B. Kirste, "ENDOR Investigations of Galvinoxyl and D-Galvinoxyl in a Smectic Mesophase," *Chem. Phys. Lett.*, 64, 63 (1979);

(x) R. Biehl, K. Möbius, S. E. O'Connor, R. I. Walter, and H. Zimmermann, "Evaluation and Assignment of Proton and Nitrogen Hyperfine Coupling Constants in the Free-Radical 1-Picryl-2,2-diphenylhydrazyl. An NMR, Electron-Nuclear Double Resonance, and Electron-Nuclear-Nuclear Triple Resonance Study." *J. Phys. Chem.*, 83, 3449 (1979).

(y) W. Lubitz, M. Plato, K. Möbius, and R. Biehl, "Alkali and H ENDOR on Aromatic Ion Pairs in Solution. An INDO Approach," *J. Phys. Chem.,* **83**, 3402 (1979).

70.   (a) Such a practice establishes the burden of identifying just what it is in one's work that is new.

(b) for context and background, see the general discussion in Ref. 1e.

71.* R. J. Kinney, W. D. Jones, and R. G. Bergman, *J. Am. Chem. Soc.,* **100**, 7902 (1978).

72.   L. Kaplan, *J. Chem. Soc., Chem. Commun.,* 106 (1969).

73.* (a) K. U. Ingold, in Organic Free Radicals, A.C.S. Symp. Ser., W. A. Pryor, Ed., Vol. 69, American Chemical Society, Washington, D.C., 1978, pp. 203–204.

(b) P. Schmid, D. Griller, and K. U. Ingold, *Int. J. Chem. Kinet.,* **11**, 333 (1979);

(c) Y. Maeda and K. U. Ingold, *J. Am. Chem. Soc.,* **101**, 4975 (1979).

74.* (a) P. Schmid and K. U. Ingold, *J. Am. Chem. Soc.,* **100**, 2493 (1978).

(b) P. Schmid and K. U. Ingold, *J. Am. Chem. Soc.,* **99**, 6434 (1977).

(c) Reference 69a, p. 192.

(d) Y. Maeda, P. Schmid, D. Griller, and K. U. Ingold, *J. Chem. Soc., Chem. Commun.,* 525 (1978).

(e) T. Doba, S. Noda, and H. Yoshida, *Bull. Chem. Soc. Jap.,* **52**, 21 (1979).

(f) T. Doba, T. Ichikawa, and H. Yoshida, *Bull. Chem. Soc. Jap.,* **50**, 3158 (1977).

(g) See also Ref. 74h.

(h) T. Doba, T. Ichikawa, and H. Yoshida, *Bull. Chem. Soc. Jap.,* **50**, 3124 (1977).

75.   Taken to be equal to the ratio of concentrations of the spin adducts as a result of assuming (1) irreversible formation of spin adduct, and (2) equal rate constants for subsequent reaction of the spin adducts. The ratio of concentrations was followed as a function of time and was extrapolated to time = 0 whenever any dependence on time of the ratio (symptomatic of 1 or 2 not holding) was observed.

76.   This is supported by the observation that the initial rate of formation of spin adducts is independent of the nature of the trap and of its concentration and is first order in peroxide at "low and intermediate concentrations," but not at high concentrations, where the rate is higher. Although it is stated, correctly, that "this phenomenon does not, of course, influence competitive kinetic data," it does bring into question the idea that the data obtained under such circumstances are the competitive kinetic data they are thought to be, as opposed to the concentrations of spin adducts having been influenced by their not having been formed exclusively by trapping of the radicals.

77.   (a) Tacitly considered to be monomeric.

(b) the equilibrium constant for dedimerization was measured, it being presumed without justification that "only the monomer will be active," that is, that only the monomer can react with a radical to give the spin adduct.

78.   (a) See also Ref. 78b and 78c.

(b) D. Rehorek, "On the Use of Nitromesitylene as a Spin Trap," *J. Prakt. Chem.,* **321**, 112 (1979).

(c) M. Kamimori, H. Sakuragi, K. Sawatari, T. Suehiro, K. Tokumaru, and M. Yoshida, "Relative Spin-trapping Ability of *N*-Benzylidene-*t*-butylamine Oxide, *N*-Benzylideneaniline Oxide, and 2,3,5,6-Tetramethylnitrosobenzene towards Phenyl and Phenylcyclohexadienyl Radicals," *Bull. Chem. Soc. Jap.,* **52**, 2339 (1979).

79.* E. G. Janzen and Y. Y. Wang, *J. Phys. Chem.,* **83**, 894 (1979).

80.   W. N. Washburn, *J. Am. Chem. Soc.,* **100**, 6235 (1978).

81.* E. J. Broomhead and K. A. McLauchlan, *J. Chem. Soc., Faraday II,* 775 (1978).

82. For reports of the radical cations of perthiacycloalkanes see, for example, K.-D. Asmus, H. A. Gillis, and G. G. Teather, *J. Phys. Chem.,* 82, 2677 (1978) and references cited therein.

83.* (a) H.-D. Beckhaus and C. Rüchardt, *Chem. Ber.,* 110, 878 (1977).
  (b) H.-D. Beckhaus, G. Hellmann, and C. Rüchardt, *Chem. Ber.,* 111, 72 (1978).
  (c) C. Rüchardt, H.-D. Beckhaus, G. Hellmann, S. Weiner, and R. Winiker, *Angew. Chem., Int. Ed. Engl.,* 16, 875 (1977).
  (d) cited in Ref. 83g and attributed, apparently in error, to Ref. 83c.
  (e) C. Rüchardt and S. Weiner, *Tetrahedron Lett.,* 1311 (1979).
  (f) H.-D. Beckhaus, G. Hellmann, C. Rüchardt, B. Kitschke, H. J. Lindner, and H. Fritz, *Chem. Ber.,* 111, 3764 (1978).
  (g) C. Rüchardt, *Zh. Vses. Khim. Ob-va,* 24, 121 (1979).
  (h) G. Hellmann, Dissertation, University of Freiburg, 1977, cited in Ref. 83g.
  (i) C. Rüchardt and R. Winiker, in press, cited in Ref. 83g.

84. See also Refs. 85 and 86.

85. (a) G. Hellmann, H.-D. Beckhaus, and C. Rüchardt, "Thermolabile Hydrocarbons, VIII. Thermolysis of 1,2-Dialkyl-1,2-diphenylethanes," *Chem. Ber.,* 112, 1808 (1979).
  (b) K.-H. Eichin, K. J. McCullough, H.-D. Beckhaus, and C. Rüchardt, "Diastereoselective Radical Recombination and Variation in Thermal Stability of Diastereomeric Hydrocarbons," *Angew. Chem., Int. Ed. Eng.,* 17, 934 (1978).
  (c) A. Haas, K. Schlosser, and S. Steenken, "C—C Bond Homolysis in (CF₃S)₃C—C-(SCF₃)₃ at Room Temperature. Thermodynamic, Kinetic, and Electron Spin Resonance Results," *J. Am. Chem. Soc.,* 101, 6282 (1979).

86. Reference 87, as related to Ref. 88.

87. (a) J. A. Gladysz, J. C. Selover and C. E. Strouse, "α-Silyloxy and α-Hydroxy Manganese Alkyls. Generation via a New Five Membered Metallocycle," *J. Am. Chem. Soc.,* 100, 6766 (1978).
  (b) D. L. Johnson and J. A. Gladysz, *J. Am. Chem. Soc.,* 101, 6433 (1979).

88. (a) H. Hillgärtner, W. P. Neumann, and B. Schroeder, "Bis(trimethyltin) Benzopinacolate, Its Reversible Radical Dissociation and Reactions," *Justus Liebigs' Ann. Chem.,* 586 (1975).
  (b) W. P. Neumann, B. Schroeder, and M. Ziebarth, "Preparation, Stability and ESR Spectroscopy of Ketyl Radicals RR'Ċ—OM(CH₃)₃," *Justus Liebigs' Ann. Chem.,* 2279 (1975).
  (c) M. Ziebarth and W. P. Neumann, "The Dissociation of O-Silylated Pinacols; Structure and Stability of the Resulting Ketyl Radicals $R^1R^2\dot{C}OSiMe_3$," *Justus Liebigs' Ann. Chem.,* 1765 (1978).

89. R. M. Marshall, J. H. Purnell, and P. D. Storey, *Proc. Roy. Soc. (Lond.),* A363, 503 (1978).

90. No mention of a reaction or compound means that no information was provided.

91. Basis of assignment of structure: (a) OK; (b) ¹H-NMR qualitatively consistent with structure; (c) mass spectrum qualitatively consistent with structure; (d) C,H analysis OK; (e) molecular weight (VP osmometry); (f) imaginative and resourceful interpretations of complex ¹H and ¹³C-NMR spectra; (g) IR; (h) an uninformative mass spectrum; (i) VPC retention time versus a reference sample on two columns; (j) "qualitative GC"; (k) chemical ionization GC/MS without reference samples; (l)

comparison of VPC retention time with that of a reference sample[91n]; (m) "identified by GC-MS analysis"; (n) agreement of boiling point with that reported for material[91o] prepared by reaction of Me_2Zn and $Bu^tCH_2CMe_2Cl$; (o) method of preparation.

92. (a) S. H. Bauer and J. Y. Beach, *J. Am. Chem. Soc.*, **64**, 1142 (1942).

(b) H. B. Bürgi and L. S. Bartell, *J. Am. Chem. Soc.*, **94**, 5236 (1972).

(c) H.-D. Beckhaus, G. Hellmann, C. Rüchardt, B. Kitschke, and H. J. Lindner, *Chem. Ber.*, **111**, 3780 (1978).

(d) W. Littke and U. Drück, *Angew. Chem., Int. Ed. Engl.*, **18**, 406 (1979).

93. (a) S. Brownstein, J. Dunogues, D. Lindsay, and K. U. Ingold, *J. Am. Chem. Soc.*, **99**, 2073 (1977).

(b) G. D. Mendenhall, *Sci. Prog. Oxf.*, **65**, 1 (1978).

(c) V. Malatesta, D. Forrest, and K. U. Ingold, *J. Am. Chem. Soc.*, **100**, 7073 (1978).

(d) H. U. Buschhaus and W. P. Neumann, *Angew. Chem., Int. Ed. Engl.*, **17**, 59 (1978).

(e) W. P. Neumann and F. Werner, *Chem. Ber.*, **111**, 3904 (1978).

(f) K. S. Chen, T. Foster, and J. K. S. Wan, *J. Chem. Soc., Perkin II*, 1288 (1979).

(g) W. P. Neumann, in *Radicaux Libres Organiques, Colloq. Int. C.N.R.S.*, **278**, 321 (1978).

94. (a) M. Ya. Botnikov, V. M. Zhulin, and I. Kh. Milyavskaya, *Izv. Akad. Nauk SSSR*, 573 (1977).

(b) R. E. Pearson and J. C. Martin, *J. Am. Chem. Soc.*, **85**, 3142 (1963).

95. (a) R. D. Gilliom and J. R. Howles, *Can. J. Chem.*, **46**, 2752 (1968).

(b) D. D. Tanner, R. Henriquez, and D. W. Reed, *Can. J. Chem.*, **57**, 2578 (1979).

(c) A. A. Zavitsas and G. M. Hanna, *J. Org. Chem.*, **40**, 3782 (1975).

(d) A. A. Zavitsas and J. A. Pinto, *J. Am. Chem. Soc.*, **94**, 7390 (1972).

(e) A. A. Zavitsas, G. Hanna, A. Arafat, J. Ogunwole, and L. R. Zavitsas, *Radicaux Libres Organiques, Coll. Int. C.N.R.S.*, **278**, 479 (1978).

96. For background, see Ref. 10b.

97.* (a) G. Foldiak and R. H. Schuler, *J. Phys. Chem.*, **82**, 2756 (1978).

(b) see also Ref. 97c.

(c) R. G. Kryger, J. P. Lorand, N. R. Stevens, and N. R. Herron, *J. Am. Chem. Soc.*, **99**, 7589 (1977).

98. J. H. Howard, in *Advances in Free Radical Chemistry*, Vol. IV, G. H. Williams, Ed., Academic, 1972, p. 49.

99.* (a) H. Paul, R. D. Small, Jr., and J. C. Scaiano, *J. Am. Chem. Soc.*, **100**, 4520 (1978).

(b) R. D. Small, Jr., J. C. Scaiano, and L. K. Patterson, *Photochem. Photobiol.*, **29**, 49 (1979).

(c) S. K. Wong, *J. Am. Chem. Soc.*, **101**, 1235 (1979).

100. (a) The possibility of Ph_2COH having been formed by reaction of the other (alkyl) radicals with Ph_2CHOH was disputed by Paul et al, who pointed out that "methyl radicals are about three orders of magnitude less reactive than *tert*-butoxy radicals and the radicals present in our system should be even less reactive than methyl."[99a] However, the determining factor here is not relative reactivities, that is, *rate constants*, but relative *rates*, which depend on reactivity *and* the steady-state concentration of the radical.

(b) The quantitative validity of their application of this assumption has been challenged[99c] with an argument, of inadequately supported strength, to the effect that (1) an estimate of $k(\cdot CH_2CMe_2OOBu^t \to Bu^tO\cdot + \text{isobutylene oxide})$, and (2) the inability to observe the ESR spectrum of $\cdot CH_2CMe_2OOBu^t$ in $(Bu^tO)_2$ combine to "demand a slower hydrogen abstraction rate from $(Bu^tO)_2$ and/or a faster decay rate of $[\cdot CH_2CMe_2OOBu^t]$" than was used in Refs. 99a and 99b.

101. The equations used do not symbolize what is desired to be said; $-d[Bu^tO\cdot]/dt$ is confused with the sum of the rates of the reactions that consume $Bu^tO\cdot$ and $d[R\cdot]/dt$ with the sum of the rates of the reactions that produce $R\cdot$. However, the operative equation actually used is, apparently accidentally, correct.

102.* (a) H. Adachi, N. Basco, and D. G. L. James, *Int. J. Chem. Kinet.*, 11, 995 (1979).

(b) P. Arrowsmith and L. J. Kirsch, *J. Chem. Soc., Faraday Trans I*, 3016 (1978).

(c) J. E. Bennett and R. Summers, *J. Chem. Soc., Perkin II*, 1504 (1977).

(d) H.-H. Schuh and H. Fischer, *Helv. Chim. Acta*, 61, 2130 (1978).

103. (a) K. D. Bartle, P. G. Butcher, C. J. Harding, and D. R. Roberts, *J. Chem. Educ.*, 55, 742 (1978).

(b) V. Ya. Katsobashvili, *React. Kinet. Catal. Lett.*, 8, 321 (1978).

104.* (a) M. Gielen and Y. Tondeur, *J. Organomet. Chem.*, 169, 265 (1979).

(b) H. Fujihara, A. Yoneda, K. Nozaki, M. Yoshihara, and T. Maeshima, Abstracts, A.C.S. Meeting, Honolulu, April 1979, p. ORGN 97.

(c) See also, H. Fujihara, K. Yamazaki, M. Yoshihara, and T. Maeshima, *J. Polymer Sci., Polymer Lett. Ed.*, 17, 507 (1979), and references cited therein.

105. (a) M. A. Churilova, A. B. Terent'ev, and R. Kh. Friedlina, *Izv. Akad. Nauk SSSR*, 121 (1977).

(b) R. G. Gasanov, A. B. Terent'ev, and R. Kh. Freidlina, *Izv. Akad. Nauk SSSR*, 542 (1977).

(c) E. M. Lipskerova, M. Ya. Mel'nikov, and N. V. Fok, *Khim. Vys. Energ.*, 13, 182 (1979).

(d) J. Czarnowski and H. J. Schumacher, *Z. Phys. Chem. (Frankf.)*, 92, 329 (1974).

(e) B. P. Roberts and J. N. Winter, *Tetrahedron Lett.*, 3575 (1979).

(f) D. D. Tanner and P. M. Rahimi, *J. Org. Chem.*, 44, 1674 (1979).

106. L. Benati, P. C. Montevecchi, and P. Spagnola, *Tetrahedron Lett.*, 815 (1978).

107. P. Beak and S. W. Mojé, *J. Org. Chem.*, 39, 1320 (1974), and references cited therein; see also the approaches of Ref. 108.

108. (a) D. H. R. Barton and S. W. McCombie, *J. Chem. Soc., Perkin I*, 1574 (1975).

(b) L. E. Khoo and H. H. Lee, *Tetrahedron Lett.*, 4351 (1968).

(c) C. Grugel and W. P. Neumann, *Justus Liebig's Ann. Chem.*, 1675 (1979).

(d) J. J. Patroni and R. V. Stick, *J. Chem. Soc., Chem. Commun.*, 449 (1978).

(e) H. Redlich, H.-J. Neumann, and H. Paulsen, *Chem. Ber.*, 110, 2911 (1977).

109. See also, S. W. Baldwin and S. A. Haut, *J. Org. Chem.*, 40, 3885 (1975).

110.* (a) N. C. Billingham, R. A. Jackson, and F. Malek, *J. Chem. Soc., Chem. Commun.*, 344 (1977).

(b) R. A. Jackson and F. Malek, *Radicaux Libres Organiques, Colloq. Int. C.N.R.S.*, 278, 525 (1978).

111. (a) J. Czarnowski and H. J. Schumacher, *Int. J. Chem. Kinet.*, 10, 111 (1978); 11, 1089 (1979).

(b) A. J. Colussi and H. J. Schumacher, *Z. Phys. Chem. N. F.*, 71, 208 (1970).

112.* P. J. Baugh, J. I. Goodall, and J. Bardsley, *J. Chem. Soc., Perkin II*, 700 (1978).

113.* (a) K.-D. Asmus, D. Bahnemann, M. Bonifačić, and H. A. Gillis, *Disc. Faraday Soc.*, **63**, 213 (1977).
  (b) Unpublished work of R. Petersen and M. C. R. Symons, reported in Ref. 113c.
  (c) M. C. R. Symons, *Disc. Faraday Soc.*, **63**, 281 (1977).
  (d) M. C. R. Symons and R. L. Petersen, *J. Chem. Soc., Faraday II*, 210 (1979).
  (e) See also references cited in Ref. 113d.
  (f) T. Berclaz, M. Geoffroy, L. Ginet, and E. A. C. Lucken, *Chem. Phys. Lett.*, **62**, 515 (1979).
  (g) M. Geoffroy, *J. Chem. Phys.*, **70**, 1497 (1979).
  (h) R. Franzi, M. Geoffroy, L. Ginet, and N. Leray, *J. Phys. Chem.*, **83**, 2898 (1979).
  (i) T. Berclaz, M. Geoffroy, and E. A. C. Lucken, *J. Magn. Res.*, **33**, 577 (1979).

114. We recognize the potential problem of electron transfer[82] or other reactions between host and guest.

115. Crown ether complexes of Br_2 have been studied as ionic brominating agents.[116a, 116b, 116e]

116. (a) K. H. Pannell and A. Mayr, *J. Chem. Soc., Chem. Commun.*, 132 (1979).
  (b) See also, Ref. 116c.
  (c) E. Shchori and J. Jagur-Grodzinski, *Isr. J. Chem.*, **10**, 959 (1972).
  (d) E. Shchori and J. Jagur-Grodzinski, *Israel J. Chem.*, **10**, 935 (1972).
  (e) For additional reports of crown/X_2 complexes (X = Cl, Br, I), see Ref. 116d, 116f, and 116g;
  (f) H. P. Hopkins, Jr., D. V. Jahagirdar, and F. J. Windler III, *J. Phys. Chem.*, **82**, 1254 (1978).
  (g) Yu. A. Serguchev and T. I. Petrenko, *Teor. Eksp. Khim.*, **13**, 705 (1977).

117.* (a) R. G. Lawler and G. T. Evans, *Ind. Chim. Belg.*, **36**, 1087 (1971).
  (b) P. W. Atkins and T. P. Lambert, *Ann. Rep. Progr. Chem.* 1975, **72A**, 88 (1976).
  (c) G. T. Evans and R. G. Lawler, *Mol. Phys.*, **30**, 1085 (1975).
  (d) A. L. Buchachenko, *Usp. Khim.*, **45**, 761 (1976).
  (e) A. L. Buchachenko, *Zh. Fiz. Khim.*, **51**, 2461 (1977).
  (f) R. Z. Sagdeev, K. M. Salikhov, and Yu. M. Molin, *Usp. Khim.*, **46**, 569 (1977).
  (g) K. M. Salikhov, F. S. Sarvarov, R. Z. Sagdeev, and Yu. N. Molin, *Kinet. Katal.*, **16**, 279 (1975).
  (h) A. L. Buchachenko, E. M. Galimov, V. V. Ershov, G. A. Nikiforov, and A. D. Pershin, *Dokl. Akad. Nauk SSSR [Phys. Chem.]*, **228**, 379 (1976).
  (i) N. J. Turro and B. Kraeutler, *J. Am. Chem. Soc.*, **100**, 7432 (1978).
  (j) N. J. Turro, B. Kraeutler, and D. R. Anderson, *J. Am. Chem. Soc.*, **101**, 7435 (1979).
  (k) R. G. Sagdeev, T. V. Leshina, M. A. Kamkha, O. I. Belchenko, Yu. N. Molin, and A. I. Rezvukhin, *Chem. Phys. Lett.*, **48**, 89 (1977).
  (l) V. A. Belyakov, V. I. Mal'tsev, E. M. Galimov, and A. L. Buchachenko, *Dokl. Akad. Nauk SSSR*, **243**, 924 (1978).
  (m) N. J. Turro, *Abstr. 178th A.C.S. Meeting*, Washington, Sept. 1979, ORGN 81.
  (n) see also a similar analysis in Ref. 117f.
  (o) Yu. N. Molin, R. Z. Sagdeev, and K. M. Salikhov, in *Soviet Scientific Reviews (B), Chemistry Reviews*, Vol. 1, M. E. Vol'pin, Ed., 1979, p. 58.
  (p) R. Haberkorn, M. E. Michel-Beyerle, and K. W. Michel, *Astron. Astrophys.*, **55**, 315 (1977).
  (q) Ref. 117o, p. 59.

(r)   A. V. Podoplelov, T. V. Leshina, Ren. Z. Sagdeev, Yu. U. Molin, and V. I. Gol'
danskii, *JETP Lett.*, 29, 380 (1979).

118.   *Compare* T. H. Maugh II, *Science*, 206, 317 (1979).

119.   S. Rummel, H. Hübner, and P. Krumbiegel, *Z. Chem.*, 7, 392 (1967).

120.   We do not agree with the comment[117f] that "since the radical recombination reaction . . . has virtually zero activation energy, the usual kinetic isotope effect is impossible."

121.   (a)   R. Haberkorn, "Theory of Magnetic Field Modulation of Radical Recombination Reactions, II. Short Time Behavior," *Chem. Phys.*, 24, 111 (1977).

(b)   W. Bube, R. Haberkorn, and M. E. Michel-Beyerle, "Magnetic Field and Isotope Effects Induced by Hyperfine Interaction in a Steady State Photochemical Experiment," *J. Am. Chem. Soc.*, 100, 5993 (1978).

(c)   W. B. Maier II, S. M. Freund, R. F. Holland, and W. H. Beattie, "Photolytic Separation of D from H in Cryogenic Solutions of Formaldehyde," *J. Chem. Phys.*, 69, 1961 (1978).

(d)   R. Z. Sagdeev, A. A. Obynochny, V. V. Pervukhin, Yu. N. Molin, and V. M. Moralyov, "Isotopic Effects of Carbon and Nitrogen in Low-Temperature Photochemical Reactions," *Chem. Phys. Lett.*, 47, 292 (1977).

(e)   See also reference 121f.

(f)   J. M. McBride and M. R. Gisler, *Mol. Cryst. Liq. Cryst.*, 52, 121 (1979), footnote 10.

(g)   A paper on "Criteria for Establishing the Existence of Nuclear Spin Isotope Effects," R. G. Lawler, *J. Am. Chem. Soc.*, 102, 430 (1980), appeared after this chapter was written.

122.   (a)   F. K. Velichko, R. A. Amriev, T. A. Pudova, and R. Kh. Freidlina, *Izv. Akad. Nauk SSSR*, 369 (1977).

(b)   W. T. Dixon, J. Foxall, G. H. Williams, D. J. Edge, B. C. Gilbert, H. Kazarians-Moghaddam, and R. O. C. Norman, *J. Chem. Soc., Perkin II*, 827 (1977).

123.   (a)   M. E. Jacox, *J. Mol. Spectrosc.*, 66, 272 (1977).

(b)   G. T. Surratt and W. A. Goddard III, *Chem. Phys.*, 23, 39 (1977).

124.   (a)   See also Ref. 124b.

(b)   J. Pacansky and H. Coufal, *J. Chem. Phys.*, 71, 2811 (1979).

125.   R. H. Staley, M. Taagepera, W. G. Henderson, I. Koppel, J. L. Beauchamp, and R. W. Taft, *J. Am. Chem. Soc.*, 99, 326 (1977).

126.   M. R. Feldman and W. C. Flythe, *J. Org. Chem.*, 43, 2596 (1978).

127.   B. Giese and K. Keller, *Chem. Ber.*, 112, 1743 (1979). We thank Professor Giese for helpful correspondence and a preprint.

128.   (a)   W. P. Unruh, T. Gedayloo, and J. D. Zimbrick, "ENDOR Measurements of Proton Tunneling in 5-Thymyl Radicals in Acidic Glasses," *J. Phys. Chem.*, 82, 2016 (1978).

(b)   P. Beckmann, "The electron-methyl group spin-spin interaction," *Mol. Phys.*, 34, 665 (1977).

(c)   M. Geoffroy, L. D. Kispert, and J. S. Hwang, "An ESR, ENDOR, and ELDOR study of tunneling rotation of a hindered methyl group in X-irradiated 2,2,5-trimethyl-1,3-dioxane-4,6-dione crystals," *J. Chem. Phys.*, 70, 4238 (1979).

(d)   J. H. Lichtenbelt and D. A. Wiersma, "Methyl Group Tunneling Rotation in the Lowest $n\pi^*$ Triplet State of Toluquinone. An Optically Detected ENDOR, LAC and CR Study," *Chem. Phys.*, 39 (1979).

(e) P. Beckmann and S. Clough, "Electron spin relaxation and tunnelling methyl groups," *J. Phys. C,* 11, 4055 (1978).

129. (a) M. Kosugi, K. Yano, M. Chiba, and T. Migita, *Chem. Lett.,* 801 (1977).

(b) M. Kosugi, H. Arai, A. Yoshino, and T. Migita, *Chem. Lett.,* 795 (1978).

(c) P. J. Kropp, J. R. Gibson, J. J. Snyder, and G. S. Poindexter, *Tetrahedron Lett.,* 207 (1978).

(d) N. J. Pienta and P. J. Kropp, *J. Am. Chem. Soc.,* 100, 655 (1978).

(e) See also L. Kaplan, *J. Am. Chem. Soc.,* 89, 4566 (1967).

(f) J. L. Charlton and G. J. Williams, *Tetrahedron Lett.,* 1473 (1977).

(g) A. E. Crease and M. D. Johnson, *J. Am. Chem. Soc.,* 100, 8013 (1978).

(h) A. Bury, C. J. Cooksey, T. Funabiki, B. D. Gupta, and M. D. Johnson, *J. Chem. Soc., Perkin II,* 1050 (1979).

(i) A. E. Crease, B. Dass Gupta, M. D. Johnson, and S. Moorhouse, *J. Chem. Soc., Dalton Trans.,* 1821 (1978).

(j) A . E. Crease, B. Dass Gupta, M. D. Johnson, E. Bialkowska, K. N. V. Duong, and A. Gaudemer, *J. Chem. Soc., Perkin I,* 2611 (1979).

130. (a) M. Gruselle and J. Y. Nedelec, *Tetrahedron,* 34, 1813 (1978).

(b) T. Kaiser and H. Fischer, *Helv. Chim. Acta,* 62, 1475 (1979).

(c) R. Foon and K. P. Schug, Paper presented to the Roy. Aust. Chem. Inst. Div. Phys. Chem. Symp., Warburton, August 1975, p. E3.1.

(d) G. A. Hill, E. Grunwald, and P. Keehn, *J. Am. Chem. Soc.,* 99, 6521 (1977).

(e) K. J. Olszyna and E. Grunwald, *J. Phys. Chem.,* 82, 2052 (1978).

(f) We thank Professor Grunwald for helpful correspondence.

131. K. Yoshino, Y. Ohkatsu, and T. Tsuruta, *Bull. Chem. Soc. Jap.,* 52, 2028 (1979).

8

NITRENES

WALTER LWOWSKI

New Mexico State University, Las Cruces, New Mexico 88003

I. INTRODUCTION

Significant progress has been made in the nitrene field during the past few years. The first vibrational, electronic, and NMR spectra have been obtained of singlet nitrenes in solution. New nitrenes have been generated. The chemistry of triplet arylnitrenes has prospered, as has the synthetic application of arylnitrenes.

Metal-nitrene complexes that transfer the R—N moiety to organic molecules have been found, and many other areas in nitrene chemistry are active. Perhaps inevitably, some nitrene myths refuse to die. Base treatment of hydroxylamine-O-sulfonic acid is still assumed to generate NH in solution (despite the work of many authors, such as Schmitz et al.[1]). Carbonylnitrenes are still automatically assumed to be intermediates in the Curtius rearrangement family [see the literature in earlier reviews[2] and other work[3]].

Although the application of nitrenes in actual syntheses is still fairly restricted in syntheses targeted to a specific substance, great strides have been made in the development of nitrene-based synthetic methods, particularly by British workers in the arylnitrene area.

Aryl- and heteroarylnitrenes have become very important in photoaffinity labeling of biochemical systems. Perhaps approximately half of the biochemical research using photoaffinity labeling employs nitrenes as the reactive intermediates.[4,5]

Recent reviews cover a number of special areas in nitrene chemistry, as seen in Table 8.1.

II. AMINONITRENES

Dervan[6-9] observed N,N-dialkylaminonitrenes in solution and recorded their infrared, electronic, and NMR spectra. N-(2,2,6,6-tetramethylpiperidyl)nitrene-(1) and N-(2,2,5,5-tetramethylpyrrolidinyl)nitrene (2) were obtained from the corresponding N-amino compounds by oxidation in ether at $-78°C$. Both nitrenes could be purified by chromatography on basic alumina. Their spectral properties are summarized in Table 8.2. The kinetics of the disappearance of 1

(1)

TABLE 8.1. Nitrene Reviews

Area	Authors	Reference
Aryl nitrenes—reactions mechanisms and synthetic uses	B. Iddon, O. Meth-Cohn, E. F. V. Scriven; H. Suschitzky and P. T. Gallagher	*Angew. Chem.* **91**, 965–982 (1979); *Angew. Chem., Int. Ed.* 18, 900–917 (1979).
Heterocycles, synthesis of five-membered	V. P. Semenov, A. N. Studenikov, and A. A. Potekhin	*Khim. Geterotsikl. Soedin,* 291–305 (1978); Plenum Press (Eng. transl.), **14**, 237 (1979).
Metal–nitrene chemistry	F. Basolo	*J. Indian Chem. Soc.* **54**, 7–12 (1977).
Transition metal complexes with R—N ligands	W. A. Nugent and B. L. Haymore	*Coordination Chem. Revs.*, **31**, 123–175 (1980).
Photoaffinity labeling	Various, especially H. Bayley and J. R. Knowles	Chapter 8, Ref 4.
Photoaffinity labeling	F. J. Darfler and A. M. Tometsko	Chapter 2 of *Chemistry and Biology of Amino Acids, Peptides and Proteins,"* Vol. 5 B. Weinstein, Ed., Dekker, New York, 1978.

TABLE 8.2. Properties of Two Dialkylaminonitrenes

Nitrene	Absorption Maxima $CH_2Cl_2-(CH_3)_2CHOH$ "nm (ϵ)"	ν_{N-N} ($^{14}N\ ^{15}N$) (cm^{-1})	^{1}H-NMR (PPM)[a,b]	E_A (unimol.), kcal/mol
2	497 (20) 487	1638 (1612)	1.05 (3) 2.32 (1)	16.8 ± 0.5 (n-hexane) 19.0 ± 0.6 (ether)
1	541 (18) 526 [in ether: 514 and 543 ($\epsilon = 13$)]	1595 (1569)	1.15 (2) 2.15 (1)	16.9 ± 0.7 (n-hexane) 20.0 ± 0.4 (ether)

[a]Entries in parentheses are relative absorption areas.
[b]For 1 ^{15}N NMR relative to HNO_3: outer N = -541 ppm, ring N = +48 ppm.[10]

and **2** were followed spectroscopically. The first-order part of the reactions predominates at higher temperatures (near 0°C) and provided the activation parameters given in Table 8.2. The assignments of structures **1** and **2** for the oxidation products were substantiated by measuring the infrared spectra of the nitrenes in which the outer nitrogen was ^{15}N instead of ^{14}N. The calculated shifts in the stretching frequencies confirm the N,N-dialkylamino-N'-nitrene structure, and the frequencies observed indicate substantial double bond character between the nitrogens.

Rees et al.[11] found a new source for phthalimidoylnitrene: dissociation of the sulfimine **3** in boiling benzene gave phthalimiodoylnitrene, which was trapped by olefins or allowed to dimerize to a mixture of cis- and trans-bisphthaloyltetrazene. With methyl methacrylate, a 65% yield of the aziridine was obtained, and the addition to cis-4-methyl-2-pentene gave only the cis- (and no trans-) aziridine.

$$\text{(structure: N–N=S(CH}_3\text{)}_2\text{)} \xrightarrow[\text{reflux}]{\text{benzene}} \text{(structure: N–N)} + S(CH_3)_2 \qquad (2)$$

3

Much earlier, Hayashi and Swern[12] investigated the generation of nitrenes from sulfimides by photolysis (rather than thermolysis), but found that much triplet nitrene was produced when this route was used to make carbonylnitrenes. It would be interesting to know whether, in the case of making carbonylnitrenes, the photolysis route is responsible for this, or whether the presence of the sulfur orbitals facilitates intersystem crossing to the triplet for the carbonylnitrene (whose ground state is the triplet), but is without such an effect with phthalimidoylnitrene (whose ground state might be the singlet).

N-Nitrenoaziridines are a new subgroup of the aminonitrenes, capable of intramolecular addition to C=C double bonds and of forming the corresponding tetrazenes;[13] (see Eq. 3). Intermolecular aziridine formation, however, has not been observed. The oxidation of cis-9-amino-9-azabicyclo[6.1.0]-(4Z)-nonene did not give tricyclic product, but 1,5-cyclooctadiene and the tetrazene.

III. ARYLNITRENES

The intricate bond reorganizations of C_6H_5N species (phenylnitrene, pyridylcarbenes, and their isomers) under flash thermolysis conditions were reviewed in Volume 1 of this series. Photolysis of matrix-isolated aryl azides and pyridyldiazo-

$$\text{PhthN—NH}_2 \xrightarrow{\text{Pb(OAc)}_4} \text{PhthN—N} \longrightarrow \text{N—NPhth} \xrightarrow{\text{N}_2\text{H}_4}$$

$$\text{N—NH}_2 \xrightarrow[-70\,^\circ\text{C}]{\text{Pb(OAc)}_4} \text{N—N} \tag{3}$$

$$\text{N—NH}_2 \xrightarrow[-70\,^\circ\text{C}]{\text{Pb(OAc)}_4} \text{N—N}^{\diagup\text{N—N}}$$

$$\text{Phth} =$$

methanes leads to a similarly intricate system of interconversions and of intermediate species. Chapman and co-workers[14-17] reported the photolyses of phenyl azide and the isomeric pyridyldiazomethanes in argon matrices at 10-12 K, using light of selected wavelength cutoffs. Infrared spectra of 3-pyridylcarbene and the 3 seven-membered ring species **5**, **10**, and **13** were obtained and used, together with the ESR spectra of the triplet nitrene and carbenes, to monitor the conversions. Warming of the matrix to 35 or 40 K led to observable thermal reactions, such as the addition of nitrogen to the triplet of **12**, re-forming **11**, as well as the addition of carbon monoxide to triplet **12**, to give 3-pyridylketene, a structure proof for **12**. For experimental reasons, the warming of the matrices takes about an hour, a time easily long enough for singlet **12** (some of which presumably was present after the irradiation) to disappear. The thermal reactions (at 40K) were monitored by ESR and IR spectroscopy. Scheme 1 presents Chapman's system of interconversions (including some thermal triplet $\rightarrow$ singlet processes).

Pending the development of more sensitive spectroscopic techniques, singlet species (exited species in the cases of arylnitrenes and pyridylcarbenes) will be hard to detect. One may resonably expect a fairly complex singlet chemistry in the matrices under Chapman's experimental conditions. Such chemistry remains to be demonstrated and might well connect with phenylnitrene conversions in solution and in flash thermolysis experiments (see Volume 1).

Singlet arylnitrene-azabicycloheptatriene interconversions have been further studied and adapted for synthetic use (see the section on solvent effects).

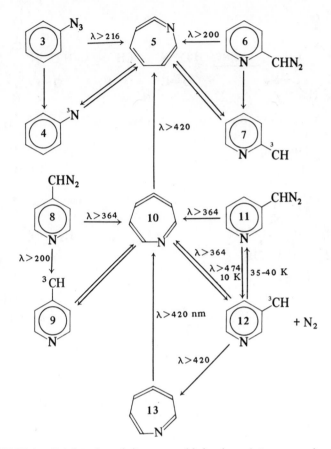

SCHEME 1.   Triplet phenylnitrene–pyridylcarbene interconversions.

Heteroaromatic systems also undergo efficient nitrene-azabicycloheterotriene isomerizations, providing good syntheses for a variety of heterocyclic amines and ring expansion products. The pyrimidine series provides good examples (Eq. 4, 5).[18, 19]

An interesting arylcarbene to arylnitrene isomerization was discovered by Dürr and Schmitz.[20] The process starts with the formation of nitrogen from diazo compound; then the system stabilizes itself further by converting from a carbene to an arylnitrene while gaining the resonance stabilization of a benzonitrile (Eq. 6).

IV. HETEROARYLNITRENES

Photolysis of the corresponding azides forms 1,3,5-triazinylnitrenes. The reactivity of these nitrenes is much greater than that of arylnitrenes.[21–24]

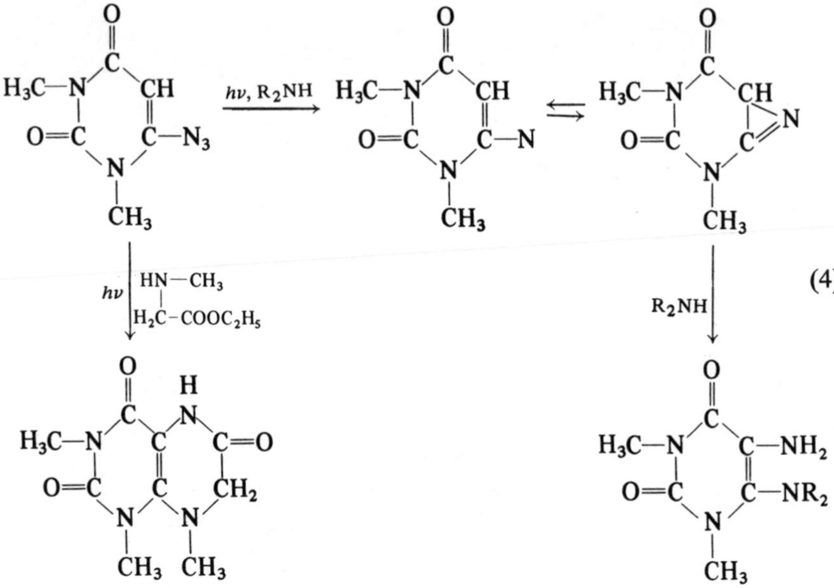

(4)

R = C₂H₅:64%

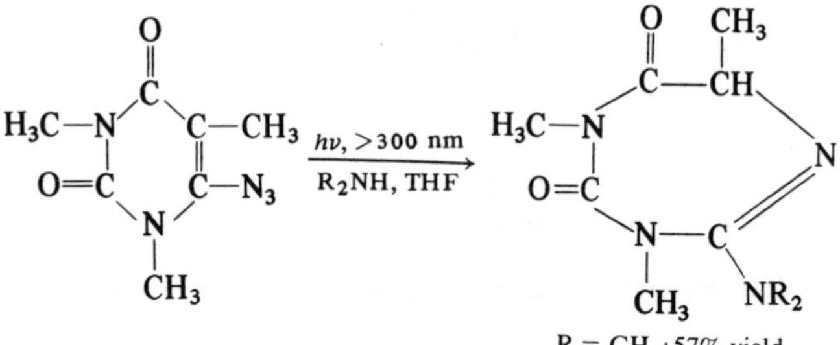

$R = C_2H_5: 64\%$

(5)

$R = CH_3: 57\%$ yield

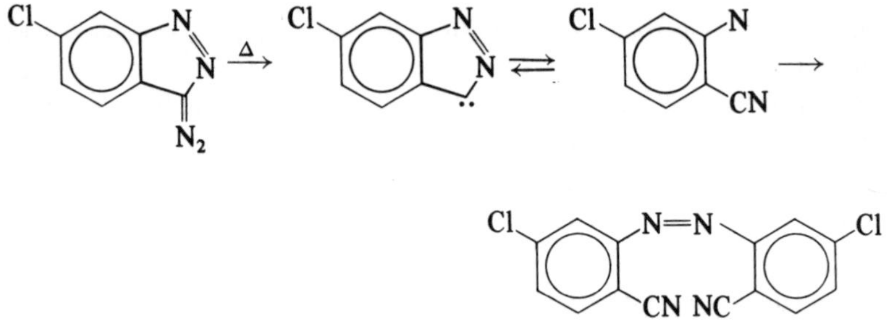

(6)

Insertion into cyclohexane C—H bonds by 2-nitreno-4,6-dimethoxy-1,3,5-triazine gives a 48% yield (based on the azide) of 2-cyclohexylamino compound. The nitrenes add to olefins to form aziridines, to sulfoxides to give sulfoximines, and to nitriles and ketones, (see Eq. 7). Thus the reactivity of triazinylnitrenes is like that of carbonylnitrenes, rather than that of typical arylnitrenes.

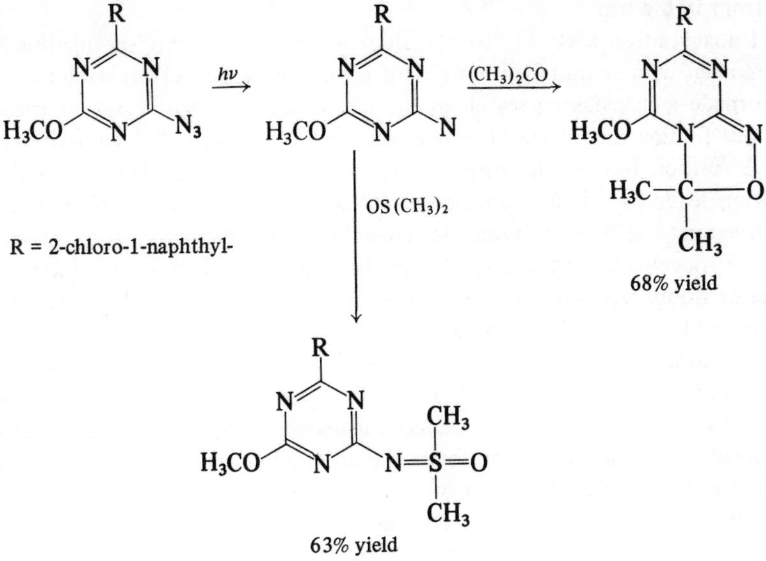

R = 2-chloro-1-naphthyl-

68% yield   (7)

63% yield

V. NEW NITRENES

Atkinson and Judkins[25] reported the first well-documented sulfenylnitrene. Oxidation with lead tetraacetate of 2,4-dinitrophenylsulfenamide gave the nitrene, which added stereospecifically to *trans*-phenyl-1-propene to give the aziridine in 64% yield (Eq. 8). Addition of the nitrene to *cis*-phenyl-1-propene,

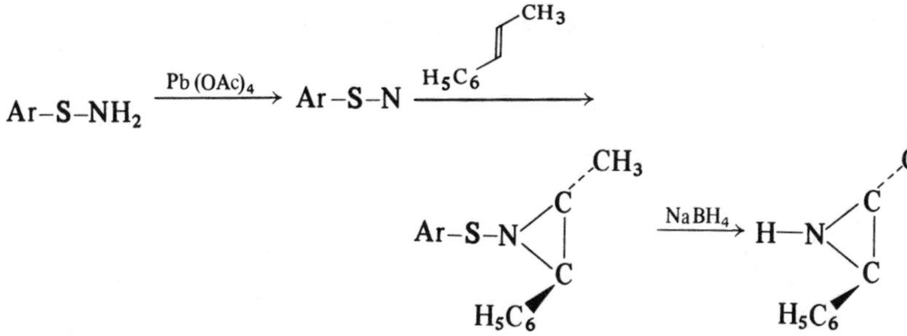

(8)

however, gave mixtures of *cis-* and *trans-*aziridines, with the ratios depending on experimental conditions. Assuming that the observed reactions are all due to (singlet and triplet) sulfenylnitrenes, the data seem to indicate that singlet and triplet 2,4-dinitrophenylsulfenylnitrenes are in equilibrium with each other. Be that as it may, the nitrene is certainly an interesting one, given the resonance interactions possible in the singlet, and the orbitals available for spin parallel electrons in the triplet.

Thioacylnitrenes, $R-C(=S)-N$, although long an object of speculation, have never been demonstrated to exist. This is not for lack of effort. Holm et al.[26] have made a painstaking search in the thermolysis of aryl-1,2,3,4-thiatriazoles without finding the nitrene. Isotopic labeling of 5-phenyl-1,2,3,4-thiatriazole in the 2-position has no discernible effect on the rate, while labeling in the 4-position leads to a kinetic isotope effect of ~ 1.04, about the value expected for breaking the $3-4$ bond in the rate-determining step, *a priori* in agreement with retrocycloaddition of N_2S. Efforts at trapping intermediates (thiobenzoyl azide or thiobenzoylnitrene) were unsuccessful. A peak in the mass spectrum of phenylthiatriazole at $m/e = 135$ was shown to be produced primarily by electron impact, rather than by a thermal reaction followed by ionization of the neutral $m/e = 135$ species. Thus the field ionization mass spectrum shows very little at $m/e = 135$. Together with an unsuccessful search for N_2S in the mass spectrum, the results of Holm argue for formation of thiobenzoyl azide (in equilibrium with the thiatriazole), followed by loss of nitrogen and atomic sulfur, to give the observed products (Eq. 9). In contrast to the thiatriazole thermolysis, photolysis does give an intermediate intervening between the thiatriazole and the final

(9)

products, benzonitrile, sulfur, and nitrogen. Holm et al.[27,28] found benzonitrile sulfide, C_6H_5CNS, in the photolyses of phenylthiatriazole and 4-phenyl-1,3,2-oxathiazolium-5-olate, a mesoionic species. The benzonitrile sulfide was trapped, and its decay was examined spectroscopically. Again, no indication of the formation of any thiobenzoylnitrene was found. This reinforces the general experience that nitrenes are easier written than formed. Some authors seem to assume the intervention of nitrenes in certain reactions simply because a nitrene looks nice on paper. Nitrenes, however, are intermediates of relatively high energy, and paths exist on many an energy surface that circumvent nitrene formation in favor of rearrangement or fragmentation processes.

There is a considerable incentive for constructing new nitrenes deliberately, so as to test hypotheses on nitrene reactivity–structure relationships, and to make available nitrenes of particular reactivities for use in synthesis. Given the large differences observed in the reactivities of, for example, dialkylamino- and sulfonylnitrenes, it is obvious that the substituent R in R-N plays a major role in determining the nitrene's properties. Furthermore, electron-attracting substituents R, such as sulfonyl, cyano, or carbonyl give nitrenes whose singlets easily insert into C—H bonds. Aryl-and alkylnitrenes do not insert into C—H bonds in their singlet states, unless fortified with electron-withdrawing substituents, such as fluorine (see Volume 1). It might then be that some net electronic effect determines nitrene reactivity. On the other hand, specific interactions of the nitrogen with nearby atoms or groups can be written in many instances, and such interactions might well influence or even govern the reactivity. Alkoxycarbonylnitrenes seem to allow such interactions (Eq. 10). Some of the structures

$$\tag{10}$$

written must be quite high in energy, but then there is no doubt that alkoxycarbonylnitrenes *are* high in energy. The ether oxygen of the alkoxycarbonylnitrene does not seem to influence its reactivity very much; all experimental observations indicate that the reactivities of alkoxycarbonyl- and of alkanoylnitrenes are very similar. To modify the reactivity, it is necessary to operate at the carbonyl group, which has been done by substituting nitrogen for the carbonyl oxygen.[29] Considerable modification of the reactivity of alkoxycarbonylnitrenes can thus be achieved. Most notably, C—H insertion can be entirely suppressed, while the addition to olefins remains unaffected or is improved. Table 8.3 shows some of

TABLE 8.3. Comparison of Ethoxycarbonylnitrene, N-Cyanoethoxy-carbimidoylnitrene, and N-Methanesulfonyl-Ethoxycarbimidoylnitrene[a]

Nitrene	C–H insertion	Yield of Azepine from Benzene	Yield of Aziridines from cis-4-Methyl-2-pentene
$C_2H_5O-C\!\!\begin{smallmatrix}\nearrow O\\ \searrow N\end{smallmatrix}$	+	~70%	58% cis 9% trans
$C_2H_5O-C\!\!\begin{smallmatrix}\nearrow NCN\\ \searrow N\end{smallmatrix}$	–	60%	85% cis 0% trans
$C_2H_5O-C\!\!\begin{smallmatrix}\nearrow N-SO_2CH_3\\ \searrow N\end{smallmatrix}$	–	No reaction	93% cis 0% trans

[a] Reference 29.

the results reported.[29] Another imidoylnitrene, $RO-C(=NH)-N$, is still less reactive.[30] It adds to dimethyl sulfoxide and reacts with suitable aryl groups intramolecularly (e.g., R = phenyl), but does little else.

VI. SOLVENT EFFECTS

Singlet nitrenes associate with unshared electron pairs, a process that may or may not lead to convalent bonding. The latter alternative prevails in chlorinated solvents, which stabilize singlet carbonylnitrenes, leading to enhanced yields of singlet nitrene products and to steric effects on the selectivity of the nitrenes (see Volume 1). More work has been done on solvent effects on carbonylnitrenes,[31-33] confirming and extending earlier conclusions.

Arylnitrenes have now been shown to undergo similar singlet stabilization by solvents containing unshared electron pairs. Tetramethylethylenediamine, used as a cosolvent in the photolyses of 1-azidonaphthalenes, enhances the yield of 1,2-diaminonaphthalenes, formed via the singlet nitrene, annelation to an azanorcaradiene, and attack by an external amine on a carbon of the three-membered ring[34] (see the section on arylnitrenes in Volume 1). Ring expansion of singlet nitrenes from o-azidobenzoic acid derivatives in methanol to give methoxy-azepines is considerably improved by using tetrahydrofuran as cosolvent.[35] The photolysis of phenylazide in the presence of methoxide in methanol gave only traces of the methoxyazepine, but using dioxane as a cosolvent improved the yield to 35%.[36] Solvents containing heavy atoms, such as bromobenzene, facilitate intersystem crossing of arylnitrenes. Thus they shift the product ratios in reactions in which singlet and triplet arylnitrenes lead to different products. See the section on arylnitrenes in Volume 1. o-Azidotriarylmethanes give, upon thermolysis, azepinoindoles (from the singlet arylnitrene) and acridanes and acridines (via the triplet nitrene), in ratios depending on the nature of the solvent used.[37] Similar results were obtained in other systems.[38]

VII. NITRENIUM IONS

Interest in nitrenium ions has been renewed as a result of the discovery of systems in which both nitrenes and their protonated forms can be demonstrated. Ogata et al.[39] had earlier demonstrated the formation of arylnitrenes in the photolysis of benzisoxazoles, which gave azepines if conducted in the presence of amines or alcohols. Reports of investigations of the same system, but in the presence of acid, are now available.[40,41] Photolyses of benzisoxazoles in concentrated hydrochloric, hydrobromic, and sulfuric acids started with the protonated heterocycle, whose predominant presence was ascertained by investigating the UV spectra of the solutions. In 96% sulfuric acid, irradiation of a number of 3-substituted-benzisoxazoles gave the same product mixture as did the corresponding 1-azido-2-acylbenzenes upon thermolysis. (Photolysis of these azides is complicated by the photochemistry of the arylketone moiety.) Photolyses of the benzisoxazoles in concentrated hydrochloric acid[40] gave results very much like those obtained using sulfuric acid, but with Cl in the place of $^-$OH (Eq. 11).

The arylnitrenium ions are not simply aryl cations with a pendant imino function as represented by the second resonance contributor in (Eq. 11), but react as electrophiles on the nitrogen. Schmid et al.[42,43] observed the attack of the nitrogen of the arylnitrenium ion from 3-methyl-2,1-benzisoxazol on benzene and toluene to form diarylamines.

Takeuchi et al.[44] have studied the reactions of protonated ethoxycarbonyl-nitrene, generated by photolysis and thermolysis of ethyl azidoformate in acetic acid. Comparison of the activation parameters of the thermolyses in acetic acid and in toluene shows that the protonation occurs after the nitrene has been formed (toluene: $k = 9.73 \times 10^{-5}$ at 109.3°C; $E_a = 32.2$ kcal/mol; $\Delta S^{\ddagger} = +4.8$ eu; acetic acid: $k = 13.0 \times 10^{-5}$ at 109.4°C; $E_a = 34.2$ kcal/mol; $\Delta S^{\ddagger} = +10.6$ eu). In pure acetic acid, the main product is ethyl acetoxycarbamate (72–75% yield in photolysis, 62–70% in thermolysis); a trace of ethyl carbamate is also

$$(11)$$

75% yield   8% yield

formed. Scheme 2 illustrates the results, including those obtained by generating the nitrenium ion in the presence of cyclohexanedione or cyclohexene.

VIII. METAL-NITRENE COMPLEXES

Many complexes containing the R—N moiety have been added recently to the already substantial number. Recent reviews by Nugent and Haymore[45] and Basolo[46] are available. The term "nitrene complex" is sometimes used indiscriminately for all complexes containing RN ligands. It would be better to reserve the term for those complexes in which the nitrogen of RN is electrophilic or electron deficient.[46] The other complexes could be named according to the nature of the nitrogen as imino (or aminato) and nitride complexes:

$$M^{n+}\!\!-\!\underset{\cdot\cdot}{N}\!-\!R \qquad M^{(n+2)+}\!\!-\!\underset{\cdot\cdot}{N}\!-\!R \qquad M\!-\!N$$

Nitreno   Imido or   Nitrido
                aminato

Metal-nitrene complexes usually disappoint organic chemists by refusing to disgorge the ligand RN as a nitrene R—N, or even to disgorge it at all, be it as a nitrene, a radical, or an ion.

Singlet nitrenes can be stabilized by π back bonding in nitrene complexes and still remain electron deficient, that is, "nitrenes," as long as charge transfer does not take place. Such transfer results in an aminato complex. Obviously, inspection of the structure of a complex is not sufficient to classify it as a nitrene versus an aminato complex. Nor is it enough to consider the mode of formation

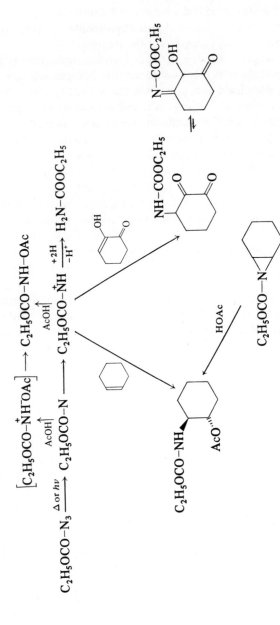

SCHEME 2. Reactions of ethoxycarbonylnitrenium ion which acetic acid, cyclohexene, and cyclohexadione.

329

(e.g., from a decomposing azide and a metal compound); electron transfer could result in an aminato complex. Hence classification should be made on the basis of the chemical (and physical) properties of the complex.

Migita et al.[47] report a reaction of methyl azidoformate with tetrakis (triphenylphosphine)palladium at room temperature in benzene solution. Nitrogen is evolved and the azide band disappears from the infrared spectrum. No further reaction is apparent when allyl ethers are added at room temperature, but heating to 80°C produces N-carbalkoxyimidates (products not formed in the decomposition of azidoformates in the presence of allyl ethers without Pd complex[48]). The nitrene–palladium complex apparently is the species attacking the allyl ether, transferring the RN moiety to give a product that is related to a plausible nitrene (C—H insertion) product by a 1,3-hydrogen shift and a tautomerization. One might name such products "nitrenoid products," to state their related but separate nature in comparison with true "nitrene products." Scheme 3 shows the reactions.

For comparison, one might note that $PdCl_2$ complexes of olefins oxidize carbamates to give products containing the $C_2H_5OCO-NH$ group, the products being formed rather selectively, as shown in Eq. 12.[49]

Over the years, many attempts to generate nitrenes from the corresponding N-halo or N-dihalo compounds have not led to useful nitrene syntheses. This problem still exists today although the decomposition of ethyl N,N-dichlorocarbamate in the presence of copper gave products that might be written as nitrene products if there were no evidence as to the products actually produced

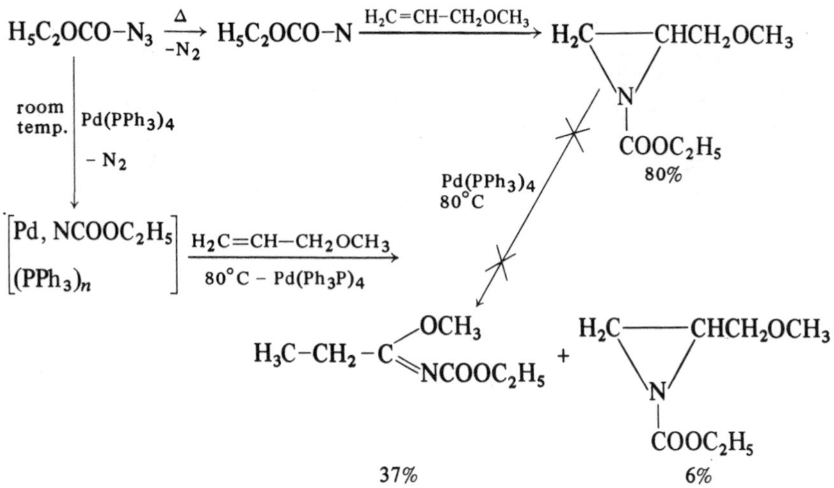

$$H_5C_2OCO-N_3 \xrightarrow[-N_2]{\Delta} H_5C_2OCO-N \xrightarrow{H_2C=CH-CH_2OCH_3} H_2C\overset{}{\triangle}CHCH_2OCH_3$$

room temp. | $Pd(PPh_3)_4$
$-N_2$

$$\begin{bmatrix} Pd, NCOOC_2H_5 \\ (PPh_3)_n \end{bmatrix} \xrightarrow[80°C - Pd(Ph_3P)_4]{H_2C=CH-CH_2OCH_3}$$

$Pd(PPh_3)_4$
80°C

$\overset{N}{\underset{COOC_2H_5}{|}}$ 80%

$$H_3C-CH_2-C\overset{OCH_3}{\underset{NCOOC_2H_5}{\diagdown}} + H_2C\overset{}{\triangle}CHCH_2OCH_3$$

$\overset{N}{\underset{COOC_2H_5}{|}}$

37%   6%

SCHEME 3.   Reactions of an allyl ether with ethoxycarbonylnitrene and with a palladium complex.

by authentic ethoxycarbonylnitrene.[50] Benzene, ethyl N,N-dichlorocarbamate, and copper powder at 50°C gave no azepine, only a 5.5% yield of ethyl N-phenyl-carbamate.[50] Apparent C—H insertion reactions were not stereospecific. Toluene gave a 21.6% yield of ethyl N-benzylcarbamate, and dioxane gave a 30% yield of ethyl N-(2-dioxanyl)carbamate. Similarly, the sodium derivative of ethyl N-chlorocarbamate, with copper powder, gave apparent C—H insertion products,[51]

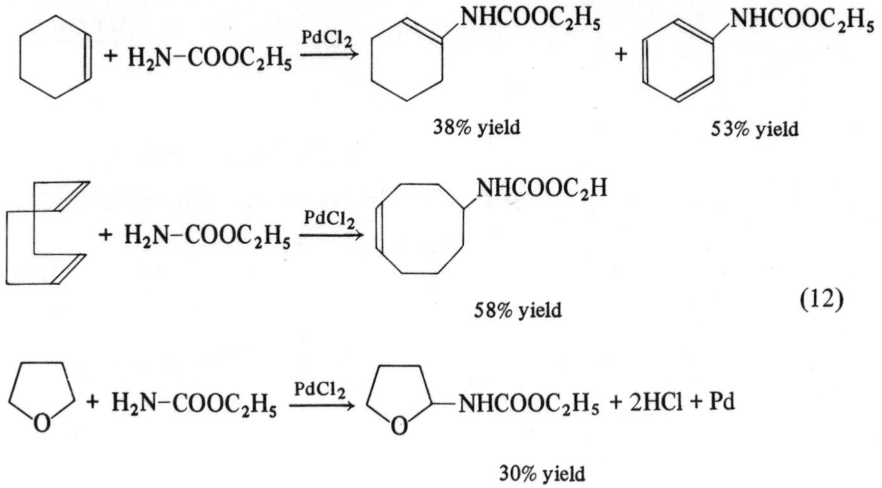

38% yield   53% yield

58% yield   (12)

30% yield

(All yields are based on PdCl$_2$.)

but not stereospecifically. Cyclohexene did not give the aziridine, only ethyl 3-cyclohexenylcarbamate, corresponding to an apparent allylic C—H insertion product. N,N-chloroarenesulfonamides behaved similarly.[52] The apparent C—H insertion products were formed nonstereospecifically. The authors[50-52] concluded that the reactions had radical mechanisms.

Metal complex nitrene transfer reagents (those complexes that transfer an RN electrophile to organic molecules) need not necessarily be RN complexes. It is quite conceivable that metal–organoazido complexes, [M(RN$_3$)], can transfer RN, with or without concurrent formation of N$_2$. We are concerned, therefore, with the mechanisms of metal nitrene complex formation, and the question of intermediate organoazido–metal complexes. It is most likely that such azido complexes are intermediates in many formations of nitrene complexes from organoazides and metal compounds. The interaction of aryl azides with certain tungsten complexes gives triazenido or nitreno complexes, depending on the particular tungsten complex used,[53] and it could be that here the former complex is a precursor of the latter. Thallium complexes can transfer the whole azido group to olefins to give N-azo-aziridine–thallium complexes. These can be ring opened to give C-azido compounds, with cleavage of the N—Tl bond,[54] as shown in Eq. 13. It would seem possible that N—N cleavage can occur in some such complexes under some conditions, with azirine formation or with the attachment to the ring nitrogen of some group R, thus creating a new group of nitrenoid reactions.

$$C_4H_9-CH=CH_2 + Tl(OCOCH_3)_{3-n}(N_3)_n \longrightarrow$$

$$C_4H_9-CH-CH_2 \xrightarrow{(CF_3CO)_2O}$$

$$N_2-Tl(OAc)_{3-n}(N_3)_{n-1}$$

$$C_4H_9-CH-CH_2N_3 + C_4H_9-CH-CH_2OCOCF_3$$
$$\quad\quad OCOCF_3 \quad\quad\quad\quad N_3$$
$$\quad\quad 73\% \quad\quad\quad\quad\quad 22\% \quad\quad (13)$$

$$+ Tl(OAc)_{3-n}(N_3)_n \longrightarrow \quad N-N{=}N-Tl(OAc)_{3-n}(N_3)_{n-1}$$

$$\downarrow H_3CCOCl$$

96%

IX. REFERENCES

1. E. Schmitz, R. Ohme, and S. Schramm, *Chem. Ber.*, 97, 2521 (1964).

2. W. Lwowski, Ed., *Nitrenes*, Wiley-Interscience, New York, 1970; especially, pp. 271ff.

3. E. Eibler and J. Sauer, *Tetrahedron Lett.*, 2659 (1974).

4. W. B. Jakoby and M. Wilchek, Ed., *Affinity Labeling, Methods Enzymol.* 46, (1977).

5. F. J. Darfler and A. M. Tometsko, in *Chemistry and Biochemistry of Amino Acids, Peptides, and Proteins*, B. Weinstein, Ed., Dekker, New York, 1978, Chap. 2.

6.* W. D. Hinsberg III and P. B. Dervan, *J. Am. Chem. Soc.*, 100, 1608 (1978).

7. P. B. Dervan and T. Uyehara, *J. Am. Chem. Soc.*, 101, 2076 (1979).

8. W. D. Hinsberg III and P. B. Dervan, *J. Am. Chem. Soc.*, 101, 6142 (1979).

9.* P. G. Schultz and P. B. Dervan, *J. Am. Chem. Soc.* 102, 878 (1980).

10. P. B. Dervan, private communication.

11. M. Edwards, T. L. Gilchrist, C. J. Harris, and C. W. Rees, *J. Chem. Res. (S)* 114 (1979).

12. Y. Hayashi and D. Swern, *J. Am. Chem. Soc.*, 95, 5205 (1973).

13.* L. Hoesch, N. Egger, and A. S. Dreiding, *Helv. Chim. Acta*, 61, 795 (1978).

14. O. L. Chapman and J. P. LeRoux, *J. Am. Chem. Soc.*, 100, 282 (1978).

15. O. L. Chapman, R. S. Sheridan, and J. P. LeRoux, *J. Am. Chem. Soc.*, 100, 6245 (1978).

16. O. L. Chapman and R. S. Sheridan, *J. Am. Chem. Soc.*, 101, 3690 (1979).

17.* O. L. Chapman, R. S. Sheridan, and J. P. LeRoux, *Rec. Trav. Chim. Pays-Bas*, 98, 334 (1979).

18.* S. Senda, K. Hirota, T. Asao, and K. Maruhashi, *J. Am. Chem. Soc.*, 100, 7661 (1978).

19. S. Senda, K. Hirota, T. Asao, and K. Maruhashi, *Tetrahedron Lett.*, 1531 (1978).

20. H. Dürr and H. Schmitz, *Chem. Ber.*, 111, 2258 (1978).

21. H. Yamada, H. Shizuka, and K. Matsui, *J. Org. Chem.*, 40, 1351 (1975).

22. R. Kayama, H. Shizuka, S. Sekiguchi, and K. Matsui, *Bull Chem. Soc. Japan*, 48, 3309 (1975).

23. T. Goka, H. Shizuka, and K. Matsui, *J. Org. Chem.*, 43, 1361 (1978).

24. T. Goka, Y. Hashida and K. Matsui, *Bull Chem. Soc. Japan*, 52, 1231 (1979).

25.* R. S. Atkinson and B. D. Judkins, *J. Chem. Soc. Chem. Commun.*, 832, 833 (1979).

26.* A. Holm, C. Larsen, and E. Larsen, *J. Org. Chem.*, 43, 4816 (1978).

27. A. Holm, N. Harris, and N. H. Toubro, *J. Am. Chem. Soc.*, 97, 6197 (1975).

28. A. Holm, J. J. Christiansen, and C. Lohse, *J. Chem. Soc., Perkin Trans. I*, 960 (1979).

29.* W. Lwowski and O. Subba-Rao, *Tetrahedron Lett.*, 727 (1980).

30. W. Lwowski, W. Reichen, and A. Mayr, unpublished results.

31. H. Takeuchi, T. Igura, M. Mitani, T. Tsuchida, and K. Koyama, *J. Chem. Soc. Perkin Trans. II*, 783 (1978).

32. H. Takeuchi, T. Nishiyama, M. Mitani, T. Tsuchida, and K. Koyama, *J. Chem. Soc., Perkin Trans. II*, 893 (1979).

33. L. D'Epifano, L. Pellicani, and P. A. Tardella, *J. Org. Chem.*, 44, 3605 (1979).

34. S. E. Carroll, B. Nay, E. F. V. Scriven, and H. Suschitzky, *Tetrahedron Lett.*, 943 (1977).

35. R. Purvis, R. K. Smalley, W. A. Strachan, and H. Suschitzky, *J. Chem. Soc., Perkin Trans., I*, 191 (1978).

36. E. F. V. Scriven and D. R. Thomas, *Chem. Ind. (Lond)*, 385 (1978).

37.* R. N. Cade, G. Jones, W. H. McKinley, and C. Price, *J. Chem. Soc., Perkin Trans. I*, 1211 (1978).

38. D. G. Hawkins, O. Meth-Cohn, and H. Suschitzky, *J. Chem. Soc., Perkin Trans. I*, 320 (1979).

39. M. Ogata, H. Matsumoto, and H. Kano, *Tetrahedron*, 25, 5205 (1969).

40. E. Giovannini, J. Rosales, and B. F. S. E. de Sousa, *Helv. Chim. Acta*, 62, 185 (1979).

41.* E. Giovannini and B. F. S. E. de Sousa, *Helv. Chim. Acta*, 62, 198 (1979).

42. T. Doppler, H. Schmid, and E.-J. Hansen, *Helv. Chim. Acta*, 62, 271 (1979).

43.* T. Doppler, H. Schmid, and E.-J. Hansen, *Helv. Chim. Acta*, 62, 304, 314 (1979).

44.* H. Takeuchi, T. Takahashi, T. Mashuda, M. Mitani, and K. Koyama, *J. Chem. Soc., Perkin Trans. II*, 1321 (1979).

45. W. A. Nugent and B. L. Haymore, *Coord. Chem. Rev.*, 31, 123 (1980).

46. F. Basolo, *J. Indian Chem. Soc.*, 54, 7 (1977).

47.* T. Migita, M. Chiba, M. Kosugi, and S. Nakaido, *Chem. Lett.*, 1403 (1978).
48.   W. Ando, H. Fuji, I. Nakamura, N. Ogino, and T. Migita, *J. Sulfur Chem.*, 8, 13 (1973).
49.   S. Ozaki and A. Tamaki, *Bull. Chem. Soc. Japan*, 51, 3391 (1978).
50.   N. Torimoto, T. Shingaki, and T. Nagai, *Bull. Chem. Soc. Japan*, 51, 2983 (1978).
51.   N. Torimoto, T. Shingaki, and T. Nagai, *J. Org. Chem.*, 44, 636 (1979).
52.   N. Torimoto and T. Shingaki, *Bull. Chem. Soc. Japan*, 50, 2780 (1977).
53.   G. L. Hillhouse and B. L. Haymore, *J. Organomet. Chem.*, 162, C-23 (1978).
54.   G. Emmer and E. Zbiral, *Justus Liebigs Ann. Chem.*, 796 (1979).

9

SILYLENES

PETER P. GASPAR

Department of Chemistry, Washington University
Saint Louis, Missouri 63130

I. INTRODUCTION

At the end of this author's last attempt in 1977 to review silylene chemistry, a forecast was made of future developments in the field.[1] An expansion of mechanistic knowledge concerning the formation and reactions of silylenes, increased use of silylenes as synthetic reagents, and the discovery of entirely new silylene reactions were confidently predicted, and have indeed occurred. Among the highlights of the recent work are the mechanistic clarification of the pyrolysis and photolysis of monosilane, both processes yielding SiH_2, the demonstration of concerted stereospecific addition by silylenes to olefins, the finding of C—H insertion and silylene-to-silylene rearrangements as entirely new reactions, and the synthesis of new classes of organosilicon compounds via silylene intermediates.

This chapter attempts to describe progress in silylene chemistry during the years 1977 to 1979. Omissions are likely to reflect the carelessness of the author rather than value judgments. Only the most essential earlier references are given, as these are included in the earlier chapter in Volume 1 of this series.[1] Several other surveys of silylene chemistry have been written recently.[2-4]

In 1980, as in 1977, the reactions of silylenes appear to resemble generally those of carbenes, but real differences are beginning to emerge. The lower reactivity, and hence greater selectivity of silylene reactions, allows processes like dimerization, almost unknown for carbenes, to occur with high yield for silylenes. Some silylene reactions, including several recently reported σ-bond insertions, have no direct parallel in carbene chemistry.

II. THE GENERATION OF SILYLENES

A. Thermolysis of Polysilanes

That workhorse method for thermal generation of silylenes, extrusion from disilanes,[1] has been supplemented in recent years by milder thermal and photo-

$$XYZSi-SiXYZ \longrightarrow XYSi: + SiXYZ_2$$

X and Y = hydrogen, halogen, alkoxy, alkyl, aryl; Z = hydrogen, halogen, alkoxy

chemical reactions, but it continues to be preparatively useful and has yielded significant mechanistic and thermochemical information.

Ring has reviewed the kinetics of thermal polysilane decomposition[5] and presented a model for the transition state shared by silylene extrusion and its microscopic inverse, insertion into a silicon–hydrogen bond.[6]

$$R^1R^2R^3Si-SiH_3 \longrightarrow \left[\begin{array}{c} H \cdots \\ R^1 \\ \diagdown Si \cdots\cdots Si \diagup H \\ R^2 \diagup \quad | \quad \diagdown H \\ R^3 \end{array} \right]^{\ddagger} \longrightarrow R^1R^2R^3\,SiH + SiH_2$$

This transition state, containing a pentacovalent sp^3d hybridized silicon atom, is in accord with the experimentally determined entropies of activation for silylene extrusion from a series of disilanes and with the ratios of preexponential factors for competitive SiH_2 insertion reactions.[6] There is still, however, considerable disagreement concerning the latter data.[1]

Doncaster and Walsh have studied the kinetics of the gas-phase pyrolysis of hexachlorodisilane in the presence of iodine and concluded that the following mechanism operates:[7]

$$Si_2Cl_6 \longrightarrow :SiCl_2 + SiCl_4$$

$$:SiCl_2 + I_2 \longrightarrow SiCl_2I_2$$

The activation parameters, for the silylene extrusion were: $\log A$ (sec^{-1}) = 13.49 ± 0.12 and $E_a = 205.9 ± 1.4$ kJ/mol. The reactions of $SiCl_2$ with $SiCl_4$, HCl, and benzene were at least 2 orders-of-magnitude slower than with I_2.

B. Silicon Atom Reactions

A remarkable series of reactions is believed to occur upon cocondensation of atomic silicon vapor with N_2 or CO at 4K.[8] From electron spin resonance and ultraviolet and infrared spectroscopic data (see below) it was concluded that triplet ground states of the novel silylenes SiN_2 and $SiCO$ were produced. Warming the matrix to 35K led to the formation of $Si(CO)_2$ by the reaction:

$$Si(CO) + CO \longrightarrow Si(CO)_2$$

If this silicon analogue of carbon suboxide survived to higher temperatures it would have many uses. SiN_2 and $SiCO$ excite great interest since both calculations and experiment suggest that all other silylenes thus far examined have singlet ground states.[1]

The formation of silylenes in the primary transformations of recoiling silicon atoms continues to be actively investigated. Although the role of [31]SiH_2 in the reactions of ^{31}Si atoms in phosphine/silane mixtures is still controversial,[9] the formation of a cyclic silylene in the addition to butadiene is suggested by the finding of a spiro-compound as a minor product.[10]

The most successful means for studying the reactions of *monomeric* SiF$_2$ has been via hot atom reactions in which nucleogenic silicon atoms abstract two fluorine atoms from their PF$_3$ precursor.[1,11-13] From such reactions, chemically stable products containing a single silylene unit predominate, since the low concentrations of intermediates preclude the dimerizations that plague thermal methods.

C. Photolysis of Silanes

a. Polysilane Photolysis

The generation of silylenes by photolysis of polysilanes was discovered a decade ago,[1] but this very useful reaction is receiving increasing attention and has been expertly reviewed.[14] This process is usually described as the extrusion of an inner silylene unit of a polysilane, but it is formally related to the thermally

induced extrusions described in Section II. A, with the migrating group Z a silyl group. The relationship is underlined by the observation that in the photolysis of an alkoxypolysilane, migration of an alkoxy group, commonly found in pyrolysis experiments, can compete with the migration of a silyl group.[15]

Processes a and b occur in the statistical ration 2:1 upon photolysis, but pyrolysis follows path b exclusively. Thus one can generate the same silylene from the same precursor by photochemical and thermal means.

Methoxytrimethylsilylsilylene may prove to be useful as a silicon atom synthon, as in the following sequence:

Branched-chain polysilanes can in general form more than one silylene upon photolysis:[14]

b. Monosilane Photolysis

While the photolysis of monosilanes is not a convenient source of silylenes for synthetic or mechanistic studies, it has received considerable attention.

Lampe and co-workers have found that decomposition of SiH_4 with 147-nm xenon resonance radiation produces SiH_2 in the major primary process, with SiH_3 formed as a minor primary product.[16]

$$SiH_4 + h\nu\ (147\ nm) \nearrow \begin{array}{l} :SiH_2 + 2H\cdot \quad \phi = 0.83 \\ \\ \cdot SiH_3 + H\cdot \quad \phi = 0.17 \end{array}$$

To account for the formation of disilane and trisilane (with quantum yields at 2 torr of 1.29 ± 0.08 and 0.46 ± 0.01, respectively) a scheme of secondary reactions involving formation of vibrationally excited disilane by both SiH_2 insertion and SiH_3 recombination was suggested. This scheme contains a new source of silylsilylene, α-elimination of molecular hydrogen from disilane (but see also below). Although this sequence accounts for the products in

$$H + SiH_4 \longrightarrow H_2 + \cdot SiH_3$$

$$2 \cdot SiH_3 \xrightarrow{\text{disproportionation}} :SiH_2 + SiH_4$$

$$2 \cdot SiH_3 \xrightarrow{\text{recombination}} SiH_3 SiH_3^*$$

$$:SiH_2 + SiH_4 \underset{\text{dissociation}}{\overset{\text{insertion}}{\rightleftharpoons}} SiH_3 SiH_3^*$$

$$SiH_3 SiH_3^* + M \longrightarrow SiH_3 SiH_3$$

$$SiH_3 SiH_3^* \longrightarrow SiH_3 SiH: + H_2$$

$$SiH_3 SiH: + SiH_4 \longrightarrow SiH_3 SiH_2 SiH_3$$

terms of known reactions of SiH_2 and $SiH_3\ddot{S}iH$, and is consistent with both pressure dependence and scavenger studies, the formation of both disilane and trisilane is reminiscent of the reactions of recoiling silicon atoms with silane[1,9] and suggests the possibility that Si or SiH intermediates may be formed in silane photolysis.

The reaction scheme proposed by Lampe illustrates a complexity in the investigation of silane decomposition—any process that converts silane to silyl radicals will also lead indirectly to silylene via both recombination and disproportionation processes. Potzinger has pointed out that the vibrationally excited disilane formed from silyl radical recombination can be collisionally quenched before dissociation to silylene, but since the disproportionation occurs via an entirely different transition state, it still persists at high pressures.[17] Potzinger and co-workers have established a disproportionation to recombination ratio of 0.7 ± 0.1 at room temperature in the reactions of hydrogen atoms with SiH_4/SiD_4 mixtures.[17] Austin and Lampe have suggested that disproportionation of silyl radicals followed by insertion of silylene is the major route to the disilane products in the hydrogen atom sensitized decomposition of monosilane.[18]

It is clear then that depending on the temperature, pressure, concentration, and nature of the coreactants,[1] the reactions initiated photochemically in SiH_4 can be enormously complex.

This complexity is apparent in the photolysis of methylated monosilanes. Formation of silylenes is reported to be an important reaction, but this can occur by various kinds of elimination, and other processes are also found. Thus Alexander and Strausz have proposed a set of primary steps in the photolysis of dimethylsilane at 147 nm, and we are confronted by an array of reactive intermediates including silylenes, silyl radicals, alkyl radicals, carbenes, and silenes.[19] Here roughly 50% of the primary reactions produce silylenes and 25% silaethylenes.

$$
\begin{array}{ll}
& \phi \\
(CH_3)_2SiD_2 + h\nu \ (147 \ nm) \longrightarrow CH_3SiD\colon + CH_3D & 0.15 \\
\qquad CH_3SiD\colon + \cdot CH_3 + \cdot D & 0.20 \\
\qquad SiD_2\colon + 2 \cdot CH_3 & 0.08 \\
\qquad CH_2 = SiD_2 + CH_4 & 0.05 \\
\qquad CH_3SiD_2H + \colon CH_2 & 0.04 \\
\qquad (CH_3)_2Si\colon + D_2 & 0.07 \\
\qquad CHSiD_2 + \cdot CH_3 + H_2 & 0.04 \\
\qquad CHSiCH_3 + D_2 + H_2 & 0.07 \\
\qquad CH_2SiCH_3 + HD + \cdot D & 0.09 \\
\qquad \cdot CH_2SiD_2CH_3 + \cdot H & 0.08 \\
\qquad CH_2 = SiDCH_3 + HD & 0.19
\end{array}
$$

For the 147-nm photolysis of tetramethylsilane Strausz and co-workers propose a mechanism in which 24% of the primary events give silylenes, 43% silyl radicals, 27% silaolefins, and 6% carbenes.[20]

$$
\begin{array}{ll}
& \phi \\
(CH_3)_4Si + h\nu \ (147 \ nm) \longrightarrow (CH_3)_3Si\cdot + \cdot CH_3 & 0.43 \\
\qquad (CH_3)_2Si\colon + 2 \cdot CH_3 & 0.24 \\
\qquad CH_2 = Si(CH_3)_2 + CH_4 & 0.17 \\
\qquad CH_2 = Si(CH_3)_2 + \cdot CH_3 + \cdot H & 0.10 \\
\qquad \colon CHSi(CH_3)_3 + H_2 & 0.02 \\
\qquad \colon CH_2 + HSi(CH_3)_3 & 0.04
\end{array}
$$

The evidence for the formation of $Me_2Si\colon$ in the photolysis of Me_4Si is rather indirect, and Tokach and Koob have omitted the processes yielding the silylene and carbenes from their proposed set of primary reactions.[99] Potzinger and co-workers, employing light of wavelength 180–190 nm, report that the formation of Me_2Si is excluded by the observation that no nonscavengeable ethane is formed, that is no molecular elimination of ethane occurs, and trapping experiments with trimethylsilane do not yield the silylene insertion product expected if radical elimination leads to Me_2Si.[100]

$$
Me_4Si + h\nu \ (180\text{-}190 \ nm) \not\longrightarrow Me_2Si\colon + CH_3CH_3
$$
$$
\not\longrightarrow Me_2Si\colon + 2 \cdot CH_3
$$
$$
\Big\downarrow HSiMe_3
$$
$$
HSiMe_2SiMe_3
$$

Two other new photolysis reactions have recently been found to give rise to silylenes. Koob and co-workers irradiated 1,1-dimethylsilacyclobutane in the gas phase.[21] The major products were 2-methyl-2-silapropene and ethylene, also formed in pyrolysis, but dimethylsilylene was an apparent minor product.

$$Me_2Si=CH_2 + CH_2=CH_2$$
$$\phi = 0.86$$

$$Me_2Si: + \quad \overset{CH_2}{\underset{CH_2-CH_2}{\diagdown}} + CH_3CH=CH_2$$
$$\phi = 0.02$$

A more efficient but equally impractical silylene source is the photolysis of silacyclopropenes.[22-24] This reaction is usually carried out in reverse, as is described later in this chapter, to produce silacyclopropenes.

$$\xrightarrow{h\nu} Me_3SiC\equiv C\phi + Me_2Si: \xrightarrow[Et_2MeSiH]{} Et_2MeSiSiMe_2H$$

$$\xrightarrow{h\nu} MeC\equiv CMe + Me_2Si: \xrightarrow[Me_3SiH]{} Me_3SiSiMe_2H$$

On the more useful side, a new photochemical silylene generator was prepared by Sakurai and co-workers by cycloaddition of cyclopropene with various siloles.[25] This is a three-step conversion of a dichlorosilane to a silylene.

R	R¹	R²
H	Me	Me
φ	Me	Me
φ	φ	Me
φ	φ	φ

D. Pyrolysis of Monosilanes

The long-standing question[1] of whether the primary step in the pyrolysis of SiH_4 is molecular elimination to SiH_2 or homolysis to SiH_3 seems finally to have been answered definitively. As we have seen in the discussion of silane

$$SiH_4 \xrightarrow{\Delta} SiH_2 + H_2$$

$$SiH_4 \xrightarrow{\Delta} SiH_3 + H$$

photolysis above, both SiH_2 and SiH_3 can react to form disilane—the major product, together with H_2, of silane pyrolysis.

Ring and co-workers have studied the decomposition of silane in a single-pulse shock tube at temperatures between 1200 and 1300K.[26] An activation energy of 56.1 ± 1.7 (later revised to 52.7 ± 1.4)[27] kcal/mol was found, and this is much lower than the 94 kcal/mol bond dissociation energy $D(SiH_3 - H)$ currently accepted for silane. Since only homogeneous and unimolecular reactions are likely under shock-tube conditions, the observed activation energy must relate to the primary step in silane decomposition, which was therefore accepted to be the molecular elimination long championed by Purnell and Walsh.[1,28] A chain length $>10^6$ was estimated to be required for a radical chain process with the observed reaction rate, but a lack of concentration dependence for the rate constant indicated that no radical chain decomposition of SiH_4 occurred—either during the shock wave or in the subsequent cooling-down period.

In the shock tube experiment no disilane, the expected product from SiH_2 and excess SiH_4, was found. This was first explained by postulating the rapid removal of SiH_2 by further decomposition.[26] Later this explanation was re-

$$SiH_2 \rightarrow SiH + H$$

jected on the thermochemical grounds, and removal of SiH_2 by dimerization was suggested.[27]

$$2SiH_2 \rightarrow [H_2 SiSiH_2]^* \rightarrow [H_3 SiSiH]^* \rightarrow Si_2 H_2 + H_2$$

Strausz and co-workers provided evidence of a different sort for molecular elimination as the mechanism for silane decomposition.[29] They carried out static pyrolyses whose products are hydrogen and disilane. Two mechanisms were considered:

Molecular elimination

$$SiH_4 \xrightarrow{\Delta} :SiH_2 + H_2$$

$$:SiH_2 + SiH_4 \rightarrow SiH_3 SiH_3$$

Homolysis followed by radical chain reactions

$$SiH_4 \xrightarrow{\Delta} \cdot SiH_3 + \cdot H$$

$$\cdot H + SiH_4 \longrightarrow H_2 + \cdot SiH_3$$

$$\cdot SiH_3 + SiH_4 \longrightarrow SiH_3 SiH_3 + \cdot H$$

$$2 \cdot SiH_3 \longrightarrow SiH_3 SiH_3$$

The occurrence of chain reactions was rendered unlikely by the observation that the presence of 10% ethylene alters neither the product yields nor the rate of silane loss. Ethylene has been shown to be an effective scavenger of silyl radicals in the presence of silane.[30]

Large quantities of added gases were found to accelerate the decomposition at partial pressures of SiH_4 of about 10^2 torr. This could be explained by the original suggestion of Purnell and Walsh that molecular elimination was occurring in the pressure falloff region of unimolecular decomposition at these pressures.[28] That this explanation is correct and also accounts for the observed reaction order of 1.5 was demonstrated by varying the partial pressure of SiH_4 at constant total pressure.[29] Ring and co-workers carried out similar experiments.[27] Under these conditions the formation of H_2 was strictly first-order, in accord with the following mechanism:

$$SiH_4 + SiH_4 \rightleftharpoons SiH_4^* + SiH_4$$

$$SiH_4 + M \rightleftharpoons SiH_4^* + M$$

$$SiH_4^* \longrightarrow :SiH_2 + H_2$$

The activation parameters found by Strausz and co-workers were in good accord with those of Ring, and Purnell and Walsh: for formation of H_2 log A $(sec^{-1}) = 13.76 \pm 0.15, E_a$ (kcal/mol) $= 55.05 \pm 0.44$; for formation of disilane $\log A = 13.51 \pm 0.29, E_a = 54.77 \pm 0.84$.

While it is now agreed that pyrolysis of SiH_4 occurs exclusively by molecular elimination, substituents exert a considerable influence on the relative importance of homolysis and molecular elimination. Neudorfl and Strausz found that both processes occur in the thermolysis of methylsilane and dimethylsilane.[31]

$$MeSiH_3 \longrightarrow MeSiH: + H_2$$

$$MeSiH_3 \xrightarrow{wall} MeSiH_2 \cdot + H \cdot$$

The heterogeneous chain initiation step is believed to be slow compared to molecular elimination, but an estimated chain length of 10^6 for reactions following homolytic cleavage would lead to a considerable contribution to product formation. The activation parameters for molecular elimination were determined: $\log A \ (\text{sec}^{-1}) = 14.95 \pm 0.11, E_a \ (\text{kcal/mol}) = 63.2 \pm 0.3$. Similar results were found for pyrolysis of dimethylsilane with activation parameters $\log A = 14.3 \pm 0.3, E_a = 68 \pm 1$. Bond dissociation energies and heats of formation were estimated from these data:

$$D(\text{MeSiH}-\text{H}) \simeq 69 \ \text{kcal/mol} \qquad \Delta H_f^0 \ (\text{MeSiH:}) \simeq 53 \ \text{kcal/mol}$$

$$D(\text{Me}_2\text{Si}-\text{H}) \simeq 74 \ \text{kcal/mol} \qquad \Delta H_f^0 \ (\text{Me}_2\text{Si:}) \simeq 44 \ \text{kcal/mol}$$

Davidson and Ring have employed low-pressure pyrolysis to study methylsilane decomposition,[32] and differ from Neudorfl and Strausz in proposing two molecular elimination processes producing silylenes in addition to a surface-initiated radical chain reaction. The formation of MeSiH was found to predominate by a factor of about 20.

$$\text{MeSiH}_3 \xrightarrow{\Delta} \text{MeSiH:} + \text{H}_2$$

$$\text{MeSiH}_3 \xrightarrow{\Delta} :\text{SiH}_2 + \text{CH}_4$$

Richardson and Simons have studied the unimolecular decomposition of dimethylsilane and ethylsilane by chemical activation through the exothermic insertion of methylene into C—H and Si—D bonds of CH_3SiD_3.[33] The follow-

$$\text{CH}_2: + \text{CH}_3\text{SiD}_3 \begin{array}{c} \nearrow \text{CH}_3\text{SiD}_2\text{CH}_2\text{D*} \\ \searrow \text{CH}_3\text{CH}_2\text{SiD}_3 * \end{array}$$

ing decomposition steps were proposed:

	$k/10^9 \ \text{sec}^{-1}$
$\text{CH}_3\text{SiD}_2\text{CH}_2\text{D*} \longrightarrow \cdot\text{CH}_3 + \text{CH}_2\text{DSiD}_2 \cdot$	0.307 ± 0.023
$\cdot\text{CH}_2\text{D} + \text{CH}_3\text{SiD}_2 \cdot$	0.308 ± 0.023
$\cdot\text{D} + \text{CH}_3\dot{\text{Si}}\text{DCH}_2\text{D}$	0.091 ± 0.006
$\text{CH}_3\text{D} + \text{CH}_2\text{DSiD:}$	0.282 ± 0.006
$\text{CH}_2\text{D}_2 + \text{CH}_3\text{SiD:}$	0.254 ± 0.007
$\text{D}_2 + \text{CH}_3\dot{\text{Si}}\text{CH}_2\text{D}$	0.75
$\text{CH}_3\text{CH}_2\text{SiD}_3 * \longrightarrow \cdot\text{D} + \text{CH}_3\text{CH}_2\text{SiD}_2 \cdot$	0.054 ± 0.003
$\text{CH}_3\text{CH}_2 \cdot + \cdot\text{SiD}_3$	0.531 ± 0.030
$\text{D}_2 + \text{CH}_3\text{CH}_2\text{SiD:}$	0.75
$\text{CH}_3\text{CH}_2\text{D} + :\text{SiD}_2$	0.434 ± 0.024
$\text{CH}_2 {=} \text{CH}_2 + \text{SiHD}_3$	1.5

Thus for chemically activated dimethylsilane, roughly two-thirds of the decompositions are believed to yield silylenes, with dimethylsilylene favored over methylsilylene. Only about one-third of the decompositions of chemically activated ethylsilane produced silylenes, with ethylsilylene favored over silylene.

III. THE REACTIONS OF SILYLENES

A. Silylene Insertion Reactions

a. Insertion into the Silicon–Silicon Bond

A reaction of silylenes only recently confirmed despite previous suggestions[1] is insertion into a silicon–silicon bond.[34] Reactions of this type were carried out in sealed tubes at 350°C and indicate considerable stability for both the disilacyclobutene reactants and the trisilacyclopentene products. It will be interesting to see whether Si–Si insertion is limited to strained ring systems.

b. Insertion into Oxygen–Hydrogen and Nitrogen–Hydrogen Bonds

Photochemically generated dimethylsilylene has been found to insert efficiently into O–H and O–D bonds of water and alcohols and into N–H bonds of primary and secondary amines.[35] No insertion was observed however into an S–H bond. These reactions provide efficient routes to derivatized silanes

$$(Me_2Si)_6 \xrightarrow{h\nu} Me_2Si:$$

$$Me_2Si + HOR \longrightarrow Me_2SiHOR$$

R	Yield (%)
$-CH_2CH_3$	87
$-CH_3$	89
$-C(CH_3)_3$	85

$$2Me_2Si + H_2O \longrightarrow (Me_2SiH)_2O$$
$$85\%$$

$$2Me_2Si + D_2O \longrightarrow (Me_2SiD)_2O$$
$$82\%$$

$$Me_2Si + HN\underset{CMe_2}{\overset{CH_2}{\diagdown |}} \longrightarrow Me_2SiH-N\underset{CMe_2}{\overset{CH_2}{\diagdown |}}$$
$$85\%$$

$$Me_2Si + HNEt_2 \longrightarrow Me_2SiHNEt_2$$
$$81\%$$

$$Me_2Si + H_2NCMe_3 \longrightarrow Me_2SiH-NHCMe_3$$
$$86\%$$

$$Me_2Si + HSCH_2CH_2CH_3 \not\longrightarrow Me_2SiH-SCH_2CH_2CH_3$$

and represent the most economical preparation to date of the Si—D bond. Mechanistically, these apparent insertions may prove to be addition reactions, with initial attack occurring on the unshared electron pairs of nitrogen and oxygen.[102] The insertion of thermally generated silylenes into O—H bonds of alcohols has been patented.[36]

c. Insertion into C—O Bonds

Another apparent insertion that may involve initial attack at oxygen is the ring expansion of furans upon reaction with thermally generated dichlorosilylene found by Chernyshev and co-workers.[37]

R^1	R^2	R^3	R^4	Yield (%)
H	H	H	H	30
H	H	H	CH_3	40
CH_3	H	H	CH_3	40
	$-(CH_2)_4-$	$-(CH_2)_4-$		70
H	H	$-(CH_2)_4-$		70

Dibenzo-p-dioxin undergoes twofold insertion.

Insertion into C—O bonds of aryloxysilanes by $SiCl_2$ has been suggested to account for the products formed from gas-phase flow pyrolysis of hexachlorodisilane in the presence of dichlorophenoxysilane.[38] One of the pathways considered requires cyclization of the phenoxysilane prior to reaction with $SiCl_2$, while the other includes a cyclization step following silylene C—O insertion:

$$Si_2Cl_6 \longrightarrow :SiCl_2 + SiCl_4$$

The initial insertion product in the latter process was isolated in 5% yield along with a 20% yield of the cyclic siloxane. Another product, $\phi SiCl_2OSiCl_3$, obtained in 10% yield, was attributed to radical side reactions. These reactions deserve further mechanistic scrutiny, as they may be complicated by other radical reactions and redistribution processes. It is somewhat surprising that insertion into the Si—O bonds, a well-known process for alkoxysilanes,[1] did not seem to occur.

d.   Reactions of Silylenes with Cyclic Siloxanes and Ethers

Insertion of a silylene into silicon-oxygen bonds of alkoxysilanes was one of the first reactions of organosilylenes to be recognized,[39] but until recently no detailed mechanism was suggested. The reactions of carbenes with cyclic ethers are believed to proceed via ylids,[40] and the reaction of a silylene with an R—O—R' system could also be expected to produce a zwitterionic intermediate because of the large driving force for formation of a silicon-oxygen bond.

Photochemically generated dimethylsilylene has been found to insert into an Si—O bond of hexamethylcyclotrisiloxane.[41] Methylphenylsilylene fails, under identical conditions, to perform this insertion,[42] indicating the delicate balance of the factors that determine the paths by which silylenes react, and how far we have yet to progress before these factors are understood.

$$Me_2Si: + \; Me_2Si \overset{O}{\underset{O}{\diagup}} \overset{\diagdown}{\underset{O}{\diagdown}} SiMe_2 \longrightarrow Me_2Si \overset{O-SiMe_2}{\underset{Me_2Si}{\diagup}} \overset{\diagdown}{\underset{O-SiMe_2}{O}} $$

When a five-membered cyclosiloxane was employed as a trapping agent, insertion by MeSiϕ, as well as by Me$_2$Si, was observed.[42] It was suggested that

$$MeS\ddot{i}R + Me_2Si \overset{\diagup\diagdown}{\underset{O}{\diagdown}} SiMe_2 \longrightarrow Me_2Si \overset{\diagup\diagdown}{\underset{MeRSi-O}{\diagdown}} SiMe_2$$

R = Me, ϕ

increased angle strain was responsible for the greater reactivity of the five-membered ring, which was found to be over 100 times as reactive as the six-membered cyclotrisiloxane toward Me$_2$Si. Other factors, electronic, steric, or thermochemical, may also be important. Certainly the detailed mechanisms for these Si—O insertions deserve further exploration. No insertion products were obtained from reactions of dimethylsilylene with octamethylcyclotetrasiloxane or hexamethylsiloxane.

$$Me_2Si: + \; Me_2Si \overset{O-SiMe_2}{\underset{Me_2Si-O}{\overset{O}{\diagdown}} \overset{SiMe_2}{O}} \overset{\times}{\nrightarrow} Me_2Si \overset{O-SiMe_2}{\underset{O-SiMe_2}{\overset{O}{\diagdown}} \overset{SiMe_2}{O}}$$

$$Me_2Si: + \; Me_3SiOSiMe_3 \overset{\times}{\nrightarrow} Me_3SiSiMe_2OSiMe_3$$

In recent experiments, Weber and co-workers have found new examples of silylene insertions into C—O bonds, and have presented possible reaction mechanisms.

Photochemically generated dimethylsilylene reacts with oxetane at 0°C to give high yields of allyloxydimethylsilane and 2,2-dimethyl-1-oxa-2-silacyclopentane.[43] No reactions were observed with unstrained aliphatic ethers, tetra-

$$Me_2Si: + \; \square_O \longrightarrow \diagup\!\!\diagdown\!\!\diagup^O{}_{SiHMe_2} + \overset{\diagup\diagdown}{\underset{O}{\diagdown}} SiMe_2$$

38%   41%

hydrofuran and diethyl ether. When the temperature was lowered to $-98°C$, only the allyloxydimethylsilane was obtained, and this observation eliminated direct insertion as the mechanism for formation of the cyclic product. It was proposed that a zwitterionic intermediate is produced by attack on the oxygen, and that this primary adduct can undergo both intramolecular rearrangement to the allyloxysilane and ring opening to a second zwitterionic species followed by cyclization. This somewhat startling ring opening to a primary carbonium

ion was suggested to explain the effects of methyl substitution on the course of the reaction. Reaction with 2,2,-dimethyloxetane gives the oxasilacyclopentane expected from fragmentation of the initial intermediate to the more stable carbonium ion. Reaction of dimethylsilylene with 2-methyloxetane leads to

13%   44%   26%

an even more complex product mixture, but the only cyclic product is that expected from fragmentation of an initially formed ylid to the more stable carbonium ion.

29%

2%   1%   26%

While these mechanisms are reasonable, more evidence is required to establish them firmly. The rearrangment of an initially formed ylid to a further zwitterionic intermediate seems unlikely when the carbonium ion formed is primary. From 3,3-dimethyloxetane, the only product obtained is 2,2,4,4-tetramethyl-1-oxa-2-silacyclopentane, but its formation via the suggested mechanism requires a substituted neopentyl carbonium ion, which would be expected to undergo rapid methyl migration.

In the solvents employed, fragmentation of the initial ylid might give rise to a diradical intermediate whose subsequent cyclization and disproportionation could yield the observed products. A similar mechanism has been suggested by Frey and Voisey for insertion of singlet CH_2 into a C—O bond of tetrahydrofuran.[40]

recombination   disproportionation

Weber sounded a useful warning against aliphatic ethers as solvents for silylene reactions. While these ethers may not give detectable reaction products, they may form addition complexes capable of delivering silylenes to various substrates with reactivities different from those of free silylenes. Thus is raised the spectre of silylenoids. Indeed Steele and Weber have found differences in substrate selectivity among pairs of alcohols when the silylene is generated in ethers rather than in cyclohexane.[44]

Zwitterionic intermediates were also suggested by Tzeng and Weber for the reactions of photogenic dimethylsilylene with α,β-unsaturated epoxides.[45] Products include 1-oxa-2-silacyclohex-4-enes, formally the result of silylene insertion accompanied by rearrangement. Deoxygenation products, 1,3-butadienes, and oligomers of dimethylsilanone are also formed. The suggested mechanism begins with attack of the silylene at the oxygen, yielding an ylid that undergoes heterolytic C—O cleavage to a zwitterionic species with a delocalized cationic center. This zwitterionic intermediate can cleave the remaining C—O bond or cyclize. Regiospecific formation of the cyclic product rules out cycloaddition

of the silanone to the butadiene as its source.

This attractive mechanistic suggestion requires careful consideration. It is not clear why a zwitterionic species is preferred over a diradical upon ring opening of the initial adduct.

Another mechanistic possibility, initial addition of the silylene to the double bond of the vinyloxirane, followed by rearrangement to the oxasilacyclohexene was considered by Goure and Barton, but rejected because oxiranes seem much more reactive toward silylene than are olefins.[101]

Goure and Barton favor ylid formation as the mechanism for the facile deoxygenation of cyclooctene oxide by dimethylsilyene. Several mechanisms were rejected, including initial insertion by Me_2Si forming a siloxetane, and silanone transfer from the ylid.

e.   Insertion into C—H Bonds

A most interesting new reaction, the insertion of silylenes into carbon-hydrogen bonds, is discussed later in connection with another newly discovered process, the rearrangement of silylenes (Section III.E).

B.   Silylene Addition Reactions

a.   Addition to Olefins

The availability of mild methods for the generation of silylenes in the photolysis of polysilanes and thermolysis of hexamethylsilirane[1] has stimulated a number of studies of silylene addition to carbon-carbon π-bonds.

Perhaps the most significant experiment was that of Tortorelli and Jones, in which it was proved that dimethylsilylene undergoes stereospecific addition to cis- and trans-2-butene.[46] In this elegant experiment, which also demonstrated that methanolysis of siliranes is stereospecific, the silylene adducts of cis- and trans-2-butene were treated with methanol. When CH_3OH is employed, the methylene group of the resulting 2-butylmethoxydimethylsilane appears in the ^{1}H NMR spectrum as a diastereotopic pair of hydrogens H_a, H_b with chemical shifts δ 1.18 and 1.55 ppm. When CH_3OD is added to the silirane, only

$$Me_2Si: + MeCH=CHMe \longrightarrow \left[\begin{array}{c} SiMe_2 \\ \diagdown \diagup \\ MeCH-CHMe \end{array} \right] \xrightarrow{MeOH} \quad Me_2SiOMe \quad H_a \\ \diagdown \diagup \quad \diagdown \\ H_b$$

one of these hydrogens is seen in the ^{1}H NMR spectrum, while the other appears in the ^{2}H NMR spectrum: for cis-2-butene D at δ 1.18 and H at 1.55, while for the trans-isomer H at δ 1.18 and D at 1.55. These observations demand that only a single stereoisomer of the silirane is formed from the cis-olefin and the other from the trans, and that both suffer stereospecific methanolysis. While not established, cis-addition is clearly indicated.

Recent results of Ishikawa, Nakagawa, and Kumada can be interpreted as confirming that silylene addition to olefins is a concerted cis-process.[47] Photochemical generation of phenyltrimethylsilylsilylene in the presence of trans-2-butene leads to the formation of only a single silacyclopropane, while two stereoisomeric silacyclopropanes are formed from addition to cis-2-butene. All three isomers give the same methanolysis product. These observations are consistent with a cis-addition mechanism.

$$(Me_3Si)_3Si\phi \xrightarrow{h\nu} \phi\ddot{S}iSiMe_3 + Me_3SiSiMe_3$$

The addition of $\phi\ddot{S}iSiMe_3$ to various olefins was examined using a low-pressure mercury lamp with Vycor filters to photolyze $(Me_3Si)_3Si\phi$. The yields of silacyclopropanes were determined by isolation of the methanolysis products.

TABLE 9.1.   Yields of Silacyclopropanes from Addition of $\phi\ddot{S}iSiMe_3$ to Olefins

Olefin	Yield of Silacyclopropane (%)
Isobutene	52
1-Butene	47
cis-2-Butene	35
trans-2-Butene	40
Trimethylvinylsilane	62

The silacyclopropanes were photochemically labile when irradiated with a quartz-filtered high-pressure mercury lamp, rearranging via both 1,2- and 1,3-hydrogen shifts, for example:

$$\begin{array}{c}\text{MeCH}\!\!-\!\!\!-\!\!\!-\!\!\text{CHMe}\\ \diagdown\text{Si}\diagup\\ \phi\diagup\;\diagdown\text{SiMe}_3\end{array} \xrightarrow{h\nu} \begin{array}{c}\text{MeC}\!\!=\!\!\text{CHMe}\\ |\\ \phi\text{SiHSiMe}_3\end{array} + \begin{array}{c}\text{MeCHCH}\!\!=\!\!\text{CH}_2\\ |\\ \phi\text{SiHSiMe}_3\end{array}$$

Addition of $\phi\ddot{S}iSiMe_3$ to conjugated dienes was also studied. 1-Silacyclopent-3-enes, the products of formal 1,4-addition, were obtained along with labile compounds believed to be the silirane 1,2-adducts on the basis of their methanolysis products. That the siliranes were intermediates in the formation of the silacyclopentenes was clearly indicated by the time dependence of the product yields. The silirane yield rises to a maximum and decreases, while the silacyclopentene yield rises monotonically, but more slowly. Thus these experiments confirm earlier suggestions[1] that the apparent 1,4-additions leading to 1-silacyclopent-3-enes really are 1,2-additions followed by rearrangement.

Addition of $\phi\ddot{S}iSiMe_3$ to 1,3-cyclooctadiene gave analogous products.

These results from Princeton and Kyoto demonstrate vividly that singlet silylenes resemble closely their carbene counterparts in their addition to olefins and conjugated dienes. Concerted 1,2-*cis*-addition seems to be the rule, and the major difference between silylene and carbene additions stems from the much lower stability of the silacyclopropane products. Thus secondary products are often isolated from silylene additions as a result of the rearrangement of the primary adducts under the influence of heat or light.

When a high-pressure mercury lamp with a quartz filter was used to generate methylphenylsilylene MeSiφ, rearrangement products from the silacyclopropane comprised the bulk of the adducts with every olefin examined.[48] With a low-pressure mercury lamp and Vycor filter, the expected methanolysis products of the silacyclopropane adducts are obtained from 1-butene, *cis*-2-butene, 1-octene, cyclohexene, and cyclooctene, with only small amounts of rearrangement products. But even with the low-pressure lamp, rearranged addition products were dominant in reactions with isobutene and tetramethylethylene.

The formation of secondary products has certainly been a complicating factor in the addition to acetylenes. Using harsh thermal methods for silylene generation, the commonly obtained aceylene adducts are 1,4-disila-2,5-cyclohexadienes, and much effort has been given to answering the question: How do two silylene and two aceylene units combine to form these products?[1]

$$2R_2Si: + 2R'C{\equiv}CR' \longrightarrow \underset{\underset{R'\quad R'}{|}}{\overset{\overset{R'\quad R'}{|}}{R_2Si \quad SiR_2}}$$

b. Addition to Acetylenes

Silacyclopropenes have long been thought to be intermediates in the formation of disilacyclohexadienes. Ishikawa and Kumada have reported a direct thermal dimerization (Eq. 59),[49] and Cornett has found that treatment of a silacyclopropene with a nucleophile at room temperature produces dimer along with polymer.[24]

$$2Me_2Si \overset{\overset{\phi}{\diagup}}{\underset{\underset{SiMe_3}{\diagdown}}{||}} \xrightarrow{\Delta} Me_2Si \overset{\phi}{\underset{Me_3Si}{\diagup}} \overset{SiMe_3}{\underset{\phi}{\diagdown}} SiMe_2$$

$$2Me_2Si \overset{\overset{Me}{\diagup}}{\underset{\underset{Me}{\diagdown}}{||}} \xrightarrow{Me_3N} Me_2Si \overset{Me}{\underset{Me}{\diagup}} \overset{Me}{\underset{Me}{\diagdown}} SiMe_2 + (-CMe{=}CMe-SiMe_2-)_n$$

Since 1976 when the first silacyclopropene was obtained from addition of a silylene to an acetylene,[1] this reaction has been widely studied using mild methods for the generation of silylenes. Ishikawa and Kumada reported in 1977 that photogenic phenyltrimethylsilylsilylene undergoes addition to a number of substituted acetylenes.[50] The silacyclopropenes were detected by conversion upon methanolysis to methoxyvinyl silanes. Photochemical rearrangement to ethynyl silanes also occurred.

R	R'
n-Bu	H
Et	Et
Me₃Si	H
Me₃Si	Me₃Si

A more extensive investigation of these reactions has recently been reported.[51] The yield of the methanolysis products from the silacyclopropene formed upon addition of $\phi\ddot{S}iSiMe_3$ to 1-hexyne passed through a maximum of about 27% with increasing irradiation time. The yield of the alkynylsilane resulting from photochemical rearrangement of the silacyclopropene increased monotonically but at a slower rate. Similar results were obtained upon addition to tert-butyl acetylene and trimethylsilylacetylene. These experiments confirm the suggestion by Haas and Ring that the formation of ethynylsilane from addition of thermally generated silylene to acetylene proceeds via the parent silacyclopropene.[52]

$$SiH_2 + HC\equiv CH \rightarrow \left[\overset{SiH_2}{HC=CH} \right] \rightarrow HC\equiv C-SiH_3$$

Ishikawa, Nakagawa, and Kumada found that addition of $\phi\ddot{S}iSiMe_3$ to internal acetylenes 3-hexyne, 1-trimethylsilyl-1-propyne, and 2,2,5,5-tetramethyl-3-hexyne led to silacyclopropenes sufficiently stable for isolation by low-pressure distillation and preparative gas chromatography.[51] Bis-(trimethylsilyl)acetylene gave a silacyclopropene that underwent thermal, as well as photochemical, rearrangement and thus could not be purified, but that reacted only slowly with methanol. The di-tert-butylacetylene adduct is so stable that after exposure to air for 5 min, 30% of the silacyclopropene is recovered.

The preparation of silacyclopropenes has been a synthetic triumph for silylene chemistry. The electronic structure and remarkable stability of silacyclopropenes are, however, still subjects of controversy (see below), as is their involvement in the formation of disilacyclohexadienes in the reactions of silylenes and acetylenes under harsh conditions.[1] Barton has shown that hexamethyl-1,2-disilacyclobut-3-ene can be converted to disilacyclohexadienes by reaction with acetylenes,[53] and Seyferth demonstrated that silacyclopropenes can be converted into disilacyclobutenes by silylene insertion.[54] Sakurai has, however, found that some disilacyclobutenes are not efficiently converted to disilacyclohexadienes in thermal reactions with acetylenes. The following reaction gives only a 1.2% yield, and the disilacyclobutene is recovered in 49% yield.[34]

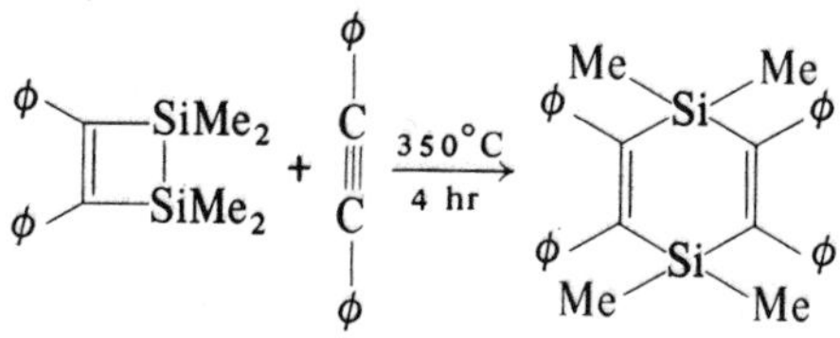

The number of known elementary reactions from which reaction sequences can be constructed for the conversion of two silylene and two acetylene molecules into one molecule of disilacyclohexadiene is increasing. It is still the case, however, that for no single set of reaction conditions is it known which sequence or sequences operate, despite the continuing efforts toward understanding the mechanisms of silylene addition.

c. Addition Reactions of Dihalosilylenes

Dihalosilylenes are quite easily accessible,[1] but have received less mechanistic attention than SiH_2 and the organosilylenes.

The addition of $SiCl_2$ to acetylenes, olefins, and 1,3-dienes resembles the corresponding reactions of SiH_2 and $SiRR'$. Chernyshev has obtained silacyclohexadienes from acetylenes, but from diphenylacetylene some disilacyclobutene was also formed.[55] Therefore a mechanism was suggested in which a 1,1-dichloro-1-silacycloprop-2-ene intermediate could be converted to a disilacyclohexadiene by either dimerization or ring expansion.

$$Cl_2Si: + CH_2{=}CH_2 \rightarrow Cl_2Si{\overset{CH_2}{\underset{CH_2}{\big<}}}_| \rightarrow Cl_2\dot{S}iCH_2\dot{C}H_2 \rightarrow Cl_2SiHCH{=}CH_2$$

Reaction of $SiCl_2$ with ethylene yielded dichlorovinylsilane as the major product (20%) accompanied by $Cl_3SiCH{=}CH_2$ (18%), Cl_3SiEt (16%), $CH_2{=}CHSiCl_2Et$ (20%), and Et_2SiCl_2 (10%). It is obvious that radical reactions are occurring, but the formation of $Cl_2SiHCH{=}CH_2$ was attributed to an addition–rearrangement mechanism.

$$Cl_3SiSiCl_3 \xrightarrow{\Delta} :SiCl_2 + SiCl_4$$

$$:SiCl_2 + RC{\equiv}CR \longrightarrow Cl_2Si\underset{CR}{\overset{CR}{\|}} \xrightarrow{\text{dimerization}} Cl_2Si\underset{\underset{R}{C}=\underset{R}{C}}{\overset{\overset{R}{C}=\overset{R}{C}}{}}SiCl_2$$

Addition of $SiCl_2$ to 1,3-butadienes led to greater than 90% yields of 1,1-dichloro-1-silacyclopent-3-enes, accompanied by small amounts of the 2-ene isomers.

R	R'
H	H
H	CH_3
CH_3	H
H	Cl

There has been a strong surge of recent interest in the chemistry of difluorosilylene.[56] In a series of pioneering studies Margrave, Timms, and co-workers found that SiF_2 is rather long lived in the gas phase, with a half-life of about 100 sec, which was not appreciably shortened in the presence of various potential reactants.[1, 57, 58] Reactions with various substrates were believed to occur on cocondensation and be initiated by dimerization of SiF_2. This view was in accord with the observation that almost all the stable reaction end products contained more than one SiF_2 unit.

Recently Seyferth has proposed a reinterpretation of difluorosilylene reaction mechanisms, suggesting that most products of SiF_2 with olefins and acetylenes involve initial formation of three-membered ring intermediates followed by dimerization or ring expansion by further reactions with SiF_2.[59] These suggestions were based on the synthesis of 1,1-difluoro-2,2,3,3-tetramethyl-1-silirane, whose reactions with olefins, dienes, and acetylenes are in accord with the suggested diradical, ring-opened intermediate. In the absence of added reaction substrates, the dimeric products are of the sort obtained from difluorosilylene.

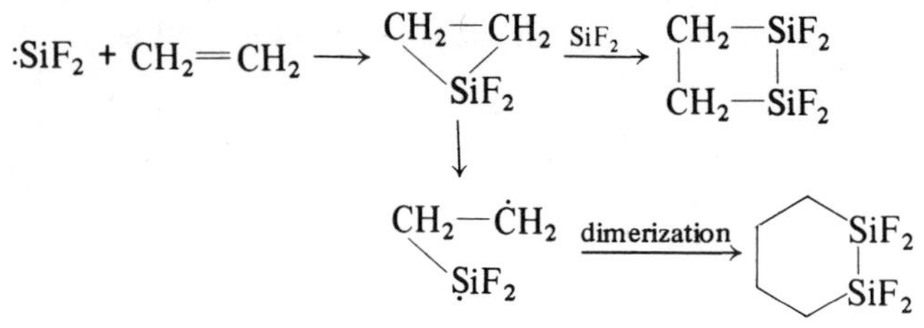

:SiF$_2$ + CH$_2$=CH$_2$ $\longrightarrow$ CH$_2$—CH$_2$ / SiF$_2$ $\xrightarrow{\text{SiF}_2}$ CH$_2$—SiF$_2$ | CH$_2$—SiF$_2$

$\downarrow$

CH$_2$—ĊH$_2$ / ṠiF$_2$ $\xrightarrow{\text{dimerization}}$ [ring] SiF$_2$ SiF$_2$

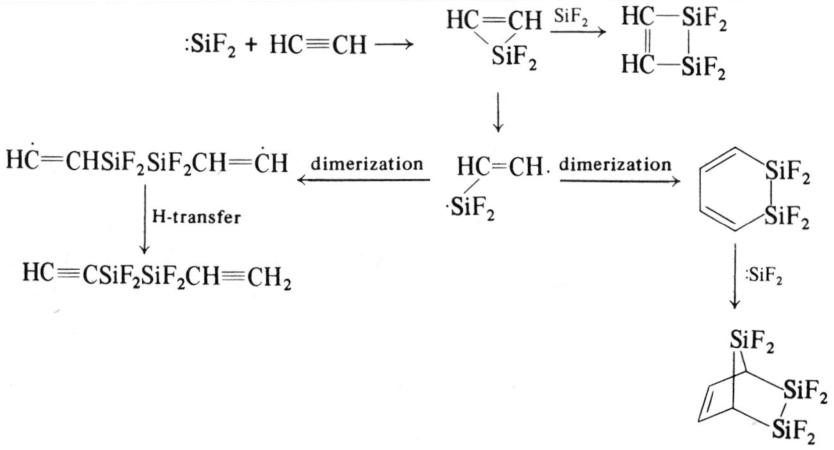

:SiF$_2$ + HC≡CH $\longrightarrow$ HC=CH / SiF$_2$ $\xrightarrow{\text{SiF}_2}$ HC—SiF$_2$ ‖ HC—SiF$_2$

$\downarrow$

HĊ=CHSiF$_2$SiF$_2$CH=ĊH $\xleftarrow{\text{dimerization}}$ HC=CH· / ṠiF$_2$ $\xrightarrow{\text{dimerization}}$ [ring] SiF$_2$ SiF$_2$

$\downarrow$ H-transfer

HC≡CSiF$_2$SiF$_2$CH=CH$_2$

$\downarrow$:SiF$_2$

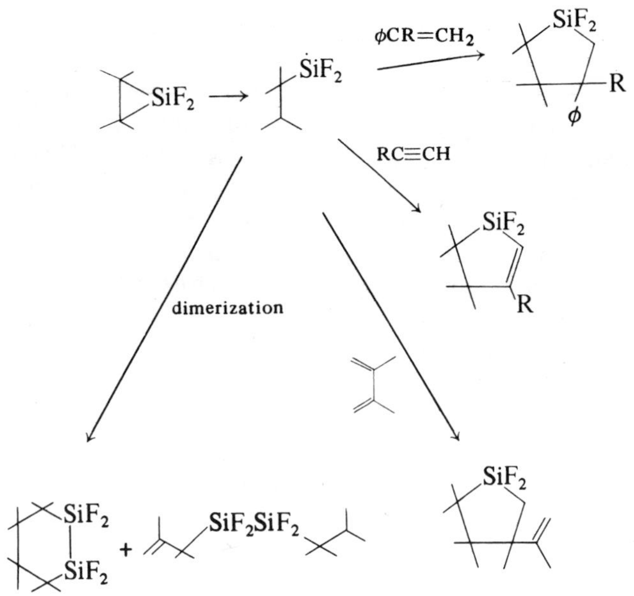

SiF$_2$ SiF$_2$ SiF$_2$ [bicyclic]

[cyclopropane]SiF$_2$ $\longrightarrow$ [ ]–SiF$_2$ · $\xrightarrow{\phi CR=CH_2}$ SiF$_2$ R ϕ

$\xrightarrow{RC≡CH}$ SiF$_2$ R

dimerization

SiF$_2$ SiF$_2$ + SiF$_2$SiF$_2$

SiF$_2$ [ring with exocyclic =CH$_2$]

360

Despite Seyferth's suggestion that SiF_2 can react directly with unsaturated substrates, the case for dimerization of SiF_2 prior to reaction with added substrates has been reiterated, together with a reminder that insertion reactions of monomeric SiF_2 also seem to occur.[60] As evidence *against* initial attack on unsaturates by monomeric SiF_2, Liu and Hwang point to the fact that all products containing two SiF_2 units have them adjacent, as expected if dimerization of SiF_2 precedes addition.[61] The fact that Seyferth obtained dimeric products with adjacent SiF_2 units from a 1,1-difluoro-1-silirane does not entirely settle the issue. The products Seyferth obtained from difluorosilirane and excess unsaturates contain one SiF_2 unit, not two as contained in the products of SiF_2 cocondensation.

It is clear that the mechanism of silirane thermolysis needs detailed elucidation. If a diradical intermediate is formed from hexamethylsilirane, as well as from the 1,1-difluoro-2,2,3,3-tetramethylsilirane, then it might act as a silylenoid, and free dimethylsilylene may not be formed.[1,62] Conversely, the diradical from the difluorosilirane may under some circumstances act as a chicken rather than an egg, and liberate SiF_2.

In favor of Seyferth's suggestion that monomeric SiF_2 can undergo addition to olefins is the finding that SiF_2 units are *not* joined to each other in the polymeric material that constitutes the major product of cocondensation of SiF_2 and propylene.[63] Although this is also the predominant structure for the polymer

$$[-\underset{\underset{CH_3}{|}}{CH}-CH_2-SiF_2-]_n$$

from SiF_2 and *cis*- and *trans*-2-butene, there is present a minor component containing $-SiF_2-SiF_2-$ units.

It has been proposed that even the apparent insertions of $(SiF_2)_n$ ($n = 1, 2, 3 \ldots$) into carbon–fluorine and carbon–chlorine bonds of haloolefins are actually additions followed by rearrangement of diradical intermediates.[61,64] Such a mechanism rationalizes both the formation of geometric isomers of the "insertion" products and the formation of cyclic products from nonhalogenated olefins. The products of cocondensation of SiF_2 and vinyl chloride[61] can be explained by a combination of this mechanism and the one suggested by Seyferth.[59]

Additions of nucleogenic $^{31}SiF_2$ (see above) to conjugated dienes have recently been reported by Tang and coworkers.[11-13] Both singlet and triplet $^{31}SiF_2$ are believed to be formed, but triplet $^{31}SiF_2$ is thought to undergo addition reactions only via donor complexes with such paramagnetic molecules as NO and O_2. Methyl groups in the diene substrate were found to hinder the addition reaction.

$$^{31}PF_3 + n \longrightarrow {}^{31}Si + p + 3\,F$$

$$^{31}Si + PF_3 \longrightarrow \longrightarrow {}^{31}SiF_2$$

R	R'
H	H
trans-Me	H
cis-Me	H
H	Me

d. Reactions of Silylenes with Carbonyl Compounds

It is only recently that the reactions of silylenes with carbonyl compounds have been investigated. Although many mechanistic questions remain unanswered, these reactions hold synthetic as well as mechanistic interest.

Ando and co-workers have obtained products from the thermal generation of dimethylsilylene in the presence of carbonyl compounds that could be attributed to the addition of the silylene to the C=O bond with the formation of oxysiliranes followed by rearrangement.[65] Aliphatic ketones afforded enol ether products also attributed to oxasilirane intermediates,[65,66] while the formation of styrene from acetophenone was attributed to carbenic fragmentation of the oxasilirane.[65] This mechanism is certainly not uniquely demanded by the data, however. Initial ylid formation leads to the same product.

Under milder conditions more direct evidence was found for the intermediacy of oxasiliranes.[66] When 2-adamantanone and 7-norbornone were allowed to react

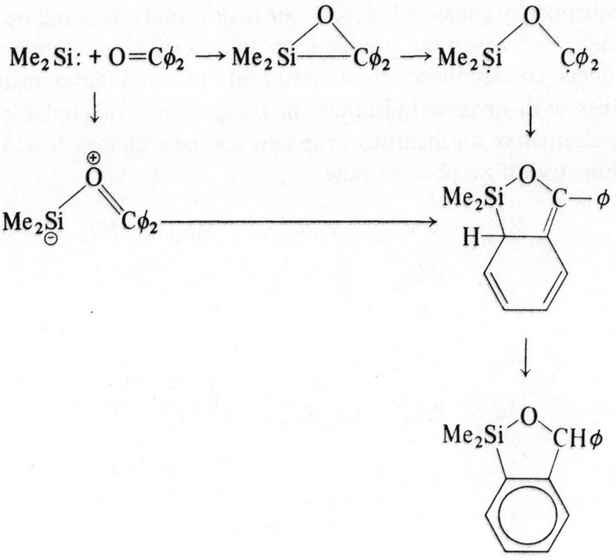

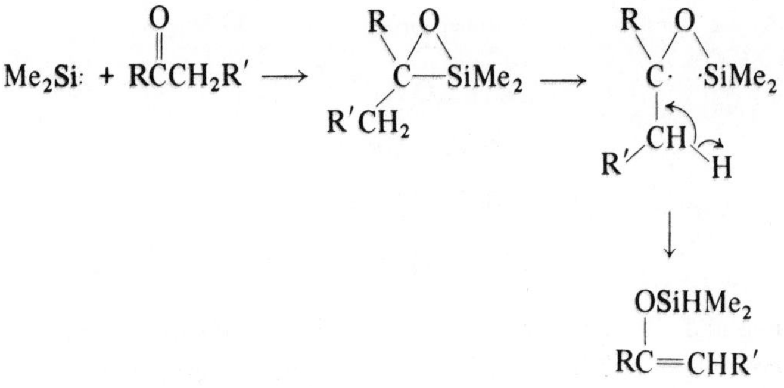

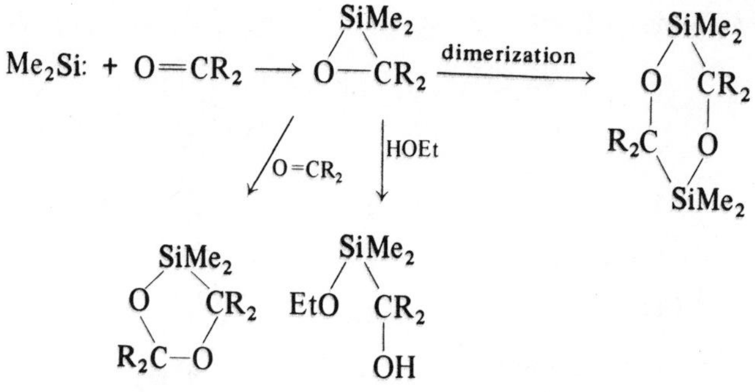

Me$_2$Si: + O=Cϕ_2 → Me$_2$Si—Cϕ_2 (epoxide, O bridge) → Me$_2$\overset{\cdot}{Si}$ \overset{O}{\underset{}{}}$ \overset{\cdot}{C}\phi_2$

Me$_2$Si$^{\ominus}$—$\overset{\oplus}{C}\phi_2$ (with O) ⟶ Me$_2$Si $\overset{O}{\underset{H}{}}$ C—ϕ (bicyclic diene)

↓

Me$_2$Si $\overset{O}{\underset{}{}}$ CHϕ (benzo-fused)

Me$_2$Si: + R$\overset{O}{\overset{\|}{C}}CH_2$R′ ⟶ $\overset{R}{\underset{R′CH_2}{}}$C—SiMe$_2$ (O epoxide) ⟶ $\overset{R\ \ O}{C}$ SiMe$_2$, R′$\overset{}{CH}$—H

↓

OSiHMe$_2$
|
RC=CHR′

Me$_2$Si: + O=CR$_2$ ⟶ O—CR$_2$ $\overset{SiMe_2}{}$ $\xrightarrow{\text{dimerization}}$ R$_2$C $\overset{SiMe_2}{\underset{O}{O}}$ CR$_2$, SiMe$_2$

↙ O=CR$_2$ ↓ HOEt

R$_2$C—O $\overset{SiMe_2}{\underset{}{}}$ O CR$_2$ EtO $\overset{SiMe_2}{}$ CR$_2$, OH

363

with photochemically generated $Me_2Si:$, the major products could be formulated as the dimers and insertion products of the oxasiliranes. Seyferth has also found products corresponding to dimerization of oxasiliranes in the reaction of phosphines with hexamethylsilirane in the presence of aliphatic ketones.[67] One of the alternative mechanisms proposed is a new silylene reaction–addition to a phosphine yielding a phosphorane.

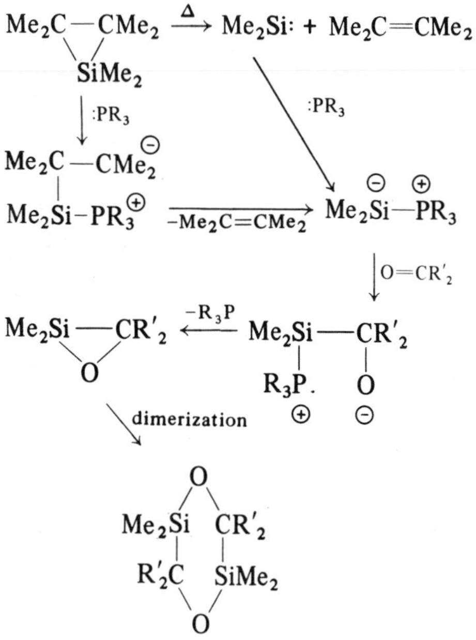

$$Me_2C-CMe_2 \xrightarrow{\Delta} Me_2Si: + Me_2C=CMe_2$$

with $SiMe_2$ bridging, reacting with $:PR_3$

$$Me_2C-CMe_2^{\ominus}$$
$$|$$
$$Me_2Si-PR_3^{\oplus} \xrightarrow[-Me_2C=CMe_2]{} Me_2Si^{\ominus}-PR_3^{\oplus}$$

with $:PR_3$ path from $Me_2C=CMe_2$

$$\downarrow O=CR'_2$$

$$Me_2Si-CR'_2 \xleftarrow{-R_3P} Me_2Si-CR'_2$$
$$\backslash O \diagup \qquad\qquad | \qquad |$$
$$\qquad\qquad\qquad R_3P^{\oplus} \quad O^{\ominus}$$

$$\diagdown \text{dimerization}$$

$$O$$
$$Me_2Si \quad CR'_2$$
$$| \qquad |$$
$$R'_2C \quad SiMe_2$$
$$O$$

Ishikawa and Kumada favored the intermediacy of carbonyl ylids in the reactions of photogenic $\phi SiSiMe_3$ with both aliphatic ketones and methyl methacrylate.[68] Alternative paths involving oxasiliranes were also given.

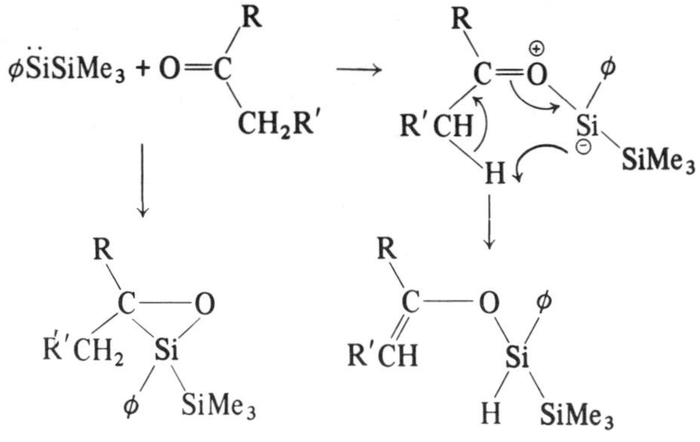

$$\phi\ddot{S}iSiMe_3 + O=C\diagup^R_{CH_2R'} \longrightarrow$$

$$R'CH \overset{R}{\underset{H}{\diagdown}} C=O^{\oplus} \quad \phi$$
$$Si^{\ominus} SiMe_3$$

$$\downarrow$$

$$R \diagdown C-O$$
$$R'CH_2 \diagup Si \diagdown$$
$$\phi \quad SiMe_3$$

$$R \diagdown C-O \quad \phi$$
$$R'CH \diagup | \quad Si \diagdown$$
$$H \quad SiMe_3$$

Ando and Ikeno have allowed dimentylsilylene to react with α-diketones and obtained 1,3-dioxa-2-silacyclopent-4-ene derivatives.[69]

$$(Me_2Si)_6 \xrightarrow{h\nu} Me_2Si:$$

$$Me_2Si: + R^1C\!\!-\!\!CR^2 \longrightarrow Me_2Si$$

R^1	R^2	Yield (%)
Me	Me	41
Et	Me	38
n-Pr	n-Pr	45
i-Pr	i-Pr	67

C. Abstraction by Silylenes

Weber and co-workers have reported the deoxygenation of dimethyl sulfoxide by reaction with silylenes. When dimethylsilylene is generated photochemically in the presence of Me_2SO, the formation of dimethylsilanone is indicated by its insertion into hexamethylcyclotrisiloxane.[41]

$$(Me_2Si)_6 \xrightarrow{h\nu} (Me_2Si)_5 + Me_2Si:$$

$$Me_2Si: + \overset{\ominus}{O}\!\!-\!\!\overset{\oplus}{S}Me_2 \longrightarrow Me_2Si\!\!=\!\!O + SMe_2$$

$$Me_2Si{=}O \ + \ Me_2Si\overset{O}{\underset{\underset{SiMe_2}{O \quad O}}{\diagup\diagdown}}SiMe_2 \ \longrightarrow \ Me_2Si\overset{O-SiMe_2}{\underset{\underset{Me_2Si-O}{O \qquad SiMe_2}}{\diagup\diagdown}}$$

Since dimethylsilanone is also formed via *non*-silylene reactions of dimethyl sulfoxide with several silylene precursors,[70-73] participation by silylenes in this oxygen abstraction cannot be determined from the product alone.

$$Me_2\overset{\oplus}{S}{-}\overset{\ominus}{O} \ + \ Me_2Si\overset{SiMe_3}{\underset{SiMe_3}{\diagup\diagdown}} \ \longrightarrow \ Me_2S \ + \ O{=}SiMe_2 \ + \ Me_3SiC{\equiv}CSiMe_3$$

$$Me_2\overset{\oplus}{S}{-}\overset{\ominus}{O} \ + \ Me_2Si\diagdown \ \longrightarrow \ Me_2S \ + \ O{=}SiMe_2 \ + \ Me_2C{=}CMe_2$$

$$Me_2S{-}O \ + \ \overset{Ar}{\underset{|}{{-}\underset{|}{Si}{-}\underset{|}{Si}{-}}} \ \overset{h\nu}{\longrightarrow} \ Me_2S \ + \ O{=}Si\diagup \ + \ Ar\underset{|}{Si}{-}$$

When heptamethyl-2-phenyltrisilane is irradiated in the presence of dimethyl sulfoxide, two different silanones are formed, one via a silylene abstraction.[74] In all these reactions the silanone is itself a reactive intermediate trapped by reaction with siloxanes, siliranes, or silirenes.

$$(Me_3Si)_2SiMe\phi \ \overset{h\nu}{\longrightarrow} \ Me_3SiSiMe_3 \ + \ Me\ddot{S}i\phi$$

$$Me\ddot{S}i\phi \ + \ \overset{\ominus}{O}{-}\overset{\oplus}{S}Me_2 \ \longrightarrow \ Me\phi Si{=}O \ + \ SMe_2$$

$$Me_2\overset{\oplus}{S}{-}\overset{\ominus}{O} \ + \ (Me_3Si)_2SiMe\phi \ \overset{h\nu}{\longrightarrow}$$

$$[Me_2S\cdots O\cdots\overset{Me}{\underset{\underset{SiMe_3}{|}}{\overset{.\phi.}{Si}}}\cdots SiMe_3]^{\ddagger} \ \longrightarrow \ Me_2S \ + \ O{=}\overset{Me}{\underset{SiMe_3}{Si}} \ + \ \phi SiMe_3$$

The deoxygenation of dimethyl sulfoxide by silylenes may well be a two-step process:

$$RR'Si{:} \ + \ \overset{\ominus}{O}{-}\overset{\oplus}{S}Me_2 \ \longrightarrow \ RR'\overset{\ominus}{Si}{-}O{-}\overset{\oplus}{S}Me_2 \ \longrightarrow \ RR'Si{=}O \ + \ SMe_2$$

D.   Silylene Dimerization

While the dimerization of SiF_2 discussed earlier in this chapter remains controversial, the dimerization of organosilylenes discovered by Conlin is now well established.[1,75] Sakurai and co-workers have obtained the expected anthracene Diels-Alder adducts of a number of disilenes following the generation of silylenes in solution by thermal extrusion.[76]

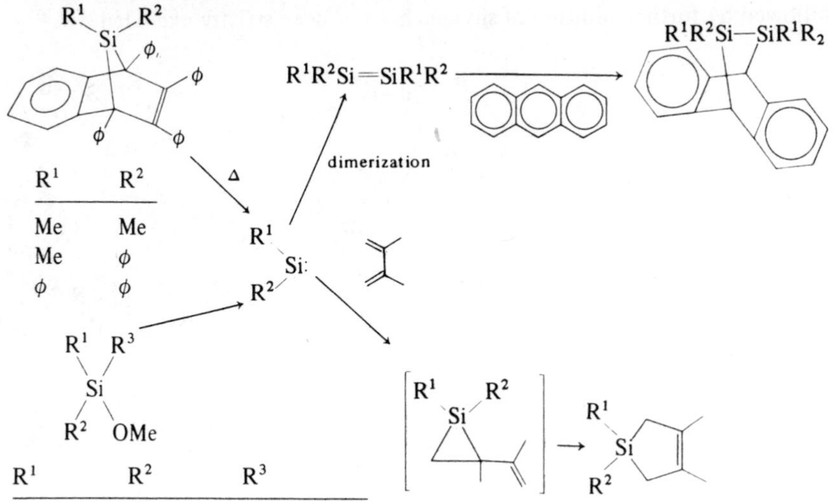

$R^1R^2Si=SiR^1R^2 \longrightarrow$

$R^1R^2Si-SiR^1R_2$

dimerization

R^1	R^2
Me	Me
Me	ϕ
ϕ	ϕ

R^1	R^2	R^3
Me	Me	$SiMe_2OMe$
Me	$SiMe_3$	$SiMe_3$
$SiMe_3$	$SiMe_3$	$SiMe_3$

The same anthracene adduct is obtained, *stereospecifically*, when the preformed disilene is produced in the presence of anthracene.[77]

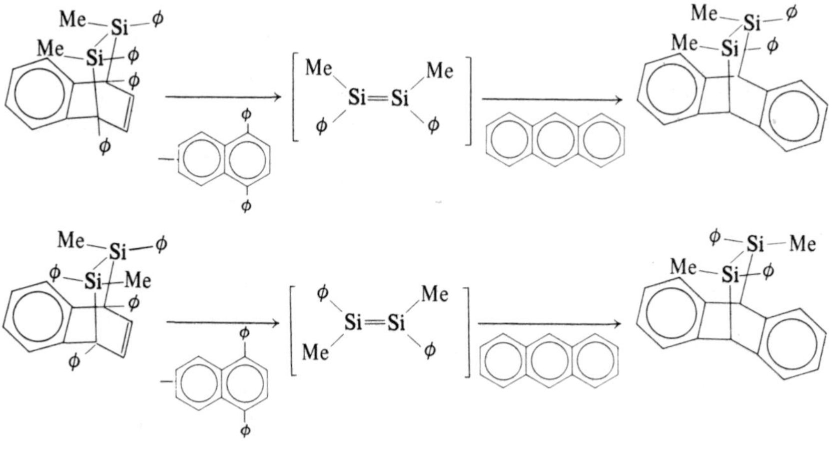

If silylene dimerization, the stereochemical integrity of the Si=Si double bond, and a concerted cycloaddition of disilene and anthracene are all accepted, the formation of an equal amount of the *cis*- and *trans*- anthracene adducts upon generation of MeSiϕ implies that no stereochemical preference exists in the dimerization of this silylene.

Of course, one could imagine an alternative mechanism to dimerization for the formation of these anthracene adducts. Addition of a silylene to anthracene followed by further addition of silylene has not been strictly excluded.

Strong evidence was found by Chen for the occurrence of dimerization when bis(trimethylsilyl)silylene is generated in solution in the absence of an added trapping reagent.[15] An oligomer is obtained that, whatever the details of its formation, requires the intermediacy of the silylene dimer, if it is conceded that trimolecular collisions of silylenes are unimportant. It is believed that the oligomer is the cyclotetrasilane indicated.

E. Silylene Rearrangements

Barton and co-workers have found that silylenes undergo rapid rearrangement in the gas phase.[78] This is one of the most interesting phenomena in the chemistry of silylenes to have been discovered in the past few years and has ramifications that will stimulate activity both on the blackboard and in the laboratory for some time to come.

When Wolff, Goure, and Barton generated $MeSiSiMe_3$ by vacuum-flow pyrolysis,[78] the products suggested that rearrangement had occurred rather than the dimerization found in solution by Sakurai.[76] To explain this result, these

$$(Me_3Si)_2SiClMe \xrightarrow[-Me_3SiCl]{\Delta} \underset{Me}{\overset{MeSi}{\diagdown}} Si: \xrightarrow{\;\;} Me_2Si \triangle SiH_2 + MeHSi \triangle SiHMe$$

$$ 28\% 15\%$$

workers proposed a novel sequence of reactions including: (1) intramolecular C—H insertion of the initial silylsilylene to give as a short-lived intermediate a highly strained disilacyclopropane; (2) competitive α-eliminations generating β-silylsilylenes; and (3) further intramolecular C—H insertions to give the observed 1,3-disilacyclobutanes. This was the first suggestion of C—H insertion by a silylene, and alkyl group shifts in the course of silylene extrusions are also unprecedented.[57]

This process involves a new silylene reaction, a silylene-to-silylene rearrangement. Carbene-to-carbene rearrangements are well known, particularly in the phenylcarbene[79,80] and vinylmethylene[80] series, and also occur via three-membered ring intermediates. In these reactions, however, intramolecular additions play the pivotal role played by insertions in the newly discovered silylene rearrangements.

$$\underset{C_bH=C_cH_2}{\overset{C_aH:}{\diagup}} \rightleftarrows \underset{C_bH-C_cH_2}{\overset{C_aH}{\diagup}} \rightleftarrows \underset{C_bH:^\diagdown C_cH_2}{\overset{C_aH}{\diagdown}}$$

The same products were obtained from gas-phase reactions of MeS̈iSiMe$_3$ as had previously been found from the extrusion of tetramethyldisilene in the absence of a trapping reagent.[75,81] Barton therefore made the daring suggestion

that in the gas phase tetramethyldisilene can rearrange to methyltrimethylsilylsilyene.

$$Me_2Si=SiMe_2 \xrightarrow{\;\;\text{hv}\;\;} Me\ddot{S}i-SiMe_3$$

This process is the reverse of the usual carbene rearrangement. There are,

however, high-energy olefin-to-carbene photochemical rearrangements.[82]

Barton's mechanism for rearrangement of tetramethyldisilene to 1,3-disilacyclobutanes offers a plausible alternative to the series of radical rearrangements and couplings first proposed by Roark and Peddle.[81] Barton provided strong evidence for the disilene-to-silylene rearrangement by trapping MeS̈iSiMe$_3$ following the generation of Me$_2$Si=SiMe$_2$. No cycloadducts of Me$_2$Si=SiMe$_2$ could be trapped upon direct generation of MeS̈iSiMe$_3$, and thus the rearrangement from the silylsilylene back to the disilene must be much slower than its inverse. Theoretical evidence is presented below for the greater stability of silylsilylene relative to disilene.

Further support for the occurrence of C—H insertion-mediated silylene rearrangements has been found by Chen et al.[83] When bis(trimethylsilyl)silylene is generated in the gas phase, the major product is 1,1,4,4-tetramethyl-1,2,4-trisilacyclopentane. This deep-seated rearrangement can be rationalized by

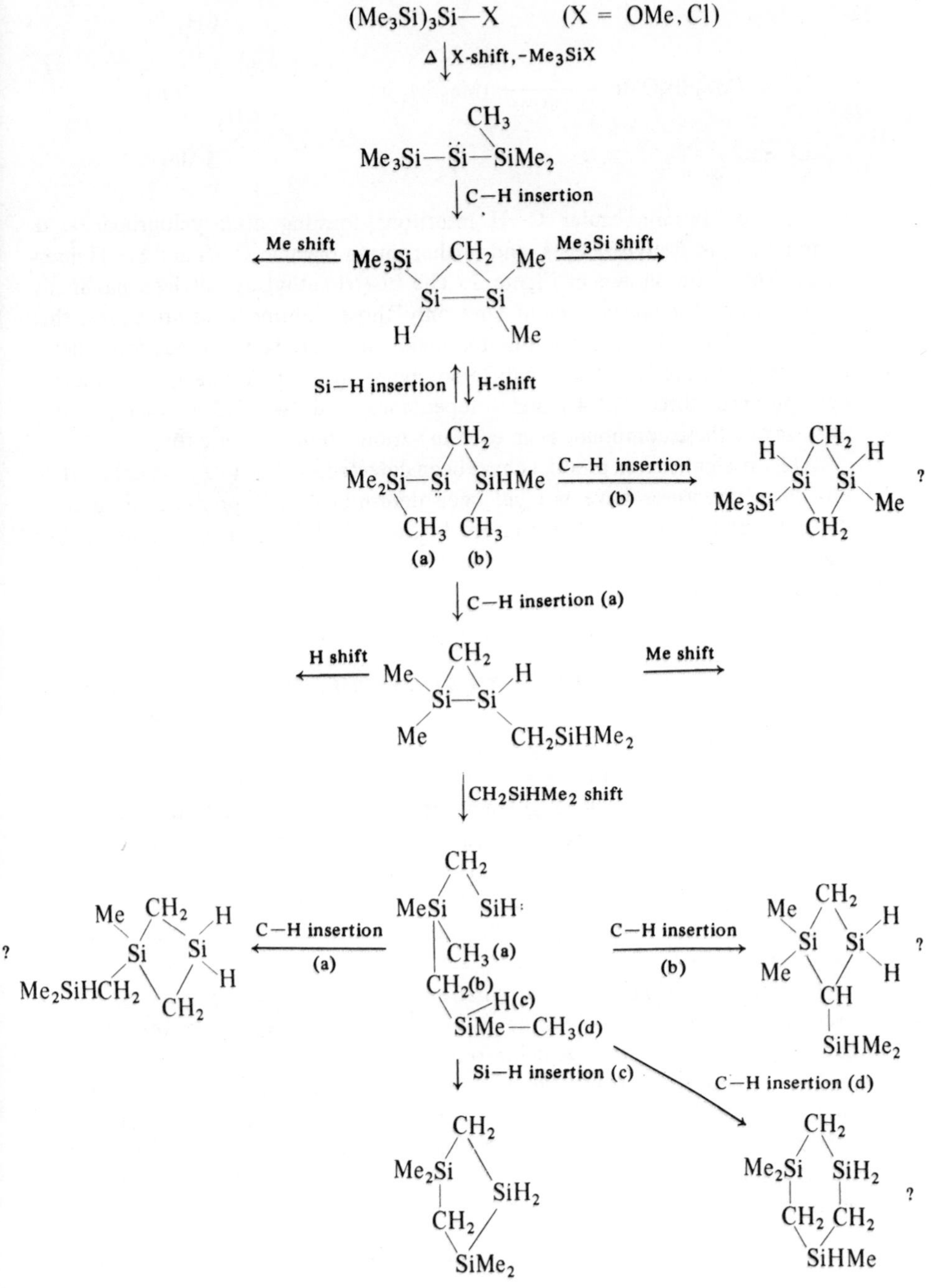

(Me$_3$Si)$_3$Si—X   (X = OMe, Cl)

Δ ↓ X-shift, –Me$_3$SiX

CH$_3$
Me$_3$Si—Sï—SiMe$_2$

↓ C—H insertion

← Me shift   Me$_3$Si   CH$_2$ Me   Me$_3$Si shift →
                      Si——Si
                  H$^{\prime}$ Me

Si—H insertion ↑↓ H-shift

CH$_2$
Me$_2$Si—Sï   SiHMe   C—H insertion   H   CH$_2$ H
         CH$_3$ CH$_3$ (b)   Me$_3$Si—Si   Si—Me   ?
          (a)   (b)   CH$_2$

↓ C—H insertion (a)

← H shift   Me   CH$_2$ H   Me shift →
                   Si——Si
               Me   CH$_2$SiHMe$_2$

↓ CH$_2$SiHMe$_2$ shift

                    CH$_2$
Me   CH$_2$ H   MeSi   SiH·   Me   CH$_2$ H
   Si   Si   C—H insertion CH$_3$(a)   C—H insertion   Si   Si   ?
? Me$_2$SiHCH$_2$ CH$_2$ (a)   CH$_2$(b)   (b)   Me   H
                              H(c)   CH
               SiMe—CH$_3$(d)   SiHMe$_2$

↓ Si—H insertion (c)   C—H insertion (d) →

        CH$_2$ CH$_2$
Me$_2$Si   SiH$_2$ Me$_2$Si   SiH$_2$
     CH$_2$ CH$_2$ CH$_2$?
        SiMe$_2$ SiHMe

FIGURE 1

371

$$(Me_3Si)_3OMe \xrightarrow[-Me_3SiOMe]{\Delta} (Me_3Si)_2Si: \xrightarrow{\Delta} \begin{array}{c} CH_2 \\ Me_2Si \quad \backslash \\ | \qquad SiH_2 \\ CH_2 \quad / \\ \backslash SiMe_2 \end{array}$$

a series of intramolecular C—H insertions forming disilacyclopropanes, α-eliminations to β-silylsilylenes, and further intramolecular C—H and Si—H insertions. These are shown in Figure 1. The bis(trimethylsilyl)silylene has many opportunities for rearrangement, and only those intermediates are shown that are required for the formation of the major product. The full reaction scheme predicts the formation of eleven different cyclic structural isomers, six 1,3-disilacyclobutanes, three 1,2,4-trisilacyclopentanes, and two 1,3,5-trisilacyclohexanes. Six of these compounds can exist in various stereoisomeric forms.

At least eight minor products have been detected, each with a yield below 10%, but their structures have not yet been determined. The product distribution depends strongly on the temperature in the 500-700°C range examined,* but not on the silylene precursor. Similar results were obtained with $(Me_3Si)_3SiCl$ as the silylene precursor.

IV. THEORETICAL CALCULATIONS

The relative stability of silylenes and isomeric π-bonded silicon compounds is a problem with which theoreticians have begun to grapple. That the interconversion of tetramethyldisilene and methyltrimethylsilylsilylene favors the silylene suggests that the silylene is more stable. While this order of stability stands in marked contrast to the thermochemistry of analogous hydrocarbons, it should not be entirely unexpected in light of the relatively low heats of formation of silylenes relative to carbenes, and the much greater stability of C=C compared to Si=C and Si=Si double bonds.

Although experimental determinations of the thermochemistry of short-lived silicon compounds continue to be difficult and hence rare, theoreticians have begun to provide the needed information.

Snyder and Wasserman have employed an *ab initio* SCF-MO calculation to predict that singlet silylsilylene $SiH_3SiH:$ is 8.6 kcal/mol more stable than planar $SiH_2=SiH_2$, which is in turn 2.5 kcal/mol more stable than triplet $SiH_3SiH:$.[84] A *trans*-bent nonplanar $SiH_2=SiH_2$ is predicted to be about 0.5 kcal/mol more stable than the planar form.

*Flash vacuum pyrolysis conditions are employed, with residence times in the hot zone less than 100 msec.

$$\angle = 12.9° \quad\ce{Si=Si}\quad \angle = 12.9°$$

(Disilene structure with H atoms, $\angle = 12.9°$ on each side)

The comparison of singlet and triplet electronic states included an estimate by Meadows and Shaefer that such a single-configuration SCF-MO wave function underestimates the stability of singlet SiH_2 relative to triplet SiH_2 by 15.4 kcal/mol.[85]

The interconversion of methylsilylene, silaethylene, and silylmethylene $H\ddot{S}i-CH_3 \rightleftharpoons H_2Si=CH_2 \rightleftharpoons H_3Si-\ddot{C}H$ has also been investigated theoretically. Gordon has studied both singlet and triplet states by a STO-4G SCF calculation.[86] The results are given in Table 9.2.

Of the three isomers, methylsilylene is predicted to be the most stable singlet ($\angle H-Si-C = 92.9°$) and silylmethylene the most stable triplet. The singlet–triplet separation for silylmethylene (about 43 kcal/mol) is predicted to be much *larger* than that for methylmethylene, so silyl substitution seems to preferentially stabilize the triplet carbene, while methyl substitution seems to preferentially stabilize the singlet carbene and silylene.

The silaethylene structure seems *not* to be most stable on either the singlet or triplet potential surface and this would suggest that isomerization from singlet $H_2Si=CH_2$ to singlet $H_3Si-\ddot{C}H$ and triplet $H_2Si=CH_2$ to triplet $H_3Si-\dot{C}H \cdot$ should be favored rather than the reverse reactions. However, the isomerization of singlet $H_3Si-\ddot{C}H$ to singlet $H_2Si=CH_2$ is predicted to be energetically downhill and has been observed experimentally in substituted form.[87,88,104,105]

$$(CH_3)_3Si-CH: \longrightarrow (CH_3)_2Si=CHCH_3$$

Strausz and co-workers have carried out open shell SCF-MO calculations on the interconversions of the triplet states of $H\ddot{S}iCH_3$, $H_2Si=CH_2$, and

TABLE 9.2.   Relative Energies Calculated for Methylsilylene, Silaethylene and Silylmethylene (kcal/mol)[86]

		Without d-Orbitals			With d-Orbitals		
		No CI	2CI	37 CI	No CI	2CI	37 CI
$H\ddot{S}i-CH_3$	Singlet	0	0	0	0		
	Triplet	21.78	25.46		24.11	28.46	
$H_2Si=CH_2$	Singlet	23.16	2.26	12.48	9.18	1.72	6.80
	Triplet	24.00	27.68		14.36	18.71	
$H_3Si-\ddot{C}H$	Singlet	66.98	64.64	70.06	50.28	49.60	54.63
	Triplet	21.68	25.36		4.88	9.23	

$H_3Si-\ddot{C}H$.[89,90] They found that triplet silaethylene is 0.8 kcal/mol less stable than its triplet silylene isomer and 4.6 kcal/mol less stable than its triplet carbene isomer. Very large barriers were found for interconversion of the triplet isomers, 83.0 kcal/mol for $H\ddot{S}i-CH_3 \rightleftarrows H_2Si=CH_2$ and 92.5 kcal/mol for $H_2Si=CH_2 \rightleftarrows H_3Si-\ddot{C}H$. Optimized geometries and absolute energies are shown in Figure 2.

SCF calculations have also been carried out on the five structural isomers of C_2SiH_4, and the following relative energies have been found:[91]

	$CH_2=Si=CH_2$	$\overset{\ddot{S}i}{\underset{CH_2-CH_2}{\diagup\diagdown}}$	$\overset{SiH_2}{\underset{CH=CH}{\frown}}$	$CH_3Si\equiv CH$	$SiH_3C\equiv CH$
Relative energies: (kcal/mol)	46	17	17	61	0

It is rather surprising that the silacyclopropylidene was predicted to be as stable as silacyclopropene, since tetramethylsilacyclopropene can be kept in

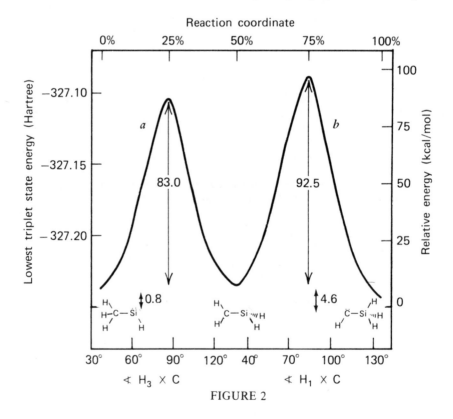

FIGURE 2

solution at room temperature for months.[92] The first excited state of silacyclo-propylidene, a triplet, was predicted to lie 34.6 kcal/mol above the ground singlet, and a first excited singlet was calculated to lie 73.5 kcal/mol above the ground singlet.[91] In contrast, Meadows and Schaefer found a 10 kcal energy difference between the ground-state singlet and lowest triplet for SiH_2.[85]

Analysis of the overlap populations indicated that there is little pseudo-aromatic delocalization of the π-electrons in silacyclopropene of the kind predicted by Jones in a CNDO semiempirical calculation.[93]

Semiempirical calculations that have proved to be extremely successful in predicting the heats of formation, molecular geometries, dipole moments, and internal rotation barriers for molecules containing first-row elements have been extended to the second row of the periodic table. Dewar obtained a value for $\Delta H_f = 60.3$ kcal/mol for SiH_2 in its ground singlet state by the MINDO/3 formalism[94] and Gordon and co-workers calculated for SiF_2 an Si—F bond length of 1.79Å and an F—Si—F bond angle of 96.2° by a minimal basis INDO calculation.[95]

A useful model has been provided by Harrison, Liedtke, and Liebman for the effects of substituents on the singlet–triplet separations of carbenes, silylenes, and isoelectronic species.[96] The authors concentrate on the differences in energy between the orbitals containing the unshared electrons—the in-plane, s-weighted σ-orbital, and the out-of-plane, pure p-orbital (1). The more electronegative is

1

the substituent Z, relative to M, the greater is the s-character of the σ-orbital, since the concomitant decrease in the s-character of the M—Z bonding MO enhances bond polarization toward Z. The increasing s-character of the σ non-bonding orbital increases the energy difference between it and the p-orbital, thus favoring the σ^2 configuration for the unshared electrons. Thus the singlet state is favored over the triplet state for which the σp promoted configuration is important. This argument rationalizes the singlet ground state for SiH_2 and the triplet ground state for CH_2.

V. PHYSICAL MEASUREMENTS

The first report of successful matrix isolation of silylenes appeared with little fanfare in 1977.[8] When silicon vapor produced by thermal evaporation at 1500-

TABLE 9.3. ESR Data of SiN_2 and SiCO in Their $^3\Sigma$ Ground States in Various Matrices at 4 K

Molecule	Matrix	x_2 or xy_2 Line (G)	y_2 Lines (G)	x_2 Line Half-width (G)	ν (GHz)	D (cm^{-1})	E (cm^{-1})	$A_\perp$ (G)
$Si^{14}N_2$	Ar	9503	($\sim$9559)	12	9.3900	2.206	(0.000 87)	60 ± 1 (^{14}N)
	Ne	9632		56	9.3890	2.276		
	N_2	9706		15	9.3880	2.343	0.001 52	
$Si^{15}N_2$	Ar	9509	9806		9.3910	2.209		7.4 ± 1 (^{15}N)
	Ne	9632			9.3890	2.276		
$^{28,29}Si^{15}N_2$	Ar				9.3895			34 ± 1 (^{29}Si)
$Si^{12}CO$	Ar	9488	($\sim$9584) 9627	5	9.3890	2.236	0.002 15	
	CO	9863		38	9.3865	2.402		
$Si^{13}CO$	Ar	9491	($\sim$9588) 9630	10	9.3900	2.237	0.002 15	<5 (^{13}C)
	Ne	9720		70	9.3910	2.333		
$^{28,29}SiCO$	Ar	9489	9627		9.3880	2.236	0.002 15	30 ± 1 (^{29}Si)

1900°C was trapped in an argon matrix containing N_2 at 4 K, a broad ESR signal was detected, and this was attributed to a triplet state of :SiN_2. A linear structure with a terminal silicon was proposed to account for the strong features of the ESR spectrum, but a small amount of bent SiN_2 was also believed present. The ESR data are given in Table 9.3. The ESR spectrum depended on the nature of the matrix, and in a N_2 matrix the slightly bent molecule predominates. Similar experiments with CO yielded ESR spectra for SiCO that were interpreted in terms of a bent triplet molecule. Various isotopic varients were examined for both SiN_2 and SiCO. From ultraviolet absorption spectra and infrared spectra of the trapped species, vibrational frequencies and force constants were deduced and these are given in Table 9.4. The small Si—N and Si—C force constants indicate quite weak bonds between the silicon atom and its ligands. The silicon analogue of carbon suboxide $Si(CO)_2$ was observed in the IR spectrum.

The occurrence of a triplet ground state for both SiN_2 and SiCO was verified by semiempirical molecular orbital calculations, and calculated spin densities were in reasonable agreement with experiment. The finding for the first time of silylenes that have triplet ground states and are unsaturated is so sensational that more observations will have to be made before these results receive the appreciation they deserve. All other silylenes thus far examined have singlet ground states.[1]

A more widely heralded report of the direct detection of a silylene concerned the spectroscopic characterization of dimethylsilylene by Drahnak, Michl, and West.[97] These workers irradiated dodecamethylcyclohexasilane in rigid matrices

TABLE 9.4.   Vibrational Frequencies and Calculated Force Constants (mdyn/Å) for SiNN and SiCO Molecules in Their Ground $^3\Sigma$ States

SiNN f_{SiN} = 2.021, f_{N-N} = 11.825
$SiCO^a$ f_{Si-C} = 5.3, f_{C-O} = 15.6, f_{SiC-CO} = 2.4

| | Frequency (cm^{-1}) | | | |
| | ν_1 | | ν_3 | |
Molecule	Obsd.	Calcd.	Obsd.	Calcd.
28-14-14	1731	1732.6	484	484.1
28-15-15	1676	1674.2	475	475.9
28-12-16	1899.3	1899.3	$(800)^a$	800.0
28-13-16	1855.3	1855.4		793.7
28-12-18	1856.4	1856.5		785.3

^{a}Si—C stretching frequency in Si^{12}C^{16}O was assumed to be 800 cm^{-1} since it was not observed but inferred from the value in the excited state.

at low temperatures, producing a yellow species with a broad electronic absorption band, λ_{max} = 453–455 nm, ϵ = 1500, and a characteristic infrared absorption at 1220 cm^{-1}. These features disappear on softening the matrix and were assigned to dimethylsilylene.

$$(Me_2Si)_6 \xrightarrow[\text{10 K, 77 K}]{h\nu \text{ (254 nm)}} (Me_2Si)_5 + Me_2Si:$$

The structural assignment was based on the position of the ultraviolet and infrared absorptions and on chemical trapping experiments. When the reaction mixtures were doped with trapping agents, namely, triethylsilane, bis(trimethylsilyl)acetylene, or cyclohexene, typical silylene products were obtained upon warming the matrix.

Although the finding of silylene products is suggestive, it does not define the structure of the yellow intermediate unequivocally. Since the work of Ishikawa and Kumada has conclusively demonstrated that photolysis of $(Me_2Si)_6$ generates free Me_2Si,[14] the finding of silylene products upon the softening of an irradiated matrix that contained initially a silylene precursor and a silylene trapping agent points to the formation and interception of a silylene. It does *not*, however, prove that the matrix-isolated, spectroscopically detected species was that silylene, nor that the silylene reacted after the matrix was softened. It does, however, seem very likely that the interpretation offered by the authors is correct. This would seem to open up not only the detailed structural characterization of organosilylenes, but also their use as preformed reagents.

Interestingly, the yellow intermediate identified as dimethylsilylene was found to be bleached by visible light. A photoinduced rearrangement to 2-silapropene was suggested as a possibility to explain the observation of an Si—H stretching band observed in the infrared spectrum.

$$\begin{array}{c} CH_3 \\ \diagdown \\ CH_3 \diagup \end{array} Si: \xrightarrow{h\nu} \begin{array}{c} CH_2 \\ \diagdown \\ CH_3 \diagup \end{array} SiH$$

An important experiment to confirm the identification of the yellow intermediate as Me_2Si: would be to determine whether photobleaching reduces the yield of silylene products found after softening of the matrix.

The structure and physical properties of silylenes have recently been reviewed by Nefedov, Kolesnikov, and Ioffe.[3] These authors have performed a valuable and heroic service by tabulating information from many sources on such properties as the heat of formation, excited state energies, bond angles and lengths, force constants, bond dissociation energies, ionization potentials, and σ–p orbital energy differences for divalent germanium, tin, and lead carbene analogues, as well as silylenes. There are already many data, but they seem to vary enormously in their accuracy, and therefore caution is advised in using them.

The cocondensation of lithium vapor and SiF_4 in an argon matrix[98] is a new method for the formation of SiF_2 for spectroscopic studies. The ν_3 and ν_1 bands in the infrared spectrum of SiF_2 were observed at 853 and 840 cm^{-1}, as was a band at 598 cm^{-1} attributed to $\cdot SiF_2^-$. An ESR signal was attributed to the $\cdot SiF_2^-$ or $\cdot (SiF_2)_n \cdot$ diradical.

VI. FUTURE DEVELOPMENTS

Silylene chemistry remains such a fast-moving field that the efforts to expand our mechanistic knowledge are still overshadowed by the discovery of new reactions. So long as this search for new products remains fruitful we can expect it to continue, but as we see in this chapter, each new reaction raises new mechanistic questions. To answer these questions is not just a matter of satisfying intellectual curiosity, nor merely a means for increasing the efficiency of preparative methods. The new phenomena of silylene chemistry represent added dimensions in the relationship between structure and reactivity that is at the heart of physical-organic chemistry. It seems likely then that the detailed elucidation of silylene reaction mechanisms will move to the center of the stage.

In the last few years mild methods for the generation of silylenes have permitted the isolation of reaction products that were previously destroyed in secondary transformations. The ability to isolate the primary products of silylene reactions has permitted these reactions to be better understood. Hence for both synthetic and mechanistic studies, continued development of new silylene precursors, particularly for photochemical reactions, is predicted.

The study of intramolecular reactions of substituted silylenes has just begun and will certainly receive much more attention. Aryl silylenes have been almost entirely neglected in the gas phase:

VII. ACKNOWLEDGMENT

The camaraderie of workers in the field and their willingness to share their latest findings is evident in this chapter. I am particularly grateful for helpful discussions, useful advice, and private communications to Professors Wataru Ando, Tom Barton, Rob Conlin, Maitland Jones, Fred Lampe, Peter Potzinger, Morey Ring, Robin Walsh, and Bob West. I thank my co-workers Yue-Shen Chen, Bruce Cohen, Becky Cornett, Adam Helfer, Jong-Pyng Hsu, Rong-juh Hwang, Eric Ma, Siu-Hong Mo, and Roger Woods for contributing good ideas and hard work to the understanding of silylenes. This chapter was prepared with financial assistance from the United States Department of Energy. This is technical report COO-1713-88.

VIII. REFERENCES

1.   P. P. Gaspar, "Silylenes," in *Reactive Intermediates*, Vol. 1, M. Jones, Jr. and R. A. Moss, Eds., Wiley, New York, 1978, pp. 229–277.

2.   E. A. Chernyshev, N. G. Komalenkova, and S. A. Bashkirova, "Silicon Analogues of Carbenes," *Usp. Khim.*, **45**, 1782 (1976).

3.   O. M. Nefedov, S. P. Kolesnikov, and A. I. Ioffe, "Group IVB Carbene Analogs– Structure and Reactivity," *Journal of Organometallic Chemistry Library*, 5, *Organometallic Chemistry Reviews*, 1977, pp. 181–218.

4.   Y.-N. Tang, "Reactions of Silicon Atoms and Silylenes," *Reactive Intermediates*, Vol. 2, R. A. Abramovitch, Ed., Plenum, New York, in the press.

5.   M. A. Ring, "Kinetics of Polysilane Decompositions," in *Homoatomic Rings, Chains and Macromolecules of Main-Group Elements*, A. L. Rheingold, Ed., Elsevier, Amsterdam, 1977, pp. 261–275.

6.*   D. P. Paquin and M. A. Ring, "Kinetics of the Decomposition of 1,1,1-Trimethyldisilane and of Trimethylsilylgermane and of Relative Rates of Silylene Insertion into Silicon-Hydrogen Bond," *J. Am. Chem. Soc.*, **99**, 1793 (1977).

7.   A. M. Doncaster and R. Walsh, "Kinetics and Mechanism of the Gas Phase Thermal Decomposition of Hexachlorodisilane in the Presence of Iodine, *J. Chem. Soc; Faraday Trans. I*, 272 (1980).

8.*   R. R. Lembke, R. F. Ferrante, and W. Weltner, Jr., "SiCO, SiN_2 and $Si(CO)_2$ Molecules: Electron Spin Resonance and Optical Spectra at 4 K," *J. Am. Chem. Soc.*, **99**, 416 (1977).

9.*   R. J. Hwang and P. P. Gaspar, "Reactions of Recoiling Silicon Atoms with Phosphine-Diene Mixtures and the Question of Silylene Intermediates, *J. Am. Chem. Soc.*, **100**, 6626 (1978).

10.   A. P. Helfer and P. P. Gaspar, to be published.

11.   O. F. Zeck, Y. Y. Su, and Y.-N. Tang, "Effect of Additives on the Reaction of Monomeric Silicon Difluoride with 1,3-Butadiene, *J. Am. Chem. Soc.*, **98**, 3474 (1976).

12.   R. A. Ferrieri, E. E. Siefert, M. J. Griffin, O. F. Zeck, and Y.-N. Tang, "Relative Reactivities of Conjugated Dienes Towards Silicon Difluoride," *J. Chem. Soc., Chem. Commun.*, 6 (1977).

13.   E. E. Siefert, R. A. Ferrieri, O. F. Zeck, and Y.-N. Tang, "Reactions of Monomeric Silicon Difluoride and Silylene with Conjugated Pentadienes," *Inorg. Chem.* **17**, 2802 (1978).

14.* M. Ishikawa and M. Kumada, "Photolysis of Organic Derivatives of Catenated Group IV Elements," *Rev. Si, Ge, Sn, Pb Compounds.*, **IV**, 7 (1979).

15.   Y.-S. Chen and P. P. Gaspar, to be published.

16.* G. G. A. Perkins, E. R. Austin and F. W. Lampe, "The 147-nm Photolysis of Monosilane," *J. Am. Chem. Soc.*, **101**, 1109 (1979).

17.   B. Reimann, A. Matten, R. Laupert, and P. Potzinger, "Zur Reaktion von Silyl-radikalen. Das Verhältnis Disproportionierung/Rekombination," *Ber. Bunsenges.* 81, 500 (1977).

18.   E. R. Austin and F. W. Lampe, "Hydrogen-Atom Sensitized Decomposition of Mono-silane," *J. Phys. Chem.,* 80, 2811 (1976).

19.* A. G. Alexander and O. P. Strausz, "Photochemistry of Silicon Compounds. 5. The 147 nm Photolysis of Dimethylsilane," *J. Phys. Chem.,* 80, 2531 (1976).

20.   L. Gammie, C. Sandorfy, and O. P. Strausz, "Photochemistry of Silicon Compounds. 6. The 147 nm Photolysis of Tetramethylsilane," *J. Phys. Chem.,* 83, 3075 (1979).

21.   S. Tokach, P. Boudjouk, and R. D. Koob, "Photolysis of 1,1 Dimethylsilacyclo-butane," *J. Phys. Chem.,* 82, 1203 (1978).

22.   M. Ishikawa, T. Fuchikami, and M. Kumada, "Photolysis of Organopolysilanes. Formation and Reactions of Substituted 1-Silacyclopropene and 1-Sila-1,2-pro-padiene," *J. Am. Chem. Soc.,* 99, 245 (1977).

23.   H. Sakurai, Y. Kamiyama, Y. Nakadaira, "Photochemical Generation of Silacyclo-propene, *J. Am. Chem. Soc.,* 99, 3879 (1977).

24.   B. J. Cornett, "Tetramethylsilacyclopropene: Its Characterization and Reactions," Doctoral Dissertation, Washington University, May 1980.

25.   H. Sakurai, Y. Kobayashi, and Y. Nakadaira, "Chemistry of Organosilicon Com-pounds XCIV. Preparations of 8-Silatricyclo[2.3.1.0^{2,4}]oct-6-enes, New Photo-chemical Silylene Generators," *J. Organomet. Chem.,* 120, C1 (1976).

26.* C. G. Newman, M. A. Ring, and H. E. O'Neal, "Kinetics of the Silane and Silylene Decompositions under Shock Tube Conditions," *J. Am. Chem. Soc.,* 100, 5945 (1978).

27.* C. G. Newman, H. E. O'Neal, M. A. Ring, F. Leska, and N. Shipley, "Kinetics and Mechanism of the Silane Decomposition," *Int. J. Chem. Kinet.,* XI, 1167 (1979).

28.   J. H. Purnell and R. Walsh, "The Pyrolysis of Monosilane," *Proc. Roy. Soc. Lond.,* Ser. A, 293, 543 (1966).

29.* P. Neudorfl, A. Jodhan, and O. P. Strausz, "Mechanism of the Thermal Decompo-sition of Monosilane," *J. Phys. Chem.,* 84, 338 (1980).

30.   K. Y. Choo and P. P. Gaspar, "Addition of Trimethylsilyl Radicals to Ethylene. A Flash Photolysis-Electron Spin Resonance Kinetic Study," *J. Am. Chem. Soc.,* 96, 1284 (1974).

31.* P. S. Neudorfl and O. P. Strausz, "Pyrolysis of Monomethyl- and Dimethyl-silanes. The Role of Molecular and Radical Processes," *J. Phys. Chem.,* 82, 241 (1978).

32.* I. M. T. Davidson and M. A. Ring, "Primary Processes in the Low-Pressure Pyrolysis of Methylsilane," *J. Chem. Soc. Farad. I,* 76, 1520 (1980).

33.* T. H. Richardson and J. W. Simons, "The decomposition Kinetics of Chemically Activated Methyl-d$_1$-methylsilane-d$_2$ and Ethylsilane-d$_3$," *Int. J. Chem. Kinet.,* X, 1055 (1978).

34.* H. Sakurai, T. Kobayashi, and Y. Nakadaira, "Chemistry of Organosilicon Com-pounds CXIX. Preparation and Some Reactions of 1,2-Disilacyclobutenes," *J. Organomet. Chem.,* 162, C43 (1978).

35.* T.-Y. Y. Gu and W. P. Weber, "Insertion of Dimethylsilylene into O—H and N—H Single Bonds," *J. Organomet. Chem.* 184, 7 (1980).

36. W. H. Atwell, "Thermolysis of Alkoxy Disilanes in the Presence of Alcohols to Provide Hydrogen-Substituted Alkoxymonosilanes," U.S. Patent 3,478,078, (November 11, 1969).

37. E. A. Chernyshev, T. L. Krasnova, V. V. Stepanov, and M. O. Labartava, "Expansion of Furan Heterocycles in Their Reactions with Hexachlorodisilane or Trichlorosilane," Zh. Obshch. Khim., 48, 2798 (1978).

38. E. A. Chernyshev, N. G. Komalenkova, S. A. Bashkirova, N. A. Batygina, and A. V. Kisin, "Reaction of Dichlorosilylene with the C—O Bond of Dichlorophenoxysilane. Formation of 1,1,3,3-Tetrachloro-2-oxa-1,3-disilaindan," Zh. Obshch. Khim., 47, 1196 (1977).

39. W. H. Atwell and D. R. Weyenberg, "Divalent Silicon Intermediates," Angew. Chem., Int. Ed. Engl., 8, 469 (1969).

40. H. M. Frey and M. A. Voisey, "Reactions of Methylene with Ethers Part 1-Dimethyl Ether, Methyl Ethyl Ether, Methyl n-propyl Ether, Methyl isopropyl Ether and Tetrahydrofuran," Trans. Faraday Soc., 64, 954 (1968).

41. H. S. D. Soysa, H. Okinoshima, and W. P. Weber, "A New Route to Dimethylsilanone [(CH$_3$)$_2$Si=O]; Dioxygenation of Dimethylsulfoxide by Dimethylsilylene," J. Organomet. Chem., 133, C17 (1977).

42.* H. Okinoshima and W. P. Weber, "Insertion of Methylphenylsilylene into Cyclic Siloxanes, Effect of Ring Size on Siloxane Reactivity," J. Organomet. Chem., 150, C25 (1978).

43.* T.-Y. Y. Gu and W. P. Weber, "Mechanism of the Reactions of Dimethylsilylene with Oxetanes," J. Am. Chem. Soc., 102, 1641 (1980).

44.* K. P. Steele and W. P. Weber, "Solvent Modified Reactivity of Dimethylsilylene," J. Am. Chem. Soc., 102, 6095 (1980).

45.* D. Tzeng and W. P. Weber, "Mechanism of Reaction of Dimethylsilylene with α,β-Unsaturated Epoxides," J. Am. Chem. Soc., 102, 1451 (1980).

46.* V. J. Tortorelli and M. Jones, Jr., "On the Stereochemistry of Addition of Dimethylsilylene to cis- and trans-Butene," J. Am. Chem. Soc., 102, 1425 (1980).

47.* M. Ishikawa, K.-I. Nakagawa, and M. Kumada, "Photolysis of Organopolysilanes. The reaction of Trimethylsilylphenylsilylene with Olefins and Conjugated Dienes," J. Organomet. Chem., 178, 105 (1979).

48.* M. Ishikawa, K.-I. Nakagawa, M. Ishigoro, and M. Kumada, "Photolysis of Organopolysilanes. The Reaction of Photochemically Generated Methylphenylsilylene with Olefins," J. Organomet. Chem., 152, 155 (1978).

49. M. Ishikawa, T. Fuchikami, and M. Kumada, "Chemistry of a 1-Silacyclopropene. Thermal Production of 1,4-Disilacyclohexadienes," J. Organomet. Chem., in press.

50. M. Ishikawa, K.-I. Nakagawa, and M. Kumada, "Photolysis of Organopolysilanes. Formation and Reactions of New Silacyclopropene Derivatives, J. Organomet. Chem., 131, C15 (1977).

51.* M. Ishikawa, K.-I. Nakagawa, and M. Kumada, "Photolysis of Oragnopolysilanes, Photochemical Formation and Reactions of 1-Trimethylsilyl-1-phenyl-1-silacyclopropene Derivatives," J. Organomet. Chem., 190, 117 (1980).

52. C. H. Haas and M. A. Ring, "Reaction of Silyl Radicals and Silylene with Acetylene and Application of Orbital Symmetry to the Pyrolysis of Silane and Disilane," Inorg. Chem., 14, 2253 (1975).

53.* T. J. Barton and J. A. Kilgour, "An Alternative Mechanism for the Formation of 1,4-Disilacyclohexa-2,5-diene from Acetylene and Silylene," *J. Am. Chem. Soc.*, 98, 7746 (1976).

54.* D. Seyferth and S. C. Vick, "The Preparation of a 1,2-Disilacyclobutane and a 1,2-Disilacyclobut-3-ene by Dimethylsilylene Insertion into the Silacyclopropane and Silacyclopropene Ring Systems. New Silacyclopropenes," *J. Organomet. Chem.*, 125, C11 (1977).

55. E. A. Chernyshev, N. G. Komalenkova, S. A. Bashkirova, and V. V. Sokolov, "Silicon-Containing Heterocyclic Compounds XXX. Thermal Reactions of Dichlorosilylene with Unsaturated Compounds," *Zh. Obshch. Khim.*, 48, 830 (1978).

56. D. L. Perry and J. L. Margrave, "The Chemistry of Silicon Difluoride," *J. Chem. Ed.*, 53, 696 (1976).

57. P. P. Gaspar and B. J. Herold, "Silicon, Germanium and Tin Structural Analogs of Carbenes," in *Carbene Chemistry*, 2nd ed., W. Kirmse, (Ed.), Academic, New York, 1971, p. 504.

58. J. L. Margrave and P. W. Wilson, "Silicon Difluoride, a Carbene Analog. Its Reactions and Properties," *Acc. Chem. Res.*, 4, 145 (1971).

59.* D. Seyferth and D. P. Duncan, "1,1-Difluoro-2,2,3,3,-tetramethyl-1-silirane: Synthesis and Novel Chemistry. Reinterpretation of Difluorosilylene Reaction Mechanisms," *J. Am. Chem. Soc.*, 100, 7734 (1978).

60.* J. L. Margrave and D. L. Perry, "Reexamination of the Evidence for Silicon-Silicon Multiply Bonded Intermediates. Possible Existence of $[F_2Si=SiF_2]$," *Inorg. Chem.*, 16, 1820 (1977).

61.* C.-S. Liu and T. I. Hwang, "Reaction of Vinyl Chloride with Difluorosilylene by Cocondensation," *J. Am. Chem. Soc.*, 101, 2996 (1979).

62.* D. Seyferth and D. C. Annarelli, "Generation of Dimethylsilylene Under Mild Conditions by the Thermolysis of Hexamethylsilirane," *J. Am. Chem. Soc.*, 97, 7162 (1975).

63.* J. C. Thompson, A. P. G. Wright, and W. F. Reynolds, "New Evidence for the Involvement of Monomeric Silicon Difluoride in Reactions with Olefins," *J. Am. Chem. Soc.*, 101, 2236 (1979).

64.* C.-S. Liu and T. I. Hwang, "Insertion *vs.* Addition of Difluorosilylene-Evidence for the Attack of Difluorosilylene on the Carbon-Carbon Bond as An Initial Step in the Insertion Reactions with *trans-* and *cis-*Difluoroethylene," *J. Am. Chem. Soc.*, 100, 2577 (1978).

65. W. Ando and M. Ikeno, "Synthesis of Silyl Enol Ether. Reaction of Silylene with Carbonyl Compounds," *Chem. Lett.*, 609 (1978).

66.* W. Ando, M. Ikeno, and A. Sekiguchi, "Chemistry of Oxasilacyclopropane. 2. Formations of Dioxasilacyclopentanes in the Reaction of Oxasilacyclopropane Derivatives with Adamantanone and Norbornone," *J. Am. Chem. Soc.*, 100, 3613 (1978).

67.* D. Seyferth and T. F. O. Lim, *J. Am. Chem. Soc.*, 100, 7074 (1978).

68.* M. Ishikawa, K.-I. Nakagawa, and M. Kumada, "Photolysis of Organopolysilanes. Reactions of Trimethylsilylphenylsilylene with Carbonyl Compounds," *J. Organomet. Chem.*, 135, C45 (1977).

69. W. Ando and M. Ikeno, "Reaction of Dimethylsilylene with α-Diketones. Formation and Reaction of 1,3-Dioxa-2-silacyclopent-4-enes," *J. Chem. Soc., Chem. Commun.*, 655 (1979).

70.   D. Seyferth, T. F. O. Lim, and D. P. Duncan, "Reactions of 1,1-Dimethyl-2,3-bis-(trimethylsilyl)-1-silirene and Hexamethylsilirane with Dimethylsulfoxide. Insertion of Dimethylsilanone into the Silirene and Silirane Rings," *J. Am. Chem. Soc.*, **100**, 1626 (1978).

71.   H. Okinoshima and W. P. Weber, "Photolysis of Aryl-Substituted Disilanes in the Presence of Dimethyl Sulfoxide," *J. Organomet. Chem.*, **149**, 279 (1978).

72.   R. E. Swaim and W. P. Weber, "Photooxidation of 1,1,1-Trimethyl-2,2,2-tri-phenyl-disilane by Dimethyl Sulfoxide," *J. Am. Chem. Soc.*, **101**, 5703 (1979).

73.   H. S. D. Soysa and W. P. Weber, "Reinvestigation of the Photolysis of Aryl-Substituted Disilanes in the Presence of Dimethyl Sulfoxide," *J. Organomet. Chem.*, **173**, 269 (1979).

74.   H. Okinoshima and W. P. Weber, "Photolysis of Heptamethyl-2-phenyltrisilane and Octamethyl-2,3-diphenyltetrasilane in the Presence of Dimethyl Sulfoxide," *J. Organomet. Chem.*, **155**, 165 (1978).

75.*   R. T. Conlin and P. P. Gaspar, "Evidence for the Dimerization of Dimethylsilylene to Tetramethyldisilene," *J. Am. Chem. Soc.*, **98**, 868 (1976).

76.*   Y. Nakadaira, T. Kobayashi, T. Otsuka, and H. Sakurai, "Efficient Trapping of Silylenes through Disilene Intermediates," *J. Am. Chem. Soc.*, **101**, 486 (1979).

77.*   H. Sakurai, Y. Nakadaira, and T. Kobayashi, "*trans-* and *cis-*1,2-Dimethyl-1,2-diphenyldisilene. Is Si=Si a True Double Bond?," *J. Am. Chem. Soc.*, **101**, 487 (1979).

78.*   W. D. Wulff, W. F. Goure, and T. J. Barton, "On the Role of Trimethylsilylmethyl-silylene in the Gas Phase Reactions of Tetramethyldisilene," *J. Am. Chem. Soc.*, **100**, 6236 (1978).

79.   M. Jones, Jr., "Gas-Phase Carbene Reactions," *Acc. Chem. Res.*, **7**, 415 (1974).

80.   C. Wentrup, "Rearrangements and Interconversions of Carbenes and Nitrenes," *Top. Curr. Chem.* **62**, 173 (1976).

81.*   D. V. Roark and G. J. D. Peddle, "Reactions of 7,8 Disilabicyclo[2.2.2] octa-2,5-dienes. Evidence for the Transient Existence of Disilene," *J. Am. Chem. Soc.*, **94**, 5837 (1972).

82.   T. R. Fields and P. J. Kropp, "Photochemistry of Alkenes III. Formation of Carbene Intermediates," *J. Am. Chem. Soc.*, **96**, 7559 (1974).

83.   Y.-S. Chen, B. H. Cohen, and P. P. Gaspar, "Rearrangement of Bis(trimethylsilyl)-silylene $(Me_3Si)_2Si$: in the Gas Phase," *J. Organometal. Chem.*, **195**, C1 (1980).

84.*   L. C. Snyder and Z. R. Wasserman, "On the Structure of the Si_2H_4 Ground State: Singlet Silysilylene," *J. Am. Chem. Soc.*, **101**, 5222 (1979).

85.*   J. H. Meadows and H. F. Schaefer III, "One- and Two-Configuration Hartree-Fock Limit Predictions for the Singlet-Triplet Separation in Methylene and Silylene," *J. Am. Chem. Soc.*, **98**, 4383 (1976).

86.*   M. S. Gordon, "The Methylsilylene-Silaethylene-Silylcarbene Isomerization," *Chem. Phys. Lett.*, **54**, 9 (1978).

87.   R. L. Kreeger and H. Schechter, "The Chemistry of Trimethylsilylcarbene," *Tetrahedron Lett.*, 2061 (1975).

88.   W. Ando, A. Sekiguchi, and T. Migita, "Photolysis of Silyldiazomethane. The Migrating Tendencies of the Groups on Silicon Atom," *Chem. Lett.*, 779 (1976).

89.* O. P. Strausz, R. K. Gosavi, G. Theodorakopoulos, and I. G. Csizmadia, "A Preliminary Investigation on the Thermodynamic Stability of Triplet Carbenoid Isomers of Silaethylene," *Chem. Phys. Lett.*, 58, 43 (1978).

90.* R. K. Gosavi, H. E. Gunning, and O. P. Strausz, "Lowest Triplet State Reaction Surface Studies for the Isomerization of Methylsilylene → Silaethylene → Silylcarbene," *Chem. Phys. Lett.*, 59, 321 (1978).

91.* J.-C. Barthelat, G. Trinquier, and G. Bertrand, "Theoretical Investigations on Some C_2SiH_4 Isomers," *J. Am. Chem. Soc.*, 101, 3785 (1979).

92.* R. T. Conlin and P. P. Gaspar, "Tetramethylsilacyclopropene," *J. Am. Chem. Soc.*, 98, 3716 (1976).

93. P. R. Jones and D. D. White, "The Aromatic Character of Silacyclopropenes," *J. Organomet. Chem.*, 154, C33 (1978).

94. M. J. S. Dewar, D. H. Lo, and C. A. Ramsden, "Ground States of Molecules. XXIX. MINDO/3 Calculations of Compounds Containing Third Row Elements," *J. Am. Chem. Soc.*, 97, 1311 (1975).

95. M. S. Gordon, M. D. Bjorke, F. J. Marsh, and M. S. Korth, "Second-Row Molecular Orbital Calculations. A Minimal Basis INDO for Na—Cl," *J. Am. Chem. Soc.*, 100, 2670 (1978).

96.* J. F. Harrison, R. C. Liedtke, and J. F. Liebman, "The Multiplicity of Substituted Acyclic Carbenes and Related Molecules," *J. Am. Chem. Soc.*, 101, 7162 (1979).

97.* T. J. Drahnak, J. Michl, and R. West, "Dimethylsilylene, $(CH_3)_2Si$," *J. Am. Chem. Soc.*, 101, 5427 (1979).

98.* D. L. Perry, P. F. Meier, R. H. Hauge, and J. L. Margrave, "Matrix-Isolation Infrared and Electron Paramagnetic Resonance Spectroscopic Studies of the Reaction of Lithium with Silicon Tetrafluoride," *Inorg. Chem.*, 17, 1364 (1978).

99. S. K. Tokach and R. D. Koob, "Photolysis of Tetramethylsilane at 147 nm. Reactivity of $(CH_3)_3Si \cdot$ and $(CH_3)_2SiCH_2$," *J. Phys. Chem.*, 83, 774 (1979).

100. E. Bastian, P. Potzinger, A. Ritter, H.-P. Schuchmann, C. von Sonntag, and G. Weddle, "The Direct Photolysis of Tetramethylsilane in the Gas and Liquid Phases," *"Ber. Bunsenges. Phys. Chem.*, 84, 56 (1980).

101.* W. F. Goure and T. J. Barton, "Reaction of Dimethylsilylene with Cyclooctene Oxide," *J. Organometal. Chem.*, 199, 33 (1980).

102. An isotope effect k_H/k_D ~2 has, however, been found recently: private communication from Professor W. P. Weber, with permission to quote.

103. W. M. Jones, "Carbene–Carbene Rearrangements in Solution," *Acc. Chem. Res.*, 10, 353 (1977).

104. O. L. Chapman, C.-C. Chang, J. Kolc, M. E. Jung, J. A. Lowe, T. J. Barton, and M. L. Tumey, "1,1,2-Trimethylsilaethylene," *J. Am. Chem. Soc.*, 98, 7844 (1976).

105. M. R. Chedekel, M. Skoglund, R. L. Kreeger, and H. Schechter, "Solid State Chemistry. Discrete Trimethylsilylmethylene," *J. Am. Chem. Soc.*, 98, 7846 (1976).

106. W. Ando, M. Ikeno, and A. Sekiguchi, "Evidence of the Formation of Oxa-silacyclopropane from the Reaction of Silylene with Ketone," *J. Am. Chem. Soc.*, 99, 6447 (1977).

107. This synthetic sequence has recently been accomplished from the silylene: Y.-S. Chen and P. P. Gaspar, to be published. For the reaction starting with a recoiling silicon atom, see above, and Ref. 10.

INDEX